Nonlinear parabolic equations and hyperbolic-parabolic coupled systems

Pitman Monographs and
Surveys in Pure and Applied Mathematics 76

Nonlinear parabolic equations and hyperbolic-parabolic coupled systems

Songmu Zheng

Fudan University

CRC Press
Taylor & Francis Group
Boca Raton London New York

CRC Press is an imprint of the
Taylor & Francis Group, an **informa** business

A CHAPMAN & HALL BOOK

First published 1995 by Longman Group Limited

Published 2019 by Chapman and Hall/CRC Press
Taylor & Francis Group
6000 Broken Sound Parkway NW, Suite 300
Boca Raton, FL 33487-2742

© 1995 by Taylor & Francis Group, LLC
CRC Press is an imprint of Taylor & Francis Group, an Informa business

First issued in paperback 2019

No claim to original U.S. Government works

ISBN-13: 978-0-367-44897-4 (pbk)
ISBN-13: 978-0-582-24488-7 (hbk)
ISSN 0269-3666

Visit the Taylor & Francis Web site at
http://www.taylorandfrancis.com

and the CRC Press Web site at
http://www.crcpress.com

British Library Cataloguing in Publication Data

A catalogue record for this book is
available from the British Library

Library of Congress Cataloging-in-Publication Data

Zheng, S. (Songmu)
Nonlinear parabolic equations and hyperbolic-parabolic coupled systems / S. Zheng.
p. cm. -- (Pitman monographs and surveys in pure and applied mathematics, ISSN 0269-3666 ;)
1. Differential equations, Parabolic. 2. Differential equations, Hyperbolic. 3. Differential equations, Nonlinear. I. Title. II. Series.
QA377.Z44 1995
515'.353--dc20
94-31194
CIP

AMS Subject Classifications: (Main) 35K, 35L, 35Q20
(Subsidiary) 82A25, 73B30, 80-xx

To Weixi and Leijun

Contents

Preface

This book is devoted to the global existence, uniqueness and asymptotic behaviour of smooth solutions to nonlinear parabolic equations and nonlinear hyperbolic–parabolic coupled systems for both small and large initial data. Most of the material in this book is based on research carried out by the author and his collaborators in recent years. The manuscript of this book has been used as lecture notes for the graduate students in Fudan University.

Nonlinear parabolic equations and nonlinear hyperbolic–parabolic coupled systems arise in the study of many physical and mechanical problems. For instance, reaction–diffusion systems describing the processes in a chemical reactor, the Cahn–Hilliard equation and coupled Cahn–Hilliard equations describing the phase separation in binary alloys, and the phase-field equations describing solid–liquid phase transitions, to name just a few, are important examples of nonlinear parabolic equations. To study the kinetics of an elastic body with heat conduction and viscosity, one is led to thermoelastic and thermoviscoelastic systems, respectively. Similar systems of partial differential equations can be derived for the kinetics of gases and liquids. One of the important features of each of these systems is that part of the system consists of nonlinear hyperbolic equations with respect to some unknown functions and the remaining part consists of nonlinear parabolic equations with respect to other unknown functions and these parts are coupled.

For any given initial data, no matter how smooth, nonlinear evolution equations, in general, have only local solutions in time and solutions will blow up in finite time unless special structure conditions are imposed. In recent years, especially starting from the mid-1970s, a great deal of effort has been made to look for global *small* smooth solutions for general nonlinear evolution equations such as nonlinear wave equations, nonlinear heat equations, etc. More precisely, assuming that the zero function satisfies nonlinear evolution equations and assuming that initial data are sufficiently small, one looks for *small* smooth solutions globally in time. Then the nonlinear evolution equations can be rewritten as the linearized evolution equations with nonlinear higher-order perturbations. The key issue is whether the local solution in time has uniform a priori estimates so that the solution can be extended locally in time and eventually globally in time. It turns out that the order of nonlinearity and the decay rate of the solution to the linearized problem are two crucial factors for global existence or blow-up in finite time for nonlinear evolution equations. Concerning the developments in this direction for nonlinear parabolic equations and nonlinear hyperbolic–parabolic coupled systems as far as the title of this book is concerned, we refer to Kawashima [1] and the references cited there, Klainerman [1], Matsumura [1-3], Matsumura & Nishida [1-4], Ponce [1], the book by Racke [1] and the references cited there, Schonbek [1], Slemrod [1], Zheng [2-4,6, 8-11], Zheng & Chen [1], Zheng & Shen [2-3], Shen & Zheng [2]. We should emphasize that the references cited in this book are not intended to be exhaustive.

Knowledge about the global existence of a small solution, however, is usually far from being enough for physical problems. Usually one looks for global solutions with *arbitrary* (not necessarily small) initial data. It turns out that the key issue is to get uniform a priori estimates of solutions by using the special structure conditions of the equations under investigation. Once the global existence and uniqueness properties are known, then interest is focused on topics related to the asymptotic behaviour of solutions: multiplicity of equilibria; convergence to an equilibrium; existence of a global attractor, inertial manifolds and inertial sets and the estimates of their fractal dimensions. Among the many references in the literature, we refer to a series of papers by Amann [1-9] for recent developments in the theory of quasilinear parabolic systems. We also refer to the books by Temam [1] and Hale [1] for the theory of infinite-dimensional dynamical systems for dissipative equations. For results on global existence for nonlinear hyperbolic–parabolic coupled systems, see e.g.,

Dafermos [4], Dafermos & Hsiao [1] and the references cited in Kawashima [1]. The material in Chapters 4 and 5 of this book presents the contribution of the author and his collaborators in this direction. The problems presented in this book have some interesting features: the phase-field equations and the coupled Cahn–Hilliard equations are not diagonal parabolic systems; instead, they are triangular parabolic systems with unusual nonlinearity; the corresponding stationary problems are nonlinear boundary value problems with nonlocal terms and constraints. It turns out that new methods have to be introduced to deal with these problems.

In what follows we briefly describe the main content of this book.

In Chapter 1, as a reference chapter, we collect concepts and some facts about Sobolev spaces, linear elliptic boundary value problems, linear and nonlinear evolution equations and infinite-dimensional dynamical systems.

In Chapter 2 we discuss decay rates of solutions to both initial value problems and initial boundary value problems for linear parabolic equations and two classes of linear hyperbolic–parabolic coupled systems: the linear one-dimensional thermoelastic and thermoviscoelastic systems. The methods we use there include the usual energy method, the spectral analysis method and a method related to a theorem on the necessary and sufficient conditions for a C_o-semigroup being exponentially stable. The last method has been systematically developed by Liu & Zheng [1–5] and Burns, Liu & Zheng [1]. We would like to mention that the exponential stability of a C_o-semigroup is also closely related to linear Gaussian quadratic optimal control for a system (for instance, see Gibson, Rosen & Tao [1]).

After the preparation in Chapter 2, Chapter 3 is mainly devoted to the study of the global existence of *small* smooth solutions to both initial value problems and initial boundary value problems for fully nonlinear parabolic equations and two classes of quasilinear hyperbolic–parabolic coupled systems which include nonlinear thermoelastic and thermoviscoelastic systems, the equations of radiation hydrodynamics, and the equations of motion of compressible viscous and heat-conductive fluids. To prove the global existence of small solutions, based on the results in Chapter 2, the weighted norm of a solution is introduced and the global iterative method or continuation argument is used. In the last section of Chapter 3, blow-up results for initial boundary value problems for a class of nonlinear parabolic equations (and also nonlinear hyperbolic equations) with *small* initial data are displayed. In this respect the question of whether solutions to the linearized problems have high enough decay rates plays a crucial role.

In Chapter 4, we discuss the global existence of solutions to the phase-field equations, the coupled Cahn–Hilliard equations and a nonlinear system of partial differential equations arising from the study of phase transitions in shape memory alloys with arbitrary initial data. Observe that several striking features appear in these systems: (i) Most of them are quasilinear and the unusual nonlinear terms such as $\frac{1}{\theta}$ and its derivatives, with θ being an unknown function, appear in the phase-field equations and coupled Cahn–Hilliard equations. (ii) Nonlinear parabolic systems (e.g., the phase-field equations and the coupled Cahn–Hilliard equations) are non-diagonal and they are not uniformly parabolic. (iii) The systems (e.g., the coupled Cahn–Hilliard equations and the system from the study of phase transitions in shape memory alloys) consist of a nonlinear fourth-order partial differential equation and a nonlinear second-order partial differential equation. It turns out that the possibility of applying the maximum principle is completely ruled out. (iv) The corresponding stationary problems are unusual nonlinear elliptic boundary value problems with non-local terms and constraints.

To obtain global existence, one has to use the special structure conditions to derive uniform a priori estimates of solutions. It is in this chapter that we focus our attention on the spirit of deriving uniform a priori estimates, including an L^∞ norm estimate of solutions.

The study of unusual stationary problems, especially in one space dimension, is given in Chapter 5. We also discuss in Chapter 5 the existence of an inertial manifold and an inertial set with finite Hausdorff dimension for a non-diagonal parabolic system: the phase-field equations. A new method, the so-called symmetrizer, is introduced.

I should like to record my appreciation of my collaborators in my mathematical career, including P. Bates, J.A. Burns, Chen Yunmei, C. Elliott, K.H. Hoffmann, W. Horn, N. Kenmochi, Z. Liu, S. Luckhaus, M. Niezgodka, A. Novick-Cohen, R. Racke, Y. Shibata and J. Sprekels for fruitful cooperation and stimulating discussions.

I would like to take this opportunity to express my sincere thanks to H. Amann for his constant interest and support. I would also like to acknowledge J.M. Ball, P. Fife, A. Friedman, K.H. Hoffmann, W. Jager, J. Moser, E. Zehnder, R. Racke and J. Sprekels for inviting me to visit their universities for extensive periods of time. Much of the work on which this book is based was carried out during these visits.

Special thanks are given to G. Roach for his interest in my research and acting as the initiator for publication of this book.

I would also like to acknowledge the NSF of China for the continuous support.

Finally, my deepest gratitude goes to my wife, Weixi Shen, also a mathematician and my collaborator in Fudan University, for her constant encouragement, advice and support in my career and also for producing the camera-ready copy of this book.

Chapter 1

Preliminaries

In this chapter we collect some basic results on function spaces and partial differential equations which will be needed in the remainder of the book. Most results are just recalled without proofs, but the relevant references are given.

1.1 Basic Facts on Sobolev Spaces

1.1.1 Sobolev Spaces $W^{m,p}(\Omega)$

Let Ω be a bounded or an unbounded domain of $I\!R^n$ with smooth boundary Γ. For $m \in I\!N$, $1 \le p \le \infty$, $W^{m,p}(\Omega)$ is defined to be the space of functions u in $L^p(\Omega)$ whose distribution derivatives of order up to m are also in $L^p(\Omega)$. Then, it is known (see, e.g., Adams [1], Lions & Magenes [1]) that $W^{m,p}(\Omega)$ is a Banach space for the norm

$$\|u\|_{W^{m,p}(\Omega)} = \left(\sum_{|\alpha| \le m} \|D^\alpha u\|_{L^p(\Omega)}^p \right)^{\frac{1}{p}} \tag{1.1.1}$$

where $\alpha = \{\alpha_1, \cdots, \alpha_n\} \in I\!N^n$, $|\alpha| = \alpha_1 + \cdots + \alpha_n$, $D^\alpha u = \frac{\partial^{\alpha_1 + \cdots + \alpha_n} u}{\partial x_1^{\alpha_1} \cdots \partial x_n^{\alpha_n}}$. When $p = 2$, we usually denote $W^{m,p}(\Omega)$ by $H^m(\Omega)$ and this is a Hilbert space for the induced inner product.

Let $C^k(\Omega)\,(k \in I\!N$ or $k = \infty)$ be the space of k times continuously differentiable functions on Ω.

1

We denote by $C_o^k(\Omega)$ the space of $C^k(\Omega)$ functions on Ω with compact support in Ω. The closure of $C_o^\infty(\Omega)$ in $W^{m,p}(\Omega)$ is denoted by $W_o^{m,p}(\Omega)$ which is a subspace of $W^{m,p}(\Omega)$.

We now recall some important properties of the Sobolev spaces $W^{m,p}(\Omega)$ (see, e.g., Adams [1]).

Theorem 1.1.1 (Density Theorem) *If Ω is a C^m domain, $m \geq 1$, $1 \leq p < \infty$, then $C^m(\bar{\Omega})$ is dense in $W^{m,p}(\Omega)$.*

Theorem 1.1.2 (Imbedding and Compactness Theorem) *Assume that Ω is a bounded domain of class C^m. Then we have*

(i) If $mp < n$, then $W^{m,p}(\Omega)$ is continuously imbedded in $L^{q^*}(\Omega)$ with $\frac{1}{q^*} = \frac{1}{p} - \frac{m}{n}$:

$$W^{m,p}(\Omega) \hookrightarrow L^{q^*}(\Omega). \tag{1.1.2}$$

Moreover, the imbedding operator is compact for any q, $1 \leq q < q^*$.

(ii) If $mp = n$, then $W^{m,p}(\Omega)$ is continuously imbedded in L^q, $\forall q$, $1 \leq q < \infty$:

$$W^{m,p}(\Omega) \hookrightarrow L^q(\Omega). \tag{1.1.3}$$

Moreover, the imbedding operator is compact, $\forall q, 1 \leq q < \infty$. If $p = 1$, $m = n$, then the above still holds for $q = \infty$.

(iii) If $k + 1 > m - \frac{n}{p} > k$, $k \in \mathrm{IN}$, then writing $m - \frac{n}{p} = k + \alpha$, $k \in \mathrm{IN}$, $0 < \alpha < 1$, $W^{m,p}(\Omega)$ is continuously imbedded in $C^{k,\alpha}(\bar{\Omega})$:

$$W^{m,p}(\Omega) \hookrightarrow C^{k,\alpha}(\bar{\Omega}), \tag{1.1.4}$$

where $C^{k,\alpha}(\bar{\Omega})$ is the space of functions in $C^k(\bar{\Omega})$ whose derivatives of order k are Hölder continuous with exponent α. Moreover, if $n = m - k - 1$, and $\alpha = 1$, $p = 1$, then (1.1.4) holds for $\alpha = 1$, and the imbedding operator is compact from $W^{m,p}(\Omega)$ to $C^{k,\beta}(\bar{\Omega})$, $\forall\, 0 \leq \beta < \alpha$.

Remark 1.1.1 *The imbedding properties (i)–(iii) are still valid for smooth unbounded domains or IR^n provided that $L^q(\Omega)$ in (1.1.3) and $C^{k,\alpha}(\bar{\Omega})$ in (1.1.4) are replaced by $L^q_{loc}(\Omega)$ and $C^{k,\alpha}(\mathcal{B})$ for any bounded domain $\mathcal{B} \subset \Omega$, respectively.*

Remark 1.1.2 *The regularity assumption on Ω can be weakened (e.g., see Adams [1]).*
When $u \in W_o^{m,p}(\Omega)$, the above imbedding properties are valid without any regularity
assumptions on Ω.

Let Ω be a smooth bounded domain of class C^m and $u \in W^{m,p}(\Omega)$. Then we can
define the trace of u on Γ which coincides with the value of u on Γ when u is a smooth
function of $C^m(\bar{\Omega})$.

Theorem 1.1.3 (Trace Theorem) *Let $\nu = (\nu_1, \cdots, \nu_n)$ be the unit outward normal*
on Γ and

$$\gamma_j u = \frac{\partial^j u}{\partial \nu^j}\bigg|_{\Gamma}, \quad \forall u \in C^m(\bar{\Omega}), \quad j = 0, \cdots, m-1. \tag{1.1.5}$$

Then the trace operator $\gamma = \{\gamma_0, \cdots, \gamma_{m-1}\}$ can be uniquely extended to a continuous
operator from $W^{m,p}(\Omega)$ to $\prod_{j=0}^{m-1} W^{m-j-\frac{1}{p},p}(\Gamma)$:

$$\gamma: \quad u \in W^{m,p}(\Omega) \mapsto \gamma u = \{\gamma_0 u, \cdots, \gamma_{m-1} u\} \in \prod_{j=0}^{m-1} W^{m-j-\frac{1}{p},p}(\Gamma). \tag{1.1.6}$$

Moreover, it is a surjective mapping.

Notice that $W^{m-j-\frac{1}{p},p}(\Gamma)$ are spaces with fractional-order derivatives. Refer to Lions
& Magenes [1] for the definition and more about that.

1.1.2   The Gagliardo–Nirenberg and Poincaré Inequalities

Throughout this book the following Gagliardo–Nirenberg interpolation inequalities
(see Nirenberg [1] and Friedman [1]) will be frequently used.

First we introduce some notation. For $p > 0$, $|u|_{p,\Omega} = \|u\|_{L^p(\Omega)}$. For $p < 0$, set
$h = \left[-\frac{n}{p}\right]$, $-\alpha = h + \frac{n}{p}$ and define

$$|u|_{p,\Omega} = \sup_{\Omega} |D^h u| \equiv \sum_{|\beta|=h} \sup_{\Omega} |D^\beta u|, \quad if \ \alpha = 0, \tag{1.1.7}$$

$$|u|_{p,\Omega} = [D^h u]_{\alpha,\Omega} \equiv \sum_{|\beta|=h} \sup_{\Omega} [D^\beta u]_\alpha$$

$$\equiv \sum_{|\beta|=h} \sup_{x,y \in \Omega} \frac{|D^\beta u(x) - D^\beta u(y)|}{|x-y|^\alpha}, \quad if \ \alpha > 0. \tag{1.1.8}$$

If $\Omega = I\!R^n$, we simply write $|u|_p$ instead of $|u|_{p,\Omega}$.

Theorem 1.1.4 *Let j, m be any integers satisfying $0 \leq j < m$, and let $1 \leq q, r \leq \infty$,
and $p \in \mathbb{R}$, $\frac{j}{m} \leq a \leq 1$ such that*

$$\frac{1}{p} - \frac{j}{n} = a(\frac{1}{r} - \frac{m}{n}) + (1 - a)\frac{1}{q}. \tag{1.1.9}$$

Then,

*(i) For any $u \in W^{m,r}(\mathbb{R}^n) \bigcap L^q(\mathbb{R}^n)$, there is a positive constant C depending only
on n, m, j, q, r, a such that the following inequality holds:*

$$|D^j u|_p \leq C |D^m u|_r^a |u|_q^{1-a} \tag{1.1.10}$$

*with the following exception: if $1 < r < \infty$ and $m - j - \frac{n}{r}$ is a nonnegative integer,
then (1.1.10) holds only for a satisfying $\frac{j}{m} \leq a < 1$.*

*(ii) For any $u \in W^{m,r}(\Omega) \bigcap L^q(\Omega)$ where Ω is a bounded domain with smooth bound-
ary, there are two positive constants C_1, C_2 such that the following inequality holds:*

$$|D^j u|_{p,\Omega} \leq C_1 |D^m u|_{r,\Omega}^a |u|_{q,\Omega}^{1-a} + C_2 |u|_{q,\Omega} \tag{1.1.11}$$

with the same exception as in (i).

*In particular, for any $u \in W_o^{m,r}(\Omega) \bigcap L^q(\Omega)$, the constant C_2 in (1.1.11) can be taken
as zero.*

The following two theorems are concerned with the useful Poincaré inequalities.

Theorem 1.1.5 *Let Ω be a bounded domain in $\mathbb{R}^n$ and $u \in H_o^1(\Omega)$. Then there is a
positive constant C depending only on Ω and n such that*

$$\|u\|_{L^2(\Omega)} \leq C \|\nabla u\|_{L^2(\Omega)}, \quad \forall u \in H_o^1(\Omega). \tag{1.1.12}$$

Theorem 1.1.6 *Let Ω be a bounded domain of C^1 in $\mathbb{R}^n$. There is a positive constant C depending only on Ω, n such that for any $u \in H^1(\Omega)$,*

$$\|u\|_{L^2(\Omega)} \leq C \left(\|\nabla u\|_{L^2(\Omega)} + \left| \int_\Omega u dx \right| \right). \tag{1.1.13}$$

1.1.3   Abstract Functions Valued in Banach Spaces

For the study of evolution equations it is convenient to introduce abstract functions
valued in Banach spaces.

Let X be a Banach space, $1 \leq p < \infty$, $-\infty \leq a < b \leq \infty$. Then $L^p((a,b);X)$ denotes the space of L^p functions from (a,b) into X. It is a Banach space for the norm

$$\|f\|_{L^p((a,b);X)} = \left(\int_a^b \|f(t)\|_X^p \, dt \right)^{\frac{1}{p}} \tag{1.1.14}$$

where the integral is understood in the Bochner sense.

For $p = \infty$, $L^\infty((a,b);X)$ is the space of measurable functions from (a,b) into X being essentially bounded. It is a Banach space for the norm

$$\|f\|_{L^\infty((a,b);X)} = \sup_{t \in (a,b)} ess\|f(t)\|_X. \tag{1.1.15}$$

Similarly, when $-\infty < a < b < \infty$ we can define Banach spaces $C^k([a,b];X)$ for the norm

$$\|f\|_{C^k([a,b];X)} = \sum_{i=0}^k \max_{t \in [a,b]} \left\| \frac{d^i f}{dt^i}(t) \right\|_X. \tag{1.1.16}$$

1.2 Linear Elliptic Equations

In this section we introduce some basic results on linear elliptic boundary value problems (refer to Nirenberg [2], Friedman [1] and Lions & Magenes [1]).

1.2.1 Boundary Value Problems

Let Ω be a domain with smooth boundary Γ. Any linear partial differential operator with, for simplicity, C^∞ coefficients $a_\alpha(x)$ in $\bar{\Omega}$ has the form

$$P(x,D) = \sum_{|\alpha| \leq \mu} a_\alpha(x) D^\alpha. \tag{1.2.1}$$

The operator P is called elliptic in $\bar{\Omega}$ if the leading homogeneous part of $P(x,\xi)$ does not vanish for $\xi \neq 0$ and $x \in \bar{\Omega}$:

$$P_\mu(x,\xi) = \sum_{|\alpha| = \mu} a_\alpha(x) \xi^\alpha \neq 0, \quad \forall\, x \in \bar{\Omega}, \ \xi \in I\!\!R^n \backslash \{0\}. \tag{1.2.2}$$

The Laplace operator Δ and the biharmonic operator Δ^2 are the most familiar elliptic operators.

It follows that for an elliptic operator P with real coefficients a_α, μ must be an even number $2m$, $m \geq 0$, $m \in \mathbb{N}$. Usually, one studies boundary value problems for elliptic operators:

$$\begin{cases} Pu = \sum_{|\alpha| \leq 2m} a_\alpha(x) D^\alpha u = f, & x \in \Omega, \\ B_j u|_\Gamma = g_j, & j = 0, \cdots m - 1, \end{cases} \qquad (1.2.3)$$

where f and g_j are given functions in $\bar{\Omega}$ and Γ, respectively, and B_j are certain partial differential operators defined on Γ and the order of each B_j is less than $2m$.

Wellposedness

When $g_j \equiv 0$, the problem (1.2.3) is said to be wellposed if

(i) *Ker* P belongs to C^∞ and $dim(Ker\, P) = \nu < \infty$.

(ii) In suitable function spaces X, Y the operator $P : X \mapsto Y$ is continuous and has closed range in Y of finite codimension ν^*, i.e, P is Fredholm.

The index of operator P is defined as follows:

$$ind\, P = \nu - \nu^*. \qquad (1.2.4)$$

The boundary operators B_j $(j = 0, \cdots, m - 1)$ for which the general boundary problems (1.2.3) are wellposed have been characterized and are known as the Lopatinsky boundary conditions. As far as spaces X, Y are concerned, $C^{2m+k+\alpha}(\bar{\Omega})$, $C^{k+\alpha}(\bar{\Omega})$ $(k \in \mathbb{N}, 0 < \alpha < 1)$ and $W^{2m+k,p}(\Omega)$, $W^{k,p}(\Omega)$ $(1 < p < \infty)$ are two suitable choices of Sobolev spaces.

Basic Results in $C^{k+\alpha}$

(i) *Fredholm*

The mapping

$$P : \left\{ u \in C^{2m+k+\alpha}(\bar{\Omega}) | B_j u = 0, on\ \Gamma, (j = 0, \cdots, m - 1) \right\} \mapsto C^{k+\alpha}(\bar{\Omega}) \qquad (1.2.5)$$

is Fredholm and its index is independent of k.

(ii) *Regularity*

If $g_j = 0$, $f \in C^{k+\alpha}(\bar{\Omega})$ and u is a weak solution of problem (1.2.3) in the distribution sense, then $u \in C^{2m+k+\alpha}(\bar{\Omega})$. Thus all functions in *Ker* P are in C^∞.

(iii) *A Priori Estimate*

There exist two positive constants C_1, C_2 independent of u such that for any $u \in C^{2m+k+\alpha}(\bar{\Omega})$ satisfying $B_j u|_\Gamma = 0$, $(j = 0, \cdots, m-1)$,

$$\|u\|_{C^{2m+k+\alpha}} \leq C_1 \|Pu\|_{C^{k+\alpha}} + C_2 \|u\|_C. \tag{1.2.6}$$

Moreover, if $Ker\ P = 0$, i.e., the uniqueness of problem (1.2.3) holds, then there is a positive constant C_3 independent of u such that

$$\|u\|_{C^{2m+k+\alpha}} \leq C_3 \|Pu\|_{C^{k+\alpha}}. \tag{1.2.7}$$

Basic Results in $W^{k,p}(\Omega)$ $(k \in I\!N,\ 1 < p < \infty)$

(i) *Fredholm*

The mapping

$$P: \left\{ u \in W^{2m+k,p}(\Omega)|\ B_j u = 0,\ on\ \Gamma,\ (j = 0, \cdots, m-1) \right\} \mapsto W^{k,p}(\Omega) \tag{1.2.8}$$

is Fredholm and its index is independent of k.

(ii) *Regularity*

If $g_j \equiv 0$, $f \in W^{k,p}(\Omega)$ and u is a weak solution of problem (1.2.3), then $u \in W^{2m+k,p}(\Omega)$. Thus all functions in $Ker\ P$ are in C^∞.

(iii) *A Priori Estimate*

There exist two positive constants C_1, C_2 independent of u such that for any $u \in W^{2m+k,p}(\Omega)$ satisfying $B_j u|_\Gamma = 0$, $(j = 0, \cdots, m-1)$ in the trace sense,

$$\|u\|_{W^{k+2m,p}} \leq C_1 \|Pu\|_{W^{k,p}} + C_2 \|u\|_{L^p}. \tag{1.2.9}$$

Moreover, if $Ker\ P = 0$, i.e., the uniqueness of problem (1.2.3) holds, then there is a positive constant C_3 independent of u such that

$$\|u\|_{W^{k+2m,p}} \leq C_3 \|Pu\|_{W^{k,p}}. \tag{1.2.10}$$

Remark 1.2.1 *For the nonhomogeneous boundary value problem (1.2.3) with $g_j \neq 0$, $(j = 0, \cdots, m-1)$ being in Sobolev spaces of fractional order, the similar wellposedness holds (see Lions & Magenes [1]).*

1.2.2   Interpolation Spaces

We recall a few facts about linear operators associated with a bilinear form and interpolation spaces associated with a positive definite operator in Hilbert spaces.

Let V, H be separable Hilbert spaces such that V is dense in H and the injection $V \hookrightarrow H$ is continuous and compact. Thus, by the Riesz representation theorem we can write

$$V \subset H \equiv H' \subset V'. \tag{1.2.11}$$

The dual product between V and V' is denoted by $\langle\,,\,\rangle$ and the inner product in H by $(\,,\,)$.

Let A be a linear continuous operator from V to V'. We can associate it with a bilinear form a on V in such a way that

$$a(u,v) = \langle Au, v \rangle, \quad \forall u, v \in V. \tag{1.2.12}$$

Suppose that a is symmetric:

$$a(u,v) = a(v,u) \tag{1.2.13}$$

and a is coercive, i.e., there exists a positive constant $\alpha > 0$ such that

$$a(u,u) \geq \alpha \|u\|_V^2, \quad \forall u \in V. \tag{1.2.14}$$

Let

$$D(A) = \{u|\, u \in V, \ Au \in H\}. \tag{1.2.15}$$

Thus by the Lax–Milgram theorem, $a(u,v)$ can be considered as an equivalent inner product. Furthermore, A is a strictly positive self-adjoint operator in H and the spectral theorem allows us to define the powers A^S of A for $s \in \mathbb{R}$. Since we assume that the injection $V \hookrightarrow H$ is compact, there exists (see Yosida [1]) a complete orthonormal basis $\{w_j\}$ of H and a sequence $\{\lambda_j\}$ such that $w_j \in D(A)$ and

$$\begin{cases} Aw_j = \lambda_j w_j, & j = 1, 2, \cdots, \\ 0 < \lambda_1 \leq \lambda_2 \leq \cdots, & \lambda_j \to \infty, \ as \ j \to \infty, \\ a(w_i, w_j) = \lambda_i \delta_{ij}, & \forall i, j \in \mathbb{N}. \end{cases} \tag{1.2.16}$$

Thus, for $s > 0$, we define

$$D(A^S) = \left\{ u \in H \,\middle|\, \sum_{j=1}^{\infty} \lambda_j^{2S} |(u, w_j)|^2 < \infty \right\}, \tag{1.2.17}$$

$$\|u\|_{D(A^S)} = \left(\sum_{j=1}^{\infty} \lambda_j^{2S} |(u, w_j)|^2 \right)^{\frac{1}{2}} \tag{1.2.18}$$

and for negative s, $D(A^S)$ is the completion of H for the norm $\left(\sum_{j=1}^{\infty} \lambda_j^{2S} |(u, w_j)|^2 \right)^{\frac{1}{2}}$.

In particular, we have $V = D(A^{\frac{1}{2}})$.

Let X and Y be two Banach spaces, with $X \subset Y$, X dense in Y, and the injection $X \hookrightarrow Y$ being continuous. In general, there are several different methods to define the intermediate spaces between X and Y. The framework described above gives a sort of definition of intermediate spaces, namely, the interpolation spaces.

Let

$$X = V = D(A^{\frac{1}{2}}), \quad Y = H = D(A^0). \tag{1.2.19}$$

$$a(u, v) = (u, v)_X. \tag{1.2.20}$$

Then interpolation spaces $[X, Y]_\theta$, $(0 \le \theta \le 1)$ are given by

$$[X, Y]_\theta = D(A^{\frac{1-\theta}{2}}), \quad \forall \theta \in [0, 1]. \tag{1.2.21}$$

The interpolation inequality

$$\|u\|_{[X,Y]_\theta} \le C(\theta) \|u\|_X^{1-\theta} \|u\|_Y^\theta, \quad \forall u \in X, \ \theta \in [0, 1] \tag{1.2.22}$$

also follows from (1.2.21).

A typical example of the above framework is

$$V = H_o^1(\Omega), \quad H = L^2(\Omega), \quad A = -\Delta \tag{1.2.23}$$

and

$$D(A) = H^2(\Omega) \bigcap H_o^1(\Omega) \tag{1.2.24}$$

with Ω being a bounded domain of C^2.

1.3   Linear Evolution Equations

In this section we recall some results on linear evolution equations of first order in
time

$$\frac{du}{dt} + Au = f, \tag{1.3.1}$$

and linear evolution equations of second order in time

$$\frac{d^2u}{dt^2} + \alpha\frac{du}{dt} + Au = f \tag{1.3.2}$$

where $\alpha \in I\!\!R$.

In particular, we will recall some results concerning linear symmetric hyperbolic system and linear parabolic system.

For linear evolution equations (1.3.1) we present the results derived from the framework of semigroup theory (see Friedman [1] and Henry [1]) and also from the framework of variational evolution equations (see Lions & Magenes [1] and Temam [1]).
The results from both settings will be needed in the remaining part of the book.

1.3.1   Variational Evolution Equations

We first consider the first-order variational evolution equations (1.3.1). Let V, H be
two separable Hilbert spaces and $a(\,,\,)$ be a bilinear form satisfying the assumptions
made in the previous section. Let A be the linear operator associated with the bilinear
form $a(\,,\,)$.

For any given $T > 0, f \in L^2([0,T]; H)$ and $u_0 \in H$ (or $u_0 \in V$), we consider the
initial value problem

$$\frac{du}{dt} + Au = f, \tag{1.3.3}$$

$$u(0) = u_0. \tag{1.3.4}$$

We have the following existence and uniqueness results (see Lions & Magenes [1] and
Temam [1]).

Theorem 1.3.1 *Suppose that V, H, a, A satisfy the assumptions in the previous section. Then for any $u_0 \in H, f \in L^2([0,T]; V')$ there exists a unique solution u of problem (1.3.3)–(1.3.4) in the sense of a distribution valued in V' such that*

$$u \in L^2([0,T]; V) \bigcap C([0,T]; H), \tag{1.3.5}$$

$$u_t \in L^2([0,T]; V'), \tag{1.3.6}$$

and

$$\|u(t)\|_H^2 + \int_0^t \|u(\tau)\|_V^2 d\tau \le \|u_0\|_H^2 + \int_0^t \|f\|_{V'}^2 d\tau, \ \ 0 \le t \le T. \tag{1.3.7}$$

Furthermore, if $u_0 \in V, f \in L^2([0,T]; H)$, then problem (1.3.3)–(1.3.4) admits a unique solution u such that

$$u \in C([0,T]; V) \bigcap L^2([0,T]; D(A)), \tag{1.3.8}$$

$$u_t \in L^2([0,T]; H), \tag{1.3.9}$$

$$\|u(t)\|_V^2 + \int_0^t \|Au\|_H^2 d\tau \le \|u_0\|_V^2 + \int_0^t \|f\|_H^2 d\tau, \ \ 0 \le t \le T. \tag{1.3.10}$$

Concerning the regularity results, we have

Theorem 1.3.2 *In addition to the assumptions in Theorem 1.3.1, we further assume that*

$$f, f_t \in L^2([0,T]; H), \tag{1.3.11}$$

$$u_0 \in D(A). \tag{1.3.12}$$

Then the unique solution of problem (1.3.3)–(1.3.4) satisfies

$$u \in C([0,T]; D(A)), \tag{1.3.13}$$

$$u_t \in L^2([0,T]; V) \bigcap C([0,T]; H), \tag{1.3.14}$$

$$u_{tt} \in L^2([0,T]; V'). \tag{1.3.15}$$

Moreover, for $t \in [0,T]$,

$$\|u(t)\|_{D(A)}^2 + \|u_t(t)\|_H^2 + \int_0^t \left(\|u_{tt}(\tau)\|_{V'}^2 + \|u_t\|_V^2 \right) d\tau$$
$$\le C_T \left(\|u_0\|_{D(A)}^2 + \|f(0)\|_H^2 + \int_0^t \|f_t\|_H^2 d\tau \right) \tag{1.3.16}$$

with C_T being a positive constant depending only on T.

A concrete example fitting the framework is

$$V = H_o^1(\Omega), \quad H = L^2(\Omega), \quad V' = H^{-1}(\Omega), \tag{1.3.17}$$

$$Au = -\sum_{i,j=1}^{n} \frac{\partial}{\partial x_i}(a_{ij}(x)\frac{\partial u}{\partial x_j}) + c(x)u, \tag{1.3.18}$$

with smooth functions $a_{ij}(x)$, $c(x)$,

$$\sum_{i,j=1}^{n} a_{ij}\xi_i\xi_j \geq \alpha \sum_{i=1}^{n} |\xi_i|^2, \quad \alpha > 0, \tag{1.3.19}$$

$$c(x) \geq 0, \tag{1.3.20}$$

$$a(u,v) = \int_\Omega \left(\sum_{i,j=1}^{n} a_{ij}\frac{\partial u}{\partial x_j}\frac{\partial v}{\partial x_i} + c(x)uv \right) dx, \quad \forall\, u,v \in H_o^1(\Omega), \tag{1.3.21}$$

$$D(A) = H^2(\Omega)\bigcap H_o^1(\Omega). \tag{1.3.22}$$

Theorem 1.3.1 and Theorem 1.3.2 show that for a given $f \in L^2([0,T]; H^{-1}(\Omega))$, $(L^2([0,T]; L^2(\Omega)))$, $u_0 \in L^2(\Omega), (H_o^1(\Omega))$, respectively, the following initial boundary value problem

$$u_t - \sum_{i,j=1}^{n} \frac{\partial}{\partial x_i}(a_{ij}(x)\frac{\partial u}{\partial x_j}) + c(x)u = f, \tag{1.3.23}$$

$$u|_\Gamma = 0, \tag{1.3.24}$$

$$u|_{t=0} = u_0(x) \tag{1.3.25}$$

admits a unique solution u,

$$u \in C([0,T]; L^2)\bigcap L^2([0,T]; H_o^1), \tag{1.3.26}$$

$$u_t \in L^2([0,T]; H^{-1}), \tag{1.3.27}$$

and

$$u \in C([0,T]; H_o^1)\bigcap L^2([0,T]; H^2), \tag{1.3.28}$$

$$u_t \in L^2([0,T]; L^2), \tag{1.3.29}$$

respectively. Moreover, if f, $f_t \in L^2([0,T]; L^2)$ and $u_0 \in H^2 \bigcap H_o^1$, then the solution u satisfies

$$u \in C([0,T]; H^2 \bigcap H_o^1), \tag{1.3.30}$$

$$u_t \in C([0,T]; L^2) \cap L^2([0,T]; H_o^1), \tag{1.3.31}$$

$$u_{tt} \in L^2([0,T]; H^{-1}). \tag{1.3.32}$$

We now turn to the second-order variational evolution equation (1.3.2). The assumptions on V, H, a and A remain the same as before. Then for the initial value problem

$$\frac{d^2 u}{dt^2} + \alpha \frac{du}{dt} + Au = f, \tag{1.3.33}$$

$$u(0) = u_0, \quad u_t(0) = u_1, \tag{1.3.34}$$

we have

Theorem 1.3.3 *Suppose that V, H, a, A satisfy the same assumptions as before. Then for $u_0 \in V$, $u_1 \in H$, $f \in L^2([0,T]; H)$, problem (1.3.33)–(1.3.34) admits a unique solution u such that*

$$u \in C([0,T]; V), \quad u_t \in C([0,T]; H). \tag{1.3.35}$$

Furthermore, for $s \geq 0$, if $u_0 \in D(A^{\frac{s+1}{2}})$, $u_1 \in D(A^{\frac{s}{2}})$, $f \in L^2([0,T]; D(A^{\frac{s}{2}}))$, then

$$u \in C([0,T]; D(A^{\frac{s+1}{2}})), \quad u_t \in C([0,T]; D(A^{\frac{s}{2}})), \tag{1.3.36}$$

$$\|u(t)\|^2_{D(A^{\frac{s+1}{2}})} + \|u_t\|^2_{D(A^{\frac{s}{2}})}$$

$$\leq C \left(\|u_0\|^2_{D(A^{\frac{s+1}{2}})} + \|u_1\|^2_{D(A^{\frac{s}{2}})} + \int_0^t \|f\|^2_{D(A^{\frac{s}{2}})} \, d\tau \right). \tag{1.3.37}$$

For the time-dependent operators $A(t)$ associated with the bilinear form $a(t; u, v)$, similar results hold under suitable assumptions (see Lions & Magenes [1] and Temam [1]).

To convert the beam equation into (1.3.2), we set

$$\Omega = (0, L), \quad H = L^2(\Omega), \quad V = H_o^2(\Omega), \quad A = \Delta^2, \quad D(A) = H^4 \cap H_o^2 \tag{1.3.38}$$

and

$$a(u, v) = \int_\Omega \Delta u \Delta v \, dx, \quad \forall u, v \in V. \tag{1.3.39}$$

Then by Theorem 1.3.3, for $u_0 \in H_o^2$, $u_1 \in L^2$, $f \in L^2([0,T];L^2)$, the following initial boundary value problem

$$\frac{d^2u}{dt^2} + \alpha\frac{du}{dt} + \Delta^2 u = f, \tag{1.3.40}$$

$$u|_\Gamma = 0, \quad \frac{\partial u}{\partial n}\bigg|_\Gamma = 0, \tag{1.3.41}$$

$$u(0) = u_0, u_t(0) = u_1 \tag{1.3.42}$$

admits a unique solution u such that

$$u \in C([0,T];H_o^2), \quad u_t \in C([0,T];L^2). \tag{1.3.43}$$

Furthermore, if $u_0 \in D(A)$, $u_1 \in H_o^2(\Omega)$, $f \in L^2([0,T];H_o^2)$, then

$$u \in C([0,T];D(A)), \quad u_t \in C([0,T];H_o^2), \quad u_{tt} \in C([0,T];L^2). \tag{1.3.44}$$

1.3.2   Setting of Semigroup Theory

Since equation (1.3.2) can be reduced to the first-order equation (1.3.1) by letting u_t be a new dependent variable, we will only discuss the setting of semigroup theory for (1.3.1).

Definition 1.3.1 *A family $\{S(t)\}$, $(0 \leq t < \infty)$ of bounded linear operators in a Banach space X is called a strongly continuous semigroup (in short, a C_o-semigroup) if*

(i) $S(t_1 + t_2) = S(t_1)S(t_2)$, $\forall t_1, t_2 \geq 0$,

(ii) $S(0) = I$,

(iii) For each $x \in X, S(t)x$ is continuous in t on $[0,\infty)$.

For such a semigroup $\{S(t)\}$, we define an operator A with domain $D(A)$ consisting of points x such that the limit

$$Ax = \lim_{h \to 0}\frac{S(h)x - x}{h}, \quad x \in D(A) \tag{1.3.45}$$

exists. We call A the infinitesimal generator of the semigroup $\{S(t)\}$. Given an operator A, if A coincides with the infinitesimal generator of $\{S(t)\}$, then we say that it generates a strongly continuous semigroup $\{S(t)\}$.

We now have (see Pazy [1]):

Theorem 1.3.4 (Hille–Yosida) *A linear (unbounded) operator A is the infinitesimal generator of a C_o-semigroup of contractions $S(t)$, $t \geq 0$, if and only if*

(i) A is closed and $\overline{D(A)} = X$.

(ii) The resolvent set $\rho(A)$ of A contains IR^+ and for every $\lambda > 0$,

$$\|(\lambda I - A)^{-1}\| \leq \frac{1}{\lambda}. \tag{1.3.46}$$

Now let X be a Hilbert space. Recall that a linear operator is dissipative if for every $x \in D(A)$, $Re(Ax, x) \leq 0$. We now have the following corollary of the Lumer–Phillips Theorem (see Pazy [1]).

Theorem 1.3.5 *Let A be a densely defined closed linear operator. If both A and its adjoint operator A^* are dissipative, then A is the infinitesimal generator of a C_o-semigroup of contractions on X.*

Definition 1.3.2 *A linear operator A in a Banach space X is sectorial if it is a closed, densely defined operator such that for some $\varphi \in (0, \frac{\pi}{2})$, some $M \geq 1$ and a real number a, the resolvent set of A contains the sector*

$$S_{a,\varphi} = \{\lambda | \varphi \leq |\arg(\lambda - a)| \leq \pi, \ \lambda \neq a\} \tag{1.3.47}$$

and

$$\|(\lambda I - A)^{-1}\| \leq \frac{M}{|\lambda - a|}, \ \forall \lambda \in S_{a,\varphi}. \tag{1.3.48}$$

It has been pointed out in Friedman [1] that many elliptic boundary value problems define sectorial operators.

Definition 1.3.3 *A strongly continuous semigroup $\{S(t)\}$ on a Banach space X is called an analytic semigroup if $t \mapsto S(t)x$ is real analytic on $0 < t < \infty$ for each $x \in X$.*

Then we have (see Henry [1])

Theorem 1.3.6 *If A is a sectorial operator, then $-A$ is the infinitesimal generator of an analytic semigroup $\{e^{-At}\}_{t \geq 0}$. Moreover,*

$$\frac{d}{dt}e^{-At} = -Ae^{-At}, \ t > 0 \tag{1.3.49}$$

and if $Re \, \lambda > a$ whenever $\lambda \in \sigma(A)$, then for $t > 0$,

$$\|e^{-At}\| \leq Ce^{-at}, \ \|Ae^{-At}\| \leq \frac{C}{t}e^{-at}, \tag{1.3.50}$$

for some constant C.

For a sectorial operator A with $Re\,\sigma(A) > 0$ we define

$$A^{-\alpha} = \frac{1}{\Gamma(\alpha)} \int_0^\infty t^{\alpha-1} e^{-At} dt, \ \ \forall\, \alpha > 0 \tag{1.3.51}$$

which is a bounded operator and one to one (see Henry [1]). Furthermore, we define

$$A^\alpha = inverse\ of\ A^{-\alpha}, \ \ \forall \alpha > 0 \tag{1.3.52}$$

$$X^\alpha = D(A^\alpha) \tag{1.3.53}$$

with the graph norm.

Theorem 1.3.7 *Suppose A is sectorial and $Re\,\sigma(A) > \delta > 0$. Then for $\alpha \geq 0$ there exists $C_\alpha < \infty$ such that*

$$\|A^\alpha e^{-At}\| \leq C_\alpha t^{-\alpha} e^{-\delta t}, \ \ t > 0. \tag{1.3.54}$$

For the initial value problem

$$\frac{du}{dt} + Au = f(t), \ \ 0 \leq t \leq T, \tag{1.3.55}$$

$$u(0) = u_0, \tag{1.3.56}$$

we have

Theorem 1.3.8 *Suppose A is a sectorial opetator on a Banach space X and f is Hölder continuous in $[0,T]$ with respect to the topology of X. Then for $u_0 \in X$, the problem (1.3.55)–(1.3.56) admits a unique solution $u \in C([0,T]; X) \cap C^1((0,T]; X)$ given by*

$$u(t) = e^{-At} u_0 + \int_0^t e^{-A(t-\tau)} f(\tau)\,d\tau. \tag{1.3.57}$$

Moreover, if $f \in C^{k+\beta}([0,T]; X)$, $(k \in \mathbb{N}, 0 < \beta < 1)$, then $u \in C^{k+1}((0,T]; X)$.

1.3.3   Hyperbolic and Parabolic Systems

In this subsection we introduce some results concerning initial value problems and initial boundary value problems for linear symmetric hyperbolic systems and linear parabolic systems.

Let us first consider the following initial value problem

$$A_1^0(x,t)\frac{\partial u}{\partial t} + \sum_{j=1}^{n} A_1^j(x,t)\frac{\partial u}{\partial x_j} = f_1(x,t), \quad (x,t) \in I\!\!R^n \times I\!\!R^+, \quad (1.3.58)$$

$$u|_{t=0} = u_0(x), \quad x \in I\!\!R^n \quad (1.3.59)$$

where u, f_1 are m-vector functions, A_1^0, A_1^j are real symmetric $m \times m$ matrices and A_1^0 is positive definite.

The system in (1.3.58) is a linear symmetric hyperbolic system. For problem (1.3.58)–(1.3.59) we have (see Kawashima [1]):

Theorem 1.3.9 *Let s, l be integers such that $s \geq [\frac{n}{2}] + 2$, $0 \leq l \leq s$. Suppose that $A_1^0, A_1^j \in C([0,T]; H^S)$, $\frac{\partial A_1^0}{\partial t}, \frac{\partial A_0^j}{\partial t} \in C([0,T]; H^{S-2}) \cap L^2([0,T]; H^{S-1})$, $f_1 \in C([0,T]; H^{l-1}) \cap L^2([0,T]; H^l)$, and $u_0 \in H^l(I\!\!R^n)$. Then problem (1.3.58)–(1.3.59) admits a unique solution $u \in C([0,T]; H^l(I\!\!R^n)) \cap C^1([0,T]; H^{l-1})$. Moreover, the following estimate holds:*

$$\|u(t)\|_{H^l}^2 \leq C_1^2 e^{C_2(Mt+M_1 t^{\frac{1}{2}})} \left(\|u_0\|_{H^l}^2 + C_2 t \int_0^t \|f_1\|_{H^l}^2 \, d\tau \right) \quad (1.3.60)$$

where

$$M = \sup_{0 \leq t \leq T} \|\{A_1^0, A_1^j\}\|_{H^S}, \quad M_1 = \left(\int_0^T \left\| \frac{\partial \{A_1^0, A_1^j\}}{\partial t} \right\|_{H^{S-1}}^2 d\tau \right)^{\frac{1}{2}} \quad (1.3.61)$$

and $C_1 > 1$ depending on O, the L^∞ norms of A_1^0, A_1^j, and $C_2 > 0$ depending on O and M.

In the above, the notation $\|\{A_1^0, A_1^j\}\|_{H^S}$ denotes the maximum of H^S norms of all the entries of A_1^0, A_1^j, etc. To prove Theorem 1.3.9, Kawashima [1] uses the results of semigroup theory by Kato [1], the standard energy method and the Friedrichs mollifier (also see Racke [1]).

We now turn to the initial boundary value problem for the first-order symmetric hyperbolic system (1.3.58). Let Ω be a bounded domain in $I\!\!R^n$ with smooth boundary Γ. In addition to the initial condition, we assume that u satisfies homogeneous boundary conditions on Γ:

$$B(t,x)u = 0 \quad (1.3.62)$$

where B is a smooth $l \times m$ matrix-valued function with rank l everywhere. The boundary condition (1.3.62) is equivalent to the condition that u is in a subspace S of dimension l in IR^m for each point on Γ.

Let

$$\beta = \sum_{j=1}^{n} A_1^j n_j \qquad (1.3.63)$$

with $n = (n_1, \cdots, n_n)$ being the unit outer normal to Γ. If β is a nonsingular matrix for $x \in \Gamma$, $t \in IR^+$, then we say that the boundary is noncharacteristic. Otherwise, we say that it is characteristic.

Following Friedrichs [1] and Lax & Phillips [1], we assume that the space S is a nonnegative subspace of IR^m with respect to β :

$$u^T \beta u \geq 0, \quad \forall u \in S. \qquad (1.3.64)$$

Furthermore, we assume that S is maximal with respect to (1.3.64). In other words, there is no subspace $\tilde{S}$ such that $\tilde{S} \supset S, \tilde{S} \neq S$ and for $u \in \tilde{S}$, (1.3.64) holds. When S is maximal with respect to (1.3.64), we usually say that the homogeneous boundary condition is admissible in the Friedrichs sense.

Let s be an integer with $s \geq [\frac{n}{2}] + 2$ and

$$u_0^{(k)} = \left.\frac{\partial^k u}{\partial t^k}\right|_{t=0} , \quad (k = 0, \cdots, s - 1) \qquad (1.3.65)$$

which can be successively obtained from the system (1.3.58) and the initial condition (1.3.57). Unlike the initial value problem, in order to obtain the existence and uniqueness of the solution $u \in \bigcap_{j=0}^{s} C^j([0,T]; H^{S-j})$ to the admissible initial boundary value problem, we require that the following compatibility conditions be satisfied:

$$\sum_{j=0}^{p} C_p^j (\partial_t^j B)(0) u_0^{(p-j)} = 0, \quad on \ \Gamma, \ \ 0 \leq p \leq s - 1. \qquad (1.3.66)$$

Then for the admissible initial boundary value for the system (1.3.58), we have (refer to Rauch & Massey III [1]):

Theorem 1.3.10 *Let s be an integer such that $s \geq [\frac{n}{2}] + 2$. Suppose A_1^0, A_1^j are real symmetric, A_1^0 is positive definite and Γ is noncharacteristic. Suppose A_1^0, A_1^j, $f_1 \in$*

$\bigcap\limits_{j=0}^{S} H^j([0,T]; H^{S-j})$, $u_0 \in H^S(\Omega)$. *Then the admissible initial boundary value problem*

for the system (1.3.58) admits a unique solution $u \in \bigcap\limits_{j=0}^{S} C^j([0,T]; H^{S-j})$. *Moreover,*

the following estimate holds for $t \in [0,T]$:

$$\sum_{j=0}^{S} \left\| \frac{\partial^j u}{\partial t^j}(t) \right\|_{H^{S-j}}^2 \leq C_1 e^{C_2 M t} \left(\sum_{k=0}^{S} \left\| u_0^{(k)} \right\|_{H^{S-k}}^2 \right.$$

$$+ \sum_{j=0}^{S-1} \left\| \frac{\partial^j f_1}{\partial t^j} \right|_{t=0} \right\|_{H^{S-1-j}}^2 + C_2 t \int_0^T \sum_{j=1}^{S} \left\| \frac{\partial^j f_1}{\partial t^j} \right\|_{H^{S-j}}^2 d\tau \right) \qquad (1.3.67)$$

where

$$M = \sum_{k=0}^{S} \int_0^t \left\| \frac{\partial^k}{\partial t^k} \{A_1^0, A_1^j\} \right\|_{H^{S-k}}^2 d\tau \qquad (1.3.68)$$

and $C_1 > 1$ *depending on* O, *the* L^∞ *norms of* A_1^0, A_1^j, *and* $C_2 > 0$ *depending on* O
and M.

Proof. The results and the proof essentially have been given in Rauch & Massey
III [1] except the explicit statement about dependence of the constant appearing in
the a priori estimate of the solution. However, it can be seen from the whole course
of the proof that in order to obtain the energy estimates, we denote the $(s-k)$th-
order tangential partial derivatives by D^{S-k} and let $D^{S-k} D_t^k$ act on the system, then
multiply it by $D^{S-k} D_t^k u$ and integrate with respect to x and t to get the estimates
of $\|D^{S-k} D_t^k u(t)\|^2$. Finally, the estimates of $\|D_n^{S-k} D^k u\|^2$, where D_n^{S-k} denotes the
$(s-k)$th-order normal derivatives, are obtained from the system (1.3.58) and the
estimates on the tangential derivatives of the solution. Thus (1.3.67) is obtained from
these estimates and the Sobolev imbedding theorem. $\qquad \square$

We now introduce some results concerning linear parabolic systems (refer to Kawashima [1]). Consider the following initial value problem for a linear parabolic system:

$$A_2^0(x,t)\frac{\partial v}{\partial t} - \sum_{j,k=1}^{n} B_2^{jk}(x,t)\frac{\partial^2 v}{\partial x_j \partial x_k} = f_2(x,t), \quad in \ I\!R^n \times I\!R^+, \qquad (1.3.69)$$

$$v|_{t=0} = v_0(x), \quad in \ I\!R^n. \qquad (1.3.70)$$

We assume that

(i) A_2^0 is a real symmetric and positive definite $m' \times m'$ matrix-valued function.

(ii) B_2^{jk} $(j, k = 1, \cdots, n)$ are $m' \times m'$ real symmetric matrix-valued functions such that

$B_2^{jk} = B_2^{kj}$ and $\sum\limits_{j,k=1}^{n} B_2^{j,k}\omega_j\omega_k$ is real symmetric positive for all $\omega = (\omega_1, \cdots, \omega_n) \in$
S^{n-1}.

(iii) v, f_2, v_0 are m'-vector functions.

(iv) A_2^0, $B_2^{jk} \in C([0,T]; H^S(\mathbb{R}^n))$, $\dfrac{\partial A_2^0}{\partial t}$, $\dfrac{\partial B_2^{jk}}{\partial t} \in C([0,T]; H^{S-2}) \cap L^2([0,T]; H^{S-1})$,

$f_2 \in L^2([0,T]; H^{l-1})$, $v_0 \in H^l(\mathbb{R}^n)$ with s, l being integers such that $s \geq [\frac{n}{2}] + 2$ and
$1 \leq l \leq s$.

Then for problem (1.3.69)–(1.3.70) we have (see Kawashima [1]):

Theorem 1.3.11 *Under the above assumptions problem (1.3.69)–(1.3.70) admits a unique solution $v \in C([0,T]; H^l) \cap L^2([0,T]; H^{l+1})$, $v_t \in L^2([0,T]; H^{l-1})$. Moreover, the following energy estimate holds for $t \in [0,T]$:*

$$\|v(t)\|_{H^l}^2 + \int_0^t \|v\|_{H^{l+1}}^2 \, d\tau$$
$$\leq C_1^2 e^{C_2(t+M_1 t^{\frac{1}{2}})} \left(\|v_0\|_{H^l}^2 + C_2 \int_0^t \|f_2\|_{H^{l-1}}^2 \, d\tau \right) \tag{1.3.71}$$

where

$$M_1 = \left(\int_0^T \left\| \frac{\partial}{\partial t} \{A_2^0, B_2^{jk}\} \right\|_{H^{S-1}}^2 \, d\tau \right)^{\frac{1}{2}} \tag{1.3.72}$$

and $C_1 \geq 1$ is a constant depending on the positivity of A_2^0 and $C_2 > 0$ is a constant depending on

$$M = \sup_{0 \leq t \leq T} \|\{A_2^0, B_2^{jk}\}\|_{H^S}. \tag{1.3.73}$$

1.4 Nonlinear Evolution Equations

In this section we briefly recall some results regarding the initial value problem for nonlinear evolution equations of the form

$$\frac{du}{dt} + Au = f(t,u), \tag{1.4.1}$$
$$u(0) = u_0. \tag{1.4.2}$$

Furthermore, a local existence and uniqueness theorem for nonlinear hyperbolic–parabolic coupled systems is given.

1.4.1 Local Existence and Uniqueness

A frequently used method of proving local existence in time is to apply the contraction mapping theorem. First one tries to find a closed subset of a suitable Banach space to work with. Then the task is to show that if the length of the time interval is sufficiently small, the nonlinear operator defined by the corresponding auxiliary linearized problem maps this subset to itself. Furthermore, this mapping is a strict contraction. Thus local existence and uniqueness follows. For problem (1.4.1)–(1.4.2) we have (see Henry [1])

Theorem 1.4.1 *Suppose A is a sectorial operator with $\operatorname{Re} \sigma(A) > 0$, $0 \le \alpha < 1$ and $f : U \mapsto X, U$ an open subset of $\mathbb{R} \times X^{\alpha}$, $f(t, u)$ is locally Hölder continuous in t, locally Lipschitz in u. Then for any $(t_0, u_0) \in U$, there exists $t^* = t^*(t_0, u_0) > 0$ such that (1.4.1) has a unique solution u on $(t^*, t_0 + t^*)$ with $u(t_0) = u_0$.*

In particular, we now establish the following

Theorem 1.4.2 *For the variational evolution equations (1.3.3)–(1.3.4) with $f = f(u)$, $f(0) = 0$ being locally Lipschitz from V to H in the following sense: $f(v) \in H$, for $v \in V$; for any $v_1, v_2 \in V$, $\|v_1\|_V \le M$, $\|v_2\|_V \le M$, there exists a constant L_M depending on M such that*

$$\|f(v_1) - f(v_2)\|_H \le L_M \|v_1 - v_2\|_V, \tag{1.4.3}$$

problem (1.3.3), (1.3.4) admits a unique local solution $u \in C([0, t^*]; V) \cap L^2([0, t^*]; D(A))$, $u_t \in L^2([0, t^*]; H)$.

Proof. Let $T > 0$ be any given constant and

$$X_T = C([0, T]; V) \cap L^2([0, T]; D(A)) \tag{1.4.4}$$

equipped with the norm

$$\|u\|_{X_T} = \left(\sup_{0 \le t \le T} \|u\|_V^2 + \int_0^T \|Au\|_H^2 \, d\tau \right)^{\frac{1}{2}}. \tag{1.4.5}$$

Let

$$S_T = \{u \in X_T, \ u(0) = u_0, \ \|u\|_{X_T} \le 2\|u_0\|_V\}. \tag{1.4.6}$$

For $v \in S_T$, we consider the auxiliary linear problem

$$\frac{du}{dt} + Au = f(v), \tag{1.4.7}$$

$$u(0) = u_0. \tag{1.4.8}$$

It easily follows from the assumptions on f and the estimate (1.3.10) that there exists
a t^* such that the nonlinear operator $\mathcal{N}$ defined by the auxiliary problem is a map
from S_{t^*} into S_{t^*}. Moreover, we can take t^* small enough, if necessary, so that $\mathcal{N}$ is
a strict contraction from S_{t^*} to S_{t^*}. Thus, the theorem follows from the contraction
mapping theorem. $\square$

We now recall a local existence and uniqueness theorem for the following initial
value problem for nonlinear hyperbolic–parabolic coupled systems:

$$A_1^0(u,v)\frac{\partial u}{\partial t} + \sum_{j=1}^{n} A_1^j(u,v)\frac{\partial u}{\partial x_j} = f_1(u,v,D_x v), \tag{1.4.9}$$

$$A_2^0(u,v)\frac{\partial v}{\partial t} - \sum_{j,k=1}^{n} B_2^{jk}(u,v)\frac{\partial^2 v}{\partial x_j \partial x_k} = f_2(u,v,D_x u, D_x v), \tag{1.4.10}$$

$$u|_{t=0} = u_0(x), \quad v|_{t=0} = v_0(x) \tag{1.4.11}$$

where $x = (x_1, \cdots, x_n) \in \mathbb{R}^n$, $t \in \mathbb{R}^+$, u, f_1, u_0 are m'-vector functions and v, f_2, v_0
are m''-vector functions; we also use the notation $D_x u = (\frac{\partial u}{\partial x_1}, \cdots, \frac{\partial u}{\partial x_n})$, $D_x v = (\frac{\partial v}{\partial x_1}, \cdots, \frac{\partial v}{\partial x_n})$. For problem (1.4.9)–(1.4.11), we make the following assumptions:

(i) Let $\mathcal{O}$ be an open convex subset in $\mathbb{R}^{m'} \times \mathbb{R}^{m''}$. A_1^0, A_2^0, A_1^j, B_2^{jk} are smooth
in $(u,v) \in \mathcal{O}$ and f_1, f_2 are smooth in $(u,v,\xi) \in \mathcal{O} \times \mathbb{R}^{nm''}$ and in $(u,v,\xi,\eta) \in \mathcal{O} \times \mathbb{R}^{n(m'+m'')}$, respectively.

(ii) A_1^0, A_2^0 are real symmetric and positive definite for $(u,v) \in \mathcal{O}$.

(iii) A_1^j are real symmetric for $(u,v) \in \mathcal{O}$.

(iv) B_2^{jk} are real symmetric and satisfy $B_2^{jk} = B_2^{kj}$ for $(u,v) \in \mathcal{O}$; $\sum_{j,k=1}^{n} B_2^{jk}\omega_j\omega_k$ is
positive definite for all $(u,v) \in \mathcal{O}$ and $\omega = (\omega_1, \cdots, \omega_n) \in S^{n-1}$.

(v) For some constant state $(\bar{u}, \bar{v}) \in \mathcal{O}$, $f_1(\bar{u}, \bar{v}, 0) = 0$, $f_2(\bar{u}, \bar{v}, 0, 0) = 0$.

Then the following local existence and uniqueness results have been proved by Kawashima [1].

Theorem 1.4.3 *Suppose that the above assumptions (i)–(v) are satisfied and* $(u_0 - \bar{u}, v_0 - \bar{v}) \in H^S(\mathbb{R}^n), (u_0, v_0) \in \mathcal{O}_1 \subset \mathcal{O}$. *Then there is a positive constant t^* de-*

pending only on $\mathcal{O}_1, d_1 = dist(\mathcal{O}_1, \partial\mathcal{O})$ and $\|\{u_0 - \bar{u}, v_0 - \bar{v}\}\|_{H^S}$ such that problem (1.4.9)–(1.4.11) admits a unique local solution (u, v) in $\mathbb{R}^n \times [0, t^*]$ such that $u - \bar{u} \in C([0, t^*]; H^S) \cap C^1([0, t^*]; H^{S-1}), v - \bar{v} \in C([0, t^*]; H^S) \cap C^1([0, t^*]; H^{S-2}) \cap L^2([0, t^*]; H^{S+1})$. Moreover,

$$\|\{u - \bar{u}, v - \bar{v}\}(t)\|_{H^S}^2 + \int_0^t \left(\|u - \bar{u}\|_{H^S}^2 + \|v - \bar{v}\|_{H^{S+1}}^2 \right) d\tau$$
$$\leq \tilde{C}^2 \|\{u_0 - \bar{u}, v_0 - \bar{v}\}\|_{H^S}^2, \quad \forall t \in [0, t^*] \tag{1.4.12}$$

where $\tilde{C} > 1$ is a constant depending on $\mathcal{O}_1, d_1$ and $\|\{u_0 - \bar{u}, v_0 - \bar{v}\}\|_{H^S}$.

We refer to Vol'pert & Hudjaev [1] for a similar local existence result.

1.4.2 Global Existence

Once a local existence theorem is established, the length of the time interval, in which the solution exists, usually depends on the norm of initial data in a suitable Banach space. Thus if, for any $T < \infty$, we have a uniform a priori estimate of the solution with respect to this norm, then the local solution can be extended step by step to $[0, \infty)$ and becomes a global one. This method is usually called *the continuation argument*. Therefore, to prove global existence, the crucial step is to get uniform a priori estimates of the solution.

If $f(t, u)$ in (1.4.1) is a mapping from $\mathbb{R}^+ \times X^\alpha$ ($\alpha \geq 0$) to X and locally Hölder continuous in t, locally Lipschitz in u, then by Zorn's Lemma there is a maximal T^* of t^* such that problem (1.4.1)–(1.4.2) admits a unique solution in $(0, T^*)$. Furthermore, (i) either $T^* = +\infty$, i.e., we have global existence; (ii) or $T^* < \infty$, and

$$\lim_{t \to T^*} \|u(t)\|_{X^\alpha} = \infty \tag{1.4.13}$$

and in this case we say that the solution u blows up in a finite time.

Theorem 1.4.4 *Suppose f satisfies the assumptions described above. Suppose that for any $T < \infty$ the solution u has a priori estimate*

$$\|u(t)\|_{X^\alpha} \leq C_T, \quad \forall t \in [0, T] \tag{1.4.14}$$

with C_T being a positive constant possibly depending on T but independent of t. Then problem (1.4.1)–(1.4.2) admits a unique global solution.

For problem (1.4.1)–(1.4.2) we have (see Henry [1])

Corollary 1.4.1 *Suppose A is a sectorial operator with $Re\,\sigma(A) > 0$ and $f : \mathbb{R}^+ \times X^\alpha \mapsto X$ locally Hölder continuous in t and locally Lipschitz in u. Suppose also that*

$$\|f(t, u)\| \le K(t)(1 + \|u(t)\|_{X^\alpha}), \ \ \forall\, t \in \mathbb{R}^+ \tag{1.4.15}$$

with $K(t)$ being continuous on $\mathbb{R}^+$. Then problem (1.4.1)–(1.4.2) admits a unique global solution.

The continuation argument also applies to general evolution equations.

1.5 Asymptotic Behaviour

In this section we recall some results about dynamical systems and asymptotic behaviour of solutions to the autonomous problem

$$\frac{du}{dt} + Au = f(u), \ \ t > 0, \tag{1.5.1}$$

$$u(0) = u_0. \tag{1.5.2}$$

In what follows we always assume that A is a sectorial operator with $Re\,\sigma(A) > 0$ in a Banach space X and $f : X^\alpha \ (0 \le \alpha < 1) \mapsto X$ is locally Lipschitz in u.

We refer the readers to Henry [1], Temam [1] and Dafermos [1] for the results in this section.

1.5.1 Global Attractor

Definition 1.5.1 *A dynamical system (nonlinear semigroup) on a complete metric space X is a nonlinear semigroup $\{S(t), t \ge 0\}$ satisfying*
(i) for each $t \ge 0, S(t)$ is continuous from X to X;
(ii) for each $x \in X, t \mapsto S(t)x$ is continuous.

For problem (1.5.1)–(1.5.2) we denote by $u(t; u_0)$ the solution at time t with initial value u_0. Then

$$S(t)u_0 = u(t; u_0), \ \ (u_0 \in X^\alpha, \ t \ge 0). \tag{1.5.3}$$

Definition 1.5.2 *Let $\{S(t); \ t \geq 0\}$ be a dynamical system on X. For any $x \in X$, the set $\gamma(x) = \{S(t)x; \ t \geq 0\} = \bigcup_{t \geq 0} S(t)x$ is called the orbit (or positive semiorbit) through x. We say x is an equilibrium point (or a stationary point) if $\gamma(x) = \{x\}$. For $x_0 \in X$, we define the ω-limit set of x_0 as*

$$\omega(x_0) = \bigcap_{s \geq 0} \overline{\bigcup_{t \geq s} S(t)x_0}. \tag{1.5.4}$$

Definition 1.5.3 *We say that a set $\mathcal{B} \subset X$ is positively invariant for the semigroup if*

$$S(t)\mathcal{B} \subseteq \mathcal{B}. \tag{1.5.5}$$

Definition 1.5.4 *An attractor is a set $\mathcal{M} \subset X$ with the following properties:*

(i) $\mathcal{M}$ is an invariant set $(S(t)\mathcal{M} = \mathcal{M}, \ \forall t \geq 0)$.

(ii) $\mathcal{M}$ has an open neighbourhood $\mathcal{O}$ such that for every $u_0 \in \mathcal{O}$, $S(t)u_0$ converges to $\mathcal{M}$ as $t \to \infty$:

$$\text{dist}(S(t)u_0, \mathcal{M}) = 0, \quad as \ t \to \infty. \tag{1.5.6}$$

If $\mathcal{M}$ is an attractor, the largest open set satisfying (ii) is called the basin of attraction of $\mathcal{M}$. We will say that $\mathcal{M}$ uniformly attracts (or simply attracts) a set $\mathcal{B} \subset \mathcal{O}$ if

$$d(S(t)\mathcal{B}, \mathcal{M}) = \sup_{x \in S(t)\mathcal{B}} \inf_{y \in \mathcal{M}} d(x, y) \to 0, \quad t \to \infty. \tag{1.5.7}$$

Definition 1.5.5 *We say that $\mathcal{M} \subset X$ is a global attractor for the semigroup $S(t)$ if $\mathcal{M}$ is a compact attractor which attracts all the bounded sets of X.*

A useful related concept is absorbing sets.

Definition 1.5.6 *Let $\mathcal{B}$ be a subset of X. We say that $\mathcal{B}$ is absorbing in X (or $\mathcal{B}$ absorbs all the bounded sets of X) if the orbit of any bounded set of X enters into $\mathcal{B}$ after a certain time which may depend on the bounded set.*

Concerning the existence of a global attractor we have (see Temam [1], Theorem 1.1. in Chapter 1)

Theorem 1.5.1 *Suppose X is a Banach space and $S(t)$ is a dynamical system and for each bounded set $\mathcal{U}$, there is a $t_1 \geq 0$ such that $\bigcup_{t \geq t_1} S(t)\mathcal{U}$ is relatively compact. Suppose that $\mathcal{B}$ is a bounded subset of X such that $\mathcal{B}$ is absorbing in X. Then the ω-limit set of $\mathcal{B}$, $\mathcal{M} = \omega(\mathcal{B})$, is the global attractor in X. Furthermore, $\mathcal{M}$ is connected.*

To apply Theorem 1.5.1, it is crucial to verify that $\bigcup_{t \geq t_1} S(t)\mathcal{U}$ is compact. For

problem (1.5.1)–(1.5.2) we have the following modification of Theorem 3.3.6 in Henry

[1].

Theorem 1.5.2 *Suppose A is a sectorial operator with $\operatorname{Re} \sigma(A) > 0$ and has compact*

resolvent and $f : X^\alpha \mapsto X$ is locally Lipschitz continuous. Furthermore, f maps

closed bounded sets in X^α into bounded sets in X. Suppose that for any bounded set

$\mathcal{U}$ and $u_0 \in \mathcal{U}, u(t; u_0)$ is uniformly bounded in X^α. Then for any $t_1 > 0$, $\bigcup_{t \geq t_1} S(t)\mathcal{U}$ is

a compact set in X^α.

Remark 1.5.1 *One can further consider the Hausdorff and fractal dimension of*

global attractors (see Temam [1] and the references cited there).

1.5.2 Convergence to Equilibrium

It is clear from the definition that the ω-limit set contains equilibria. In applications

we often encounter a special class of system, called the gradient system, for which the

ω-limit set consists of equilibria and the structure of the flow on the global attractor

can be described in some detail.

Definition 1.5.7 *A dynamical system $\{S(t); t \geq 0\}$ on X is said to be a gradient*

system if

(i) $\forall x \in X$, $\bigcup_{t \geq 0} S(t)x$ is precompact.

(ii) There exists a Lyapunov function $F : X \mapsto \mathbb{R}$ such that:

(ii_1) $F(x)$ is a continuous function and is bounded from below.

(ii_2) $F(x) \to \infty$, as $\|x\|_X \to \infty$.

(ii_3) $F(S(t)x)$ is nonincreasing in t for each $x \in X$.

(ii_4) If x is such that $S(t)x$ is defined for $t \in \mathbb{R}^+$ and $F(S(t)x) = F(x)$, for $t \in \mathbb{R}^+$,

then x is an equilibrium point.

For the gradient system we have (see Hale [1])

Theorem 1.5.3 *If $S(t)$ is a gradient system, then for each $x \in X$, the ω-limit set*

$\omega(x)$ consists of equilibria.

It follows from Theorem 1.5.1 and Theorem 1.5.3 that if the set of equilibrium points is discrete, then for each x, $\omega(x)$ must be only one equilibrium point. Therefore, we have

Corollary 1.5.1 *Let $S(t)$ be a gradient system and let $\mathcal{E}$ be the set of equilibrium points of $S(t)$. Suppose $\mathcal{E}$ is a discrete set. Then for each x, $S(t)x$ converges to an equilibrium point as time goes to infinity.*

To give a more detailed description of global attractors, we first introduce the concept of stable and unstable sets (see Temam [1]).

Definition 1.5.8 *Let $S(t)$ be a nonlinear continuous semigroup on a Banach space X and $\mathcal{B} \subset X$ be an invariant set. The stable set $\mathcal{M}_-(\mathcal{B})$ of $\mathcal{B}$ is the set of points u_* which belong to a complete orbit $\{u(t),\ t \in \mathbb{R}\}$, i.e., $\exists\, t_0 \in \mathbb{R}, u(t_0) = u_*$ and such that*

$$d(u(t), \mathcal{B}) \to 0, \text{ as } t \to +\infty. \tag{1.5.8}$$

The unstable set $\mathcal{M}_+(\mathcal{B})$ of $\mathcal{B}$ is the set of points u_* which belong to a complete orbit $\{u(t),\ t \in \mathbb{R}\}$ and such that

$$d(u(t), \mathcal{B}) \to 0, \quad \text{as } t \to -\infty. \tag{1.5.9}$$

Then we have (see Temam [1], Theorem 4.1 in Chapter 7)

Theorem 1.5.4 *Let $S(t)$ be a nonlinear semigroup on a Banach space X such that the mapping $\{t, u_0\} \mapsto S(t)u_0$ from $\mathbb{R}^+ \times X$ into X is continuous. We assume that $S(t)$ possesses a Lyapunov function F and a global attractor $\mathcal{M}$. Let $\mathcal{E}$ be the set of equilibria, i.e., fixed points of the semigroup. Then*

$$\mathcal{M} = \mathcal{M}_+(\mathcal{E}). \tag{1.5.10}$$

Furthermore, if $\mathcal{E}$ is discrete, then

$$\mathcal{M} = \bigcup_{z \in \mathcal{E}} \mathcal{M}_+(z). \tag{1.5.11}$$

1.5.3 Inertial Manifolds and Inertial Sets

In general the global attractors are very complicated fractals and convergence of the orbit to the global attractor could be very slow. But for some dissipative dynamical

systems there exist inertial manifolds and inertial sets to which the orbit converges
exponentially fast (see Temam [1] and the references cited there; Eden, Foias, Nicolaenko & Temam [1], [2]).

Definition 1.5.9 *Let H be a Hilbert space and $S(t)$ be a dynamical system on H.
An inertial manifold of the system is a finite-dimensional Lipschitz manifold $\mathcal{M}$ with
the following properties:*

(i) $\mathcal{M}$ is positively invariant for the semigroup.

(ii) $\mathcal{M}$ attracts exponentially all the orbits.

In what follows we recall the results on the existence of an inertial manifold for
the system $S(t)$ associated with

$$\frac{du}{dt} + Au + F(u) = 0, \tag{1.5.12}$$

$$u(0) = u_0. \tag{1.5.13}$$

The framework we work with is the following:

(i) H is a Hilbert space and A is a linear closed unbounded positive self-adjoint
operator densely defined in H. We also assume that A is an isomorphism from $D(A)$
onto H and A^{-1} is compact. Then A has eigenpairs $\{\lambda_j\}$ and $\{w_j\}$ such that

$$Aw_j = \lambda_j w_j, \quad j = 1, 2, \cdots, \tag{1.5.14}$$

$$0 < \lambda_1 \le \lambda_2 \le \cdots, \quad \lambda_j \to \infty, \text{ as } j \to \infty. \tag{1.5.15}$$

As shown in Section 1.2 we can define A^α for $\alpha \in \mathbb{R}$.

(ii) Concerning the nonlinear operator F, we assume that for $\alpha \in \mathbb{R}$, F is Lipschitz
on the bounded sets of $D(A^\alpha)$ with values in $D(A^{\alpha-\frac{1}{2}})$:

$$\|A^{\alpha-\frac{1}{2}}F(u) - A^{\alpha-\frac{1}{2}}F(v)\| \le C_M \|A^\alpha(u-v)\|, \ \forall u, v \in D(A^\alpha), \ \|A^\alpha u\| \le M, \ \|A^\alpha v\| \le M. \tag{1.5.16}$$

(iii) Concerning problem (1.5.12)–(1.5.13), we assume that for some $\alpha \in \mathbb{R}$ and for
every $u_0 \in D(A^\alpha)$, there is a unique global solution u defined on $\mathbb{R}^+$ such that

$$u \in C([0,\infty); D(A^\alpha)) \bigcap L^2([0,T]; D(A^{\alpha+\frac{1}{2}})), \ \forall T > 0. \tag{1.5.17}$$

Furthermore, the associated nonlinear semigroup $S(t)$ is continuous in $D(A^\alpha)$, $\forall t \ge 0$,
and it has an absorbing set $\mathcal{B}_o$ in $D(A^\alpha)$ contained in the ball of $D(A^\alpha)$ centred at 0

of radius $\frac{\rho}{2}$.

(iv) Let $\theta(s)$ be a cut-off function such that

$$\theta(s) = 1, \quad for\ 0 \le s \le 1, \tag{1.5.18}$$

$$\theta(s) = 0, \quad for\ s \ge 2, \tag{1.5.19}$$

and let $\theta_\rho(s) = \theta(\frac{s}{\rho})$. Let

$$F_\theta(u) = \theta_\rho(\|A^\alpha u\|)F(u), \quad \forall u \in D(A^\alpha). \tag{1.5.20}$$

We assume that the ball of $D(A^\alpha)$ centred at 0 of radius ρ is absorbing for

$$\frac{du}{dt} + Au + F_\theta(u) = 0, \tag{1.5.21}$$

$$u(0) = u_0. \tag{1.5.22}$$

(v) For some N large enough,

$$\|A^{\alpha+\frac{1}{2}}q\|^2 = (A^{\alpha+1}q, A^\alpha q) \ge \Lambda\|A^\alpha q\|^2, \ \forall q \in QD(A^\alpha), \tag{1.5.23}$$

$$\|A^{\alpha+\frac{1}{2}}p\|^2 = (A^{\alpha+1}p, A^\alpha p) \le \lambda\|A^\alpha p\|^2, \ \forall p \in PD(A^\alpha) \tag{1.5.24}$$

where $\lambda = \lambda_N$, $\Lambda = \lambda_{N+1}$, P is the projection in H onto the space spanned by $w_1, \cdots w_n$ and $Q = I - P$. Furthermore, we assume that Λ is large enough, i.e.,

$$\Lambda > M_2^2 \left(\frac{l+1}{l} + 4K_4 + 11 \right)^{\frac{1}{2}}, \tag{1.5.25}$$

and

$$\frac{\Lambda - \lambda}{\Lambda^{1/2} + \lambda^{1/2}} > 2M_2 \frac{l+1}{l} \tag{1.5.26}$$

where $M_2, 0 < l < \frac{1}{8}$ and K_4 are certain positive constants.

Remark 1.5.2 *Condition (1.5.26) is usually called the spectral gap condition.*

We now can state a theorem concerning the existence of an inertial manifold (see Temam [1] and Foias, Sell & Temam [1]).

Theorem 1.5.5 *Suppose the above assumptions (i)–(v) are satisfied. Then there exists an inertial manifold of finite dimension. More precisely, there is a finite-dimensional manifold $\mathcal{M}$ such that $\mathcal{M}$ is Lipschitz and $\mathcal{M}$ attracts all the orbits of (1.5.12) and (1.5.21). Moreover, $\mathcal{M}$ is positively invariant for $S_\theta(t)$ associated with (1.5.21) and $S(t)(\mathcal{M} \cap \mathcal{B}_o) \subset \mathcal{M} \cap \mathcal{B}_o$ for all $t \ge 0$.*

To weaken the crucial spectral gap condition (1.5.26), we work with a slightly different setting: assumption (ii) is replaced by

(ii)' F is Lipschitz on the bounded sets of $D(A^\alpha)$ with values in $D(A^{\alpha-\gamma})$, for some $\alpha \in \mathbb{R}$ and some γ, $0 \leq \gamma < \frac{1}{2}$, and (1.5.16) is replaced by a similar one with $\alpha - \frac{1}{2}$ replaced by $\alpha - \gamma$. The spectral gap condition (1.5.26) is replaced by

$$\frac{\Lambda - \lambda}{\Lambda^\gamma + \lambda^\gamma} > 8M_2 \frac{l+1}{l}. \tag{1.5.27}$$

Theorem 1.5.6 *Suppose the assumptions of Theorem 1.5.5 with (ii) replaced by (ii)' and (1.5.26) by (1.5.27), respectively, are satisfied. Then the conclusions of Theorem 1.5.5 are still valid.*

Nevertheless, the spectral gap condition is a severe restriction on applications of Theorem 1.5.5 and Theorem 1.5.6, in particular, to problems in higher space dimensions.

Recently, Eden, Foias, Nicolaenko & Temam [1], [2] introduced a notion of inertial sets which is similar to the notion of inertial manifolds except that an inertial set is, in general, not a manifold but a fractal set. In other words, an inertial set is a set of finite fractal dimension which attracts all solutions at an exponential rate. More precisely, let H be a separable Hilbert space and $\mathcal{B}$ a compact subset of H. Let $\{S(t)\}_{t \geq 0}$ be a nonlinear continuous semigroup that leaves $\mathcal{B}$ positively invariant and let $\mathcal{A} = \cap\{S(t)\mathcal{B};\ t \geq 0\}$, i.e., $\mathcal{A}$ is the global attractor for $\{S(t)\}_{t \geq 0}$ on $\mathcal{B}$.

Definition 1.5.10 *A set $\mathcal{M}$ is called an inertial set for $(\{S(t)\}_{t \geq 0},\ \mathcal{B})$ if*

(i) $\mathcal{A} \subseteq \mathcal{M} \subseteq \mathcal{B}$,

(ii) $S(t)\mathcal{M} \subseteq \mathcal{M}$ for every $t \geq 0$,

(iii) for every u_0 in $\mathcal{B}$, $dist_H(S(t)u_0,\ \mathcal{M}) \leq C_1 exp\{-C_2 t\}$ for all $t \geq 0$ with $C_1,\ C_2$ being positive constants independent of u_0, and

(iv) $\mathcal{M}$ has finite fractal dimension $d_F(\mathcal{M})$.

A result on the existence of inertial sets has been given in Eden, Foias, Nicolaenko & Temam [1]. Instead of the spectral gap condition for the existence of inertial manifolds, it requires the squeezing property defined as follows.

Definition 1.5.11 *A continuous semigroup $\{S(t)\}_{t \geq 0}$ on a separable Hilbert space H is said to satisfy the squeezing property on $\mathcal{B}$ if there exists $t_* > 0$ such that $S_* = S(t_*)$*

satisfies the following condition: there exists an orthogonal projection P of rank N_0 such that if for every u and v in $\mathcal{B}$

$$\|P(S_*u - S_*v)\| \leq \|(I - P)(S_*u - S_*v)\| \tag{1.5.28}$$

then

$$\|S_*u - S_*v\| \leq \frac{1}{8}\|u - v\|. \tag{1.5.29}$$

The following is the result by Eden, Fioas, Nicolaenko & Temam [1]:

Theorem 1.5.7 *If $(\{S(t)\}_{t\geq 0}, \mathcal{B})$ satisfies the squeezing property on $\mathcal{B}$ and if $S_* = S(t_*)$ is Lipschitz on $\mathcal{B}$ with Lipschitz constant L, then there exists an inertial set $\mathcal{M}$ for $(\{S(t)\}_{t\geq 0}, \mathcal{B})$ such that*

$$d_F(\mathcal{M}) \leq N_0 \max\left(1, \frac{\ln(16L + 1)}{\ln 2}\right) \tag{1.5.30}$$

and

$$dist_H(S(t)u_0, \mathcal{M}) \leq C_1 exp\{\frac{-C_2}{t_*}t\} \tag{1.5.31}$$

where $d_F(\mathcal{M})$ is the fractal dimension of $\mathcal{M}$.

These results will be applied in Chapter 5 to the phase-field equations.

1.6 Notation and Some Useful Inequalities

Throughout this book we use the following common notation.

1. In addition to the notation $\frac{\partial^k}{\partial t^k}$, $\frac{\partial^\alpha}{\partial x_1^{\alpha_1} \cdots \partial x_n^{\alpha_n}}$, we also use D_t^k, D^α to denote the corresponding partial derivatives, i.e., $D_t^k = \frac{\partial^k}{\partial t^k}$, $D^\alpha = \frac{\partial^\alpha}{\partial x_1^{\alpha_1} \cdots \partial x_n^{\alpha_n}}$ with an exception in Section 3.4. The subscripts t and x are often used to denote the partial derivatives with respect to t and x, respectively, i.e., $u_{tt} = \frac{\partial^2 u}{\partial t^2}$, $v_{xx} = \frac{\partial^2 v}{\partial x^2}$, etc.

2. We simply denote by $\|\cdot\|$ the L^2 norm of a function.

3. We often use $C, C_i (i \in I\!N)$ to denote a universal constant which may vary in different places.

In addition to the Nirenberg and Poincaré inequalities, the following elementary inequalities are very useful and will be frequently referred to in the remainder of the book (refer to Beckenbach & Bellman [1] for the proofs):

1. The Young inequality

Let a, b and ε be positive constants and $p,\, q \geq 1$, $\frac{1}{p} + \frac{1}{q} = 1$. Then

$$ab \leq \frac{\varepsilon^p a^p}{p} + \frac{b^q}{q\varepsilon^q}. \tag{1.6.1}$$

2. The Jensen inequality

Let $\varphi(u) : u \in [\alpha, \beta] \mapsto I\!R$ be a convex function. Suppose that $f : t \in [a, b] \mapsto [\alpha, \beta]$, and $P(t)$ are continuous functions with $P(t) \geq 0$, $P(t) \not\equiv 0$. Then the following inequality holds:

$$\varphi\left(\frac{\int_a^b f(t)P(t)\,dt}{\int_a^b P(t)\,dt} \right) \leq \frac{\int_a^b \varphi(f(t))P(t)\,dt}{\int_a^b P(t)\,dt}. \tag{1.6.2}$$

3. The Gronwall inequality

Suppose that a, b are nonnegative constants and $u(t)$ is a nonnegative integrable function. Suppose that the following inequality holds for $0 \leq t \leq T$:

$$u(t) \leq a + b \int_0^t u(s)\,ds. \tag{1.6.3}$$

Then for $0 \leq t \leq T$,

$$u(t) \leq ae^{bt}. \tag{1.6.4}$$

Chapter 2

Decay of Solutions to Linear Evolution Equations

In this chapter we establish the results on decay rates of solutions to both initial value problems and initial boundary value problems for linear parabolic equations and two classes of linear hyperbolic–parabolic coupled systems: linear one-dimensional thermoelastic systems and thermoviscoelastic systems.

In Section 2.1.1 the results on decay rates of solution to the initial value problem for the heat equation are obtained by using the Poisson formula and the Young inequality. It turns out that the decay rates depend on the space dimension n.

In Section 2.1.2 we are concerned with the initial boundary value problem for linear parabolic equations. Under the assumption that the elliptic operator in linear parabolic equations is self-adjoint and its first eigenvalue is strictly positive, we obtain the exponential decay of the solution to the linear parabolic equation using the results in semigroup theory. It is noteworthy that the decay rates do not depend on the space dimension n for the initial boundary value problem with bounded domains.

Getting decay rates of solutions to one-dimensional linear thermoelastic systems and thermoviscoelastic systems, which are important examples of linear hyperbolic–parabolic coupled systems, is more difficult.

In Sections 2.2.1 and 2.3.1 the initial value problems for linear one-dimensional thermoelastic systems and thermoviscoelastic systems are considered. We should mention that the linearized systems for the equations of radiation hydrodynamics and for the equations of motion of compressible viscous and heat-conductive fluids in one space dimension are just the same as the linear one-dimensional thermoelastic and thermoviscoelastic systems, respectively. Using the Fourier transform method and making delicate spectral analysis we obtain the decay rates of solutions to both systems. It is noteworthy that the decay rates are the same as in the heat equation. In other words, as far as the decay rate is concerned, for linear one-dimensional thermoelastic and thermoviscoelastic systems, the parabolic part in the systems is dominant. However, we should point out that it is not the case for higher-dimensional problems (see Racke [1–2] and the references cited there).

In Sections 2.2.2 and 2.3.2 the initial boundary value problems for both linear thermoelastic and thermoviscoelastic systems are considered. These systems can be equivalently converted to first-order evolution equations. In Section 2.2.2 we obtain the exponential stability of the corresponding C_o-semigroup using a theorem by Huang [1] and the PDE method. This technique recently developed by Liu and Zheng [1–5] (also see Burns, Liu & Zheng [1]) was also used to deal with a higher-dimensional problem: the initial boundary value problem for the Kirchhoff plate with thermal or viscous damping. However, concerning the initial boundary value problems for linear thermoelastic system in higher space dimensions the situation is more complicated and in general, as Dafermos [1] analysed, one cannot expect to obtain exponential decay.

In Section 2.3.2 we display another important method, namely the energy method, to obtain the exponential decay of solutions. Usually the results obtained by the energy method are such that the solution decays exponentially in the higher-order Sobolev spaces, a weaker statement than the exponential stability of the C_o-semigroup.

In Section 2.3.2 we establish a theorem (Theorem 2.3.2) to show that if the infinitesimal generator is invertible, then the exponential stability of the C_o-semigroup is equivalent to the exponential decay of the solution in $D(A)$ which is usually obtained by the energy method.

The results in this chapter will be used in Chapter 3 to obtain the global existence of solutions to the corresponding nonlinear evolution equations with small initial data.

2.1 Linear Parabolic Equations

2.1.1 Initial Value Problems

Consider the following initial value problem for the heat equation

$$\begin{cases} u_t - \Delta u = 0, & x \in I\!R^n, \ t > 0 \\ u|_{t=0} = \varphi(x), & x \in I\!R^n. \end{cases} \tag{2.1.1}$$

It is well known that for any $\varphi \in L^p(I\!R^n)$, $1 \le p \le \infty$, the function $u(x,t)$ given by the following Poisson formula:

$$u(x,t) = (4\pi t)^{-\frac{n}{2}} \int_{I\!R^n} e^{\frac{-|x-y|^2}{4t}} \varphi(y) dy \overset{\Delta}{=} \int_{I\!R^n} K(x-y,t)\varphi(y) dy \tag{2.1.2}$$

is C^∞ in $x \in I\!R^n$, $t > 0$, and satisfies the heat equation.

Moreover, we have

Lemma 2.1.1 *Suppose $\varphi \in L^p(I\!R^n)$, $1 \le p \le \infty$. Then we have for $t > 0$*

$$\|u(t)\|_{L^p} \le \|\varphi\|_{L^p}, \tag{2.1.3}$$

$$\|D^k u(t)\|_{L^q} \le C t^{-\left(\frac{n}{2r} + \frac{k}{2}\right)} \|\varphi\|_{L^p} \tag{2.1.4}$$

with $\frac{1}{q} = \frac{1}{p} - \frac{1}{r}, 1 \le r, q \le \infty$ and C being a positive constant depending only on p, q, r and k.

Proof. We notice that $K(x-y,t) = K(y-x,t)$ and

$$\int_{I\!R^n} K(x-y,t) dy = \int_{I\!R^n} K(x-y,t) dx = 1. \tag{2.1.5}$$

By the Hölder inequality, we have for $\frac{1}{p} + \frac{1}{p'} = 1$

$$|u(x,t)| \le \int_{I\!R^n} K(x-y,t)|\varphi(y)| dy$$

$$\le \left(\int_{I\!R^n} K(x-y,t)|\varphi(y)|^p dy \right)^{\frac{1}{p}} \left(\int_{I\!R^n} K(x-y,t) dy \right)^{\frac{1}{p'}}$$

$$= \left(\int_{I\!R^n} K(x-y,t)|\varphi|^p dy \right)^{\frac{1}{p}}. \tag{2.1.6}$$

Thus

$$\int_{\mathbb{R}^n} |u(x,t)|^p dx \le \int_{\mathbb{R}^n_x} \int_{\mathbb{R}^n_y} K(x-y,t)|\varphi(y)|^p dy dx$$

$$= \int_{\mathbb{R}^n_y} \left(\int_{\mathbb{R}^n_x} K(x-y,t) dx \right) |\varphi(y)|^p dy = \|\varphi\|_{L^p}^p \tag{2.1.7}$$

and (2.1.3) follows.

To prove (2.1.4), we use the following well-known Young inequality (see Reed & Simon [1]):

$$\|f * g\|_{L^q} = \| \int_{\mathbb{R}^n} f(x-y)g(y) dy \|_{L^q} \le \|f\|_{L^{r'}} \|g\|_{L^p} \tag{2.1.8}$$

with $\frac{1}{p} + \frac{1}{r'} = \frac{1}{q} + 1$, $1 \le p, q, r' \le \infty$. A straightforward calculation easily shows that

$$\|K(t)\|_{L^{r'}} \le C \left(\int_{\mathbb{R}^n} \frac{1}{t^{\frac{nr'}{2}}} e^{-\frac{|x|^2 r'}{4t}} dx \right)^{\frac{1}{r'}}$$

$$\le C t^{-\frac{n}{2}(1-\frac{1}{r'})} = C t^{-\frac{n}{2r}} \tag{2.1.9}$$

with $\frac{1}{r} + \frac{1}{r'} = 1$. Similarly, we have

$$\|D_{x_i} K(t)\|_{L^{r'}} \le C \left(\int_{\mathbb{R}^n} \frac{1}{t^{\frac{nr'}{2}}} \left(\frac{|x_i|}{t} \right)^{r'} e^{-\frac{|x|^2 r'}{4t}} dx \right)^{\frac{1}{r'}}$$

$$\le C t^{-\frac{n}{2r} - \frac{1}{2}} \tag{2.1.10}$$

and, in general,

$$\|D^k K(t)\|_{L^{r'}} \le C t^{-(\frac{n}{2r} + \frac{k}{2})}. \tag{2.1.11}$$

Combining (2.1.11) with the Young inequality yields (2.1.4). Thus the proof is completed. $\square$

Lemma 2.1.2 *Let $\varphi(x) \in C_o^\infty(\mathbb{R}^n)$ and k be a positive integer. Then $u(x,t)$ given by (2.1.2) satisfies the following estimates:*

$$\|D^k u\|_{L^1} \le C t^{-\frac{k}{2}} \|\varphi\|_{L^1}, \quad t > 0, \tag{2.1.12}$$

$$\|D^k u\| \le C t^{-\frac{n+2k}{4}} \|\varphi\|_{L^1}, \quad t > 0, \tag{2.1.13}$$

$$\|D^k u\|_{L^\infty} \le C t^{-\frac{n+k}{2}} \|\varphi\|_{L^1}, \quad t > 0, \tag{2.1.14}$$

$$\|D^k u\|_{L^1} \le C(1+t)^{-\frac{k}{2}} (\|\varphi\|_{L^1} + \|D^k \varphi\|_{L^1}), \quad t \ge 0, \tag{2.1.15}$$

$$\|D^k u\| \leq C(1+t)^{-\frac{n+2k}{4}}(\|\varphi\|_{L^1} + \|D^k \varphi\|), \quad t \geq 0, \qquad (2.1.16)$$

$$\|D^k u\| \leq C(1+t)^{-\frac{n+2k}{4}}\|\varphi\|_{W^{n+k,1}}, \quad t \geq 0, \qquad (2.1.17)$$

$$\|D^k u\| \leq C(1+t)^{-\frac{k}{2}}\|\varphi\|_{H^k}, \quad t \geq 0, \qquad (2.1.18)$$

$$\|D^k u\|_{L^\infty} \leq C(1+t)^{-\frac{n+k}{2}}(\|\varphi\|_{L^1} + \|D^k \varphi\|_{H^{[\frac{n}{2}]+1}}), \quad t \geq 0, \qquad (2.1.19)$$

$$\|D^k u\|_{L^\infty} \leq C(1+t)^{-\frac{n+k}{2}}\|\varphi\|_{W^{k+n,1}}, \quad t \geq 0, \qquad (2.1.20)$$

$$\|D^k u\|_{L^\infty} \leq C(1+t)^{-\frac{n+2k}{4}}\|\varphi\|_{H^{k+[\frac{n}{2}]+1}}, \quad t \geq 0. \qquad (2.1.21)$$

Proof. (2.1.12)–(2.1.14) directly follow from (2.1.4) and (2.1.15)–(2.1.21) follow from (2.1.3), (2.1.4) and the Sobolev imbedding theorem. □

2.1.2 Initial Boundary Value Problems

Consider the following initial boundary value problem:

$$u_t - \sum_{i,j=1}^{n} \frac{\partial}{\partial x_i}(a_{ij}(x)\frac{\partial u}{\partial x_j}) + c(x)u = 0, \quad (x,t) \in \Omega \times (0,\infty), \qquad (2.1.22)$$

$$u|_\Gamma = 0, \quad t > 0, \qquad (2.1.23)$$

$$u|_{t=0} = u_0(x), \quad x \in \Omega \qquad (2.1.24)$$

where $\Omega \subset \mathbb{R}^n$, for simplicity, is a bounded domain with C^∞ boundary Γ and $a_{i,j}(x), c(x) \geq 0$ are C^∞ functions in $\bar{\Omega}$ satisfying

$$\sum_{i,j=1}^{n} a_{i,j}(x)\xi_i\xi_j \geq \mu \sum_{i,j=1}^{n} \xi_i^2, \quad \mu > 0, \quad x \in \bar{\Omega}. \qquad (2.1.25)$$

Let

$$Au = -\sum_{i,j=1}^{n} \frac{\partial}{\partial x_i}(a_{ij}(x)\frac{\partial u}{\partial x_j}) + c(x)u \qquad (2.1.26)$$

with

$$D(A) = H^2 \cap H_o^1. \qquad (2.1.27)$$

Therefore, for $u \in D(A) \subset L^2$

$$(Au, u) = \int_\Omega \left(\sum_{i,j=1}^{n} a_{i,j}\frac{\partial u}{\partial x_j}\frac{\partial u}{\partial x_i} + c(x)u^2 \right) dx \geq 0. \qquad (2.1.28)$$

Thus $-A$ is a dissipative operator in L^2 and generates a C_o-semigroup $S(t)$. Moreover, as mentioned in Chapter 1, A is a sectorial operator and $S(t)$ is an analytic semigroup. For $s \in \mathbb{R}$, we can define $D(A^S)$.

Lemma 2.1.3 *Let $s \geq 0$ be an integer and $\lambda_1 > 0$ be the first eigenvalue of A. Then, for any $u_0 \in D(A^{S+\frac{1}{2}}) \subset H^{2S+1}$, there is a positive constant C depending only on μ, the coefficients and Ω such that*

$$\sum_{k=0}^{S} \|D_t^k u(t)\|_{H^{2(S-k)+1}} \leq Ce^{-\lambda_1 t}\|u_0\|_{H^{2S+1}}. \tag{2.1.29}$$

Proof. It follows from (2.1.19) that

$$\frac{1}{2}\frac{d}{dt}\|u\|_{D(A^{\frac{1}{2}})}^2 + \|Au\|^2 = 0. \tag{2.1.30}$$

Hence,

$$\frac{1}{2}\frac{d}{dt}\|u\|_{D(A^{\frac{1}{2}})}^2 + \lambda_1\|u\|_{D(A^{\frac{1}{2}})}^2 \leq 0, \tag{2.1.31}$$

$$\|u\|_{D(A^{\frac{1}{2}})}^2 \leq e^{-2\lambda_1 t}\|u_0\|_{D(A^{\frac{1}{2}})}^2. \tag{2.1.32}$$

Let

$$u_k = D_t^k u, \quad 0 \leq k \leq s. \tag{2.1.33}$$

Then u_k satisfies the prolonged system

$$\frac{du_k}{dt} + Au_k = 0, \tag{2.1.34}$$

$$u_k(0) = (-A)^k u_0. \tag{2.1.35}$$

In the same manner as before, we obtain for $0 \leq k \leq s$

$$\|D_t^k u(t)\|_{D(A^{\frac{1}{2}})} = \|u_k(t)\|_{D(A^{\frac{1}{2}})} \leq e^{-\lambda_1 t}\|A^k u_0\|_{D(A^{\frac{1}{2}})}. \tag{2.1.36}$$

Since $u_k = D_t^k u$ satisfies

$$Au_k = -D_t^{k+1}u, \quad (x,t) \in \Omega \times (0,\infty), \tag{2.1.37}$$

$$u_k|_\Gamma = 0, \quad t > 0. \tag{2.1.38}$$

For any fixed t, applying the regularity results for the above elliptic boundary value problem stated in Chapter 1, we obtain for $k = s - 1$,

$$\|u_{S-1}(t)\|_{H^3} \leq C\|D_t^S u\|_{H^1} \leq Ce^{-\lambda_1 t}\|A^S u_0\|_{H^1} \leq Ce^{-\lambda_1 t}\|u_0\|_{H^{2S+1}}, \tag{2.1.39}$$

and for $k = s - 2, \ldots, 0$ successively we have

$$\|u_k(t)\|_{H^{2(s-k)+1}} \leq C\|D_t^{k+1}u\|_{H^{2(s-k-1)+1}} \leq Ce^{-\lambda_1 t}\|u_0\|_{H^{2s+1}}. \tag{2.1.40}$$

Thus the proof is complete. $\square$

Remark 2.1.1 *For other initial boundary value problems with the Neumann or Robin boundary conditions the same conclusion holds provided that the corresponding first eigenvalue of A is positive.*

2.2 Linear Thermoelastic Systems

In this section we consider the following one-dimensional thermoelastic system

$$u_{tt} - \alpha^2 u_{xx} + \gamma_1 \theta_x = 0, \tag{2.2.1}$$

$$\theta_t + \gamma_2 u_{xt} - k\theta_{xx} = 0 \tag{2.2.2}$$

with $\alpha > 0, k > 0, \gamma_1, \gamma_2$ being positive constants. Physically, the function u represents the displacement of a rod and the function θ represents the temperarure. The first equation (2.2.1) is a hyperbolic equation for u and the second equation (2.2.2) is a parabolic equation for θ. They are coupled to each other. It is well known that for the linear wave equation the total energy is conserved for all time. On the other hand, as shown in the previous section, the energy function of solutions to the heat equation has a certain decay rate. In this section we show that for this particular hyperbolic–parabolic coupled system, namely the one-dimensional thermoelastic system, the solution has the same decay rate as for the heat equation. On the other hand we want to mention that this is not true for the higher-dimensional thermoelastic system (see Racke [1,2]). The material of this section is based on Zheng & Shen [3,4]. We also refer to Kawashima [1] and Matsumura [2].

2.2.1 Initial Value Problems

We first consider the initial value problem for (2.2.1)–(2.2.2):

$$u|_{t=0} = \varphi(x), \quad u_t|_{t=0} = \psi(x), \quad \theta|_{t=0} = \theta_0(x), \quad x \in \mathbb{R}. \tag{2.2.3}$$

Let

$$u_1 = \alpha u_x, \quad u_2 = u_t, \quad v = \sqrt{\frac{\gamma_2}{\gamma_1}}\,\theta. \tag{2.2.4}$$

Then equations (2.2.1)–(2.2.2) are reduced to the following first-order evolution system

$$u_{1t} - \alpha u_{2x} = 0, \tag{2.2.5}$$

$$u_{2t} - \alpha u_{1x} + \beta v_x = 0, \tag{2.2.6}$$

$$v_t + \beta u_{2x} - k v_{xx} = 0 \tag{2.2.7}$$

with

$$\beta = \gamma_1\sqrt{\frac{\gamma_2}{\gamma_1}} = \gamma_2\sqrt{\frac{\gamma_1}{\gamma_2}}. \tag{2.2.8}$$

The initial conditions (2.2.3) turn out to be

$$u_1|_{t=0} = u_1^0(x) = \alpha\varphi_x, \quad u_2|_{t=0} = u_2^0(x) = \psi, \quad v|_{t=0} = v^0(x) = \sqrt{\frac{\gamma_2}{\gamma_1}}\,\theta_0. \tag{2.2.9}$$

Introduce

$$U = (u_1,\, u_2,\, v)^T. \tag{2.2.10}$$

Then (2.2.5)–(2.2.7) and (2.2.9) can be rewritten as

$$U_t = AU, \tag{2.2.11}$$

$$U|_{t=0} = U_0(x) = (u_1^0,\, u_2^0,\, v^0)^T \tag{2.2.12}$$

with

$$A = \begin{pmatrix} 0 & \alpha D & 0 \\ \alpha D & 0 & -\beta D \\ 0 & -\beta D & k D^2 \end{pmatrix}. \tag{2.2.13}$$

Here we have used the notation $D^i = \dfrac{\partial^i}{\partial x^i}$.

In the following Theorems 2.2.1–2.2.3 we give the decay rates of L^p norms ($p = 1, 2, \infty$) of the solution U.

Theorem 2.2.1 *Suppose $U_0(x) \in W^{2,1}(\mathrm{IR})$. Then for $t \geq 0$, the solution $U(x,t)$ to problem (2.2.11)–(2.2.12) satisfies the following estimates:*

$$\|U(t)\|_{L^1} \leq C\|U_0\|_{L^1}, \tag{2.2.14}$$

$$\|DU(t)\|_{L^1} \le C(1+t)^{-\frac{1}{2}}\|U_0\|_{W^{1,1}}, \tag{2.2.15}$$

$$\|U(t)\|_{L^\infty} \le C(1+t)^{-\frac{1}{2}}\|U_0\|_{W^{1,1}}, \tag{2.2.16}$$

$$\|U(t)\| \le C(1+t)^{-\frac{1}{4}}\|U_0\|_{W^{1,1}}, \tag{2.2.17}$$

$$\|D^2U(t)\|_{L^1} \le C(1+t)^{-1}\|U_0\|_{W^{2,1}}, \tag{2.2.18}$$

$$\|DU(t)\|_{L^\infty} \le C(1+t)^{-1}\|U_0\|_{W^{2,1}}. \tag{2.2.19}$$

From now on we denote by C a positive constant depending only on the coefficients α, β and k

Proof. The basic strategy of the proof is to use the Fourier transform with respect to x.

Throughout this book we denote the Fourier transform of a function (or a vector function) f by $\hat{f}(\xi)$ or $\mathcal{F}(f)$:

$$\hat{f}(\xi) = \mathcal{F}(f) = \int_{-\infty}^{\infty} e^{-ix\xi} f(x)dx. \tag{2.2.20}$$

We also denote the inverse Fourier transform by $\mathcal{F}^{-1}$.

Taking the Fourier transform for (2.2.11)–(2.2.12), we obtain

$$\hat{U}_t = A\hat{U}, \tag{2.2.21}$$

$$\hat{U}|_{t=0} = \hat{U}_0 \tag{2.2.22}$$

with

$$A = \begin{pmatrix} 0 & i\alpha\xi & 0 \\ i\alpha\xi & 0 & -i\beta\xi \\ 0 & -i\beta\xi & -k\xi^2 \end{pmatrix}. \tag{2.2.23}$$

Let λ_i $(i = 1, 2, 3)$ be the characteristic roots of $det(\lambda I - A) = 0$. Then they satisfy the following equation

$$\lambda^3 + k\xi^2\lambda^2 + (\alpha^2 + \beta^2)\xi^2\lambda + k\alpha^2\xi^4 = 0. \tag{2.2.24}$$

The solution to equation (2.2.24) can be explicitly given by

$$
\begin{cases}
\lambda_1(\xi) = P + Q - \frac{k}{3}\xi^2, \\[2mm]
\lambda_2(\xi) = -\frac{1}{2}(P+Q) + \frac{\sqrt{3}}{2}(P-Q)i - \frac{k}{3}\xi^2, \\[2mm]
\lambda_3(\xi) = -\frac{1}{2}(P+Q) - \frac{\sqrt{3}}{2}(P-Q)i - \frac{k}{3}\xi^2
\end{cases}
\tag{2.2.25}
$$

with

$$
P = \left(-\frac{q}{2} + \sqrt{\Delta}\right)^{\frac{1}{3}}, \quad Q = \left(-\frac{q}{2} - \sqrt{\Delta}\right)^{\frac{1}{3}}
\tag{2.2.26}
$$

and

$$
\Delta = \frac{q^2}{4} + \frac{p^3}{27}, \quad p = \frac{3(\alpha^2 + \beta^2)\xi^2 - k^2\xi^4}{3},
\tag{2.2.27}
$$

$$
\tag{2.2.28}
$$

$$
q = \frac{2k^3\xi^6 + (27\alpha^2 - 9(\alpha^2 + \beta^2))k\xi^4}{27}.
\tag{2.2.29}
$$

A straightforward calculation shows (see Zheng & Shen [3], also see Matsumura [2], Kim [1])

(i) $\lambda_i(0) = 0$ and $\lambda_i(\xi)\,(i = 1, 2, 3)$ possess the following properties:
As $\xi \to 0$,

$$
\begin{cases}
\lambda_1 = -\frac{k\alpha^2}{\alpha^2 + \beta^2}\xi^2 + O(\xi^3), \\[3mm]
\lambda_2 = -\frac{k\beta^2}{2(\alpha^2 + \beta^2)}\xi^2 + i\sqrt{\alpha^2 + \beta^2}\,\xi + O(\xi^3), \\[3mm]
\lambda_3 = \bar{\lambda}_2 = -\frac{k\beta^2}{2(\alpha^2 + \beta^2)}\xi^2 - i\sqrt{\alpha^2 + \beta^2}\,\xi + O(\xi^3),
\end{cases}
\tag{2.2.30}
$$

$$
\left|\frac{d\lambda_1}{d\xi}\right| = O(\xi), \quad \left|\frac{d\lambda_2}{d\xi}\right| = O(1), \quad \left|\frac{d\lambda_3}{d\xi}\right| = O(1).
\tag{2.2.31}
$$

As $\xi \to \infty$,

$$
\begin{cases}
\lambda_1 = -k\xi^2 + \frac{\beta^2}{k} + \frac{\alpha_1}{k^3}\xi^{-2} + O(\xi^{-3}), \\[3mm]
\lambda_2 = -\frac{\beta^2}{2k} - \frac{\alpha_1}{3k^3}\xi^{-2} + O(\xi^{-4}) + i(\alpha\xi + \frac{\alpha_2}{k^2}\xi^{-1} + O(\xi^{-3})), \\[3mm]
\lambda_3 = -\frac{\beta^2}{2k} - \frac{\alpha_1}{3k^3}\xi^{-2} + O(\xi^{-4}) - i(\alpha\xi + \frac{\alpha_2}{k}\xi^{-1} + O(\xi^{-3})),
\end{cases}
\tag{2.2.32}
$$

$$\frac{d\lambda_1}{d\xi} + 2k\xi = O(\xi^{-3}), \quad \frac{d\lambda_2}{d\xi} - i\alpha = O(\xi^{-2}), \quad \frac{d\lambda_3}{d\xi} + i\alpha = O(\xi^{-2}) \qquad (2.2.33)$$

where

$$\alpha_1 = (45\alpha^2 - \beta^2)(\alpha^2 - 3\beta^2), \quad \alpha_2 = \frac{-88\alpha^4 + 28\alpha^2\beta^2 - \beta^4}{8\alpha}. \qquad (2.2.34)$$

(ii) $\lambda_j \neq \lambda_k, (j \neq k)$, except for at most two values of ξ with $|\xi| > 0$.

Furthermore, we claim that

(i) For any ξ, $|\xi| > 0$, $Re\,\lambda_i(\xi) < 0\,(i = 1, 2, 3)$.

(ii) There are positive constants δ_1, δ_2 and $C_i\,(i = 1, 2, 3, 4)$ depending on δ_1, δ_2 such that for $i = 1, 2, 3$,

$$\begin{cases} as \ |\xi| \leq \delta_1, & -C_2|\xi|^2 \leq Re\,\lambda_i(\xi) \leq -C_1|\xi|^2, \\ as \ \delta_1 \leq |\xi| \leq \delta_2, & Re\,\lambda_i(\xi) \leq -C_3, \\ as \ |\xi| \geq \delta_2, & Re\,\lambda_i(\xi) \leq -C_4. \end{cases} \qquad (2.2.35)$$

To prove claim (i), owing to (2.2.30), (2.2.32), it suffices to show that for any $\xi \neq 0$, the characteristic equation (2.2.24) has no pure imaginary root. If it is not true, then there is a real number η such that $i\eta$ satisfies (2.2.24). It turns out that

$$\begin{cases} \eta^3 - (\alpha^2 + \beta^2)\xi^2\eta = 0, \\ k\xi^2\eta^2 - k\alpha^2\xi^4 = 0. \end{cases} \qquad (2.2.36)$$

But (2.2.36) can hold only for $\xi = 0$, a contradiction.

Claim (ii) follows from claim (i), (2.2.30) and (2.2.32). Thus, except for at most two values of ξ, $|\xi| > 0$, the solution $\widehat{U}$ to the initial value problem (2.2.21)–(2.2.22) can be expressed as

$$\widehat{U}(\xi, t) = \widehat{G}(\xi, t)\widehat{U}_0(\xi) \qquad (2.2.37)$$

where $\widehat{G}$ is a 3×3 matrix function with

$$\widehat{G}_{11} = \frac{\lambda_2\lambda_3(\lambda_1 + k\xi^2)}{k\xi^2(\lambda_1 - \lambda_2)(\lambda_1 - \lambda_3)}e^{\lambda_1 t} + \frac{\lambda_1\lambda_3(\lambda_2 + k\xi^2)}{k\xi^2(\lambda_2 - \lambda_1)(\lambda_2 - \lambda_3)}e^{\lambda_2 t}$$
$$+ \frac{\lambda_1\lambda_2(\lambda_3 + k\xi^2)}{k\xi^2(\lambda_3 - \lambda_1)(\lambda_3 - \lambda_2)}e^{\lambda_3 t}, \qquad (2.2.38)$$

$$\widehat{G}_{12} = \widehat{G}_{21} = \frac{i\alpha\xi(\lambda_1 + k\xi^2)}{(\lambda_1 - \lambda_2)(\lambda_1 - \lambda_3)}e^{\lambda_1 t} + \frac{i\alpha\xi(\lambda_2 + k\xi^2)}{(\lambda_2 - \lambda_1)(\lambda_2 - \lambda_3)}e^{\lambda_2 t}$$

$$+\frac{i\alpha\xi(\lambda_3 + k\xi^2)}{(\lambda_3 - \lambda_1)(\lambda_3 - \lambda_2)}e^{\lambda_3 t}, \tag{2.2.39}$$

$$\widehat{G}_{13} = \widehat{G}_{31} = \frac{\alpha\beta\xi^2}{(\lambda_1 - \lambda_2)(\lambda_1 - \lambda_3)}e^{\lambda_1 t} + \frac{\alpha\beta\xi^2}{(\lambda_2 - \lambda_1)(\lambda_2 - \lambda_3)}e^{\lambda_2 t}$$
$$+\frac{\alpha\beta\xi^2}{(\lambda_3 - \lambda_1)(\lambda_3 - \lambda_2)}e^{\lambda_3 t}, \tag{2.2.40}$$

$$\widehat{G}_{22} = \frac{\lambda_1(\lambda_1 + k\xi^2)}{(\lambda_1 - \lambda_2)(\lambda_1 - \lambda_3)}e^{\lambda_1 t} + \frac{\lambda_2(\lambda_2 + k\xi^2)}{(\lambda_2 - \lambda_1)(\lambda_2 - \lambda_3)}e^{\lambda_2 t}$$
$$+\frac{\lambda_3(\lambda_3 + k\xi^2)}{(\lambda_3 - \lambda_1)(\lambda_3 - \lambda_2)}e^{\lambda_3 t}, \tag{2.2.41}$$

$$\widehat{G}_{23} = \widehat{G}_{32} = \frac{-i\beta\lambda_1\xi}{(\lambda_1 - \lambda_2)(\lambda_1 - \lambda_3)}e^{\lambda_1 t} + \frac{-i\beta\lambda_2\xi}{(\lambda_2 - \lambda_1)(\lambda_2 - \lambda_3)}e^{\lambda_2 t}$$
$$+\frac{-i\beta\lambda_3\xi}{(\lambda_3 - \lambda_1)(\lambda_3 - \lambda_2)}e^{\lambda_3 t}, \tag{2.2.42}$$

$$\widehat{G}_{33} = \frac{-\lambda_1(\lambda_2 + k\xi^2)(\lambda_3 + k\xi^2)}{k\xi^2(\lambda_1 - \lambda_2)(\lambda_1 - \lambda_3)}e^{\lambda_1 t} + \frac{-\lambda_2(\lambda_1 + k\xi^2)(\lambda_3 + k\xi^2)}{k\xi^2(\lambda_2 - \lambda_1)(\lambda_2 - \lambda_3)}e^{\lambda_2 t}$$
$$+\frac{-\lambda_3(\lambda_1 + k\xi^2)(\lambda_2 + k\xi^2)}{k\xi^2(\lambda_3 - \lambda_1)(\lambda_3 - \lambda_2)}e^{\lambda_3 t}. \tag{2.2.43}$$

Moreover, $\widehat{G}_{ij}(\xi, t)$ are C^∞ functions in $\mathbb{R} \times (0, \infty)$.

In order to obtain the L^1 norm estimate of U, we use the technique shown by Lemma 2.2.1 and Lemma 2.2.2. Suppose that $\widehat{f}(\xi, t)$, the Fourier transform of $f(x, t)$, is expressed as the product of the Fourier transforms of two functions $\widehat{g}$ and $\widehat{f}_0$:

$$\widehat{f}(\xi, t) = \widehat{g}(\xi, t)\widehat{f}_0(\xi). \tag{2.2.44}$$

Then we have

Lemma 2.2.1 *Suppose that* $\widehat{g}, \frac{\partial \widehat{g}}{\partial \xi}, \xi\widehat{g}, \frac{\partial}{\partial \xi}(\xi\widehat{g}) \in L^2(\mathbb{R}_\xi)$, $f_0 \in L^1(\mathbb{R}_x)$. *Then for* $t \geq 0$ *the following estimates hold:*

$$\|f(t)\|_{L^1} \leq \left(\sqrt{2T}\|\widehat{g}\|_{L^2(\mathbb{R}_\xi)} + \sqrt{\frac{2}{T}}\left\|\frac{\partial}{\partial \xi}\widehat{g}\right\|_{L^2(\mathbb{R}_\xi)}\right)\|f_0\|_{L^1}, \tag{2.2.45}$$

$$\|Df(t)\|_{L^1} \leq \left(\sqrt{2T}\|\xi\widehat{g}\|_{L^2(\mathbb{R}_\xi)} + \sqrt{\frac{2}{T}}\left\|\frac{\partial}{\partial \xi}(\xi\widehat{g})\right\|_{L^2(\mathbb{R}_\xi)}\right)\|f_0\|_{L^1} \tag{2.2.46}$$

with T *being an arbitrary positive constant.*

Proof. By the basic property of the Fourier transform we have

$$f(x,t) = g(x,t) * f_0(x) = \int_{\mathbb{R}} g(x-y,t)f_0(y)dy. \tag{2.2.47}$$

Thus

$$\|f(t)\|_{L^1} \le \|g(t)\|_{L^1}\|f_0\|_{L^1}, \tag{2.2.48}$$

$$\|Df(t)\|_{L^1} \le \|Dg(t)\|_{L^1}\|f_0\|_{L^1}. \tag{2.2.49}$$

On the other hand, we have

$$\|g(t)\|_{L^1} = \int_{|x| \le T} |g(x,t)|dx + \int_{|x| \ge T} \frac{1}{|x|}|xg(x,t)|dx$$

$$\le \sqrt{2T}\left(\int_{|x| \le T}|g(x,t)|^2dx\right)^{\frac{1}{2}} + \left(\int_{|x| \ge T}\frac{1}{|x|^2}dx\right)^{\frac{1}{2}}\left(\int_{|x| \ge T}|xg|^2dx\right)^{\frac{1}{2}}$$

$$\le \sqrt{2T}\|\hat{g}(t)\|_{L^2(\mathbb{R}_\xi)} + \sqrt{\frac{2}{T}}\left\|\frac{\partial}{\partial \xi}\hat{g}(t)\right\|_{L^2(\mathbb{R}_\xi)}. \tag{2.2.50}$$

Combining (2.2.48) and (2.2.50) yields (2.2.45). The proof of (2.2.46) can be carried out in the same manner. Thus the proof of Lemma 2.2.1 is completed. $\square$

Corollary 2.2.1 *Suppose that there are functions $\hat{h}(\xi,t)$, $\widehat{M}(\xi,t)$ such that $\hat{g} - \hat{h}$, $\frac{\partial}{\partial \xi}(\hat{g}-\hat{h})$, $\xi\hat{g} - \widehat{M}$, $\frac{\partial}{\partial \xi}(\xi\hat{g} - \widehat{M}) \in L^2(\mathbb{R}_\xi)$ and $\|\mathcal{F}^{-1}(\hat{h}\hat{f}_0)\|_{L^1}$, $\|\mathcal{F}^{-1}(\widehat{M}\hat{f}_0)\|_{L^1} < \infty$. Then the following estimates hold:*

$$\|f\|_{L^1} \le \left(\sqrt{2T}\|\hat{g} - \hat{h}\|_{L^2} + \sqrt{\frac{2}{T}}\left\|\frac{\partial}{\partial \xi}(\hat{g} - \hat{h})\right\|_{L^2}\right)\|f_0\|_{L^1}$$

$$+\|\mathcal{F}^{-1}(\hat{h}\hat{f}_0)\|_{L^1}, \tag{2.2.51}$$

$$\|Df\|_{L^1} \le \left(\sqrt{2T}\|\xi\hat{g} - \widehat{M}\|_{L^2} + \sqrt{\frac{2}{T}}\left\|\frac{\partial}{\partial \xi}(\xi\hat{g} - \widehat{M})\right\|_{L^2}\right)\|f_0\|_{L^1}$$

$$+\|\mathcal{F}^{-1}(\widehat{M}\hat{f}_0)\|_{L^1}. \tag{2.2.52}$$

This can be seen by setting $\hat{f} = (\hat{g} - \hat{h})\hat{f}_0 + \hat{h}\hat{f}_0$, $\xi\hat{f} = (\xi\hat{g} - \widehat{M})\hat{f}_0 + \widehat{M}\hat{f}_0$. $\square$

Lemma 2.2.2 *Suppose that $u_0 \in W^{1,1}(\mathbb{R})$ and $\alpha, \beta, k, \gamma, \mu$ are constants with $\beta \ne 0, \mu, k > 0$. Then for $t \ge 0$ the following estimates hold:*

$$\|\mathcal{F}^{-1}(e^{-\frac{\beta^2}{2k}t}\cos\alpha\xi t\,\hat{u}_0(\xi))\|_{L^1} \le Ce^{-Ct}\|u_0\|_{L^1}, \tag{2.2.53}$$

$$\|\mathcal{F}^{-1}(e^{-\frac{\beta^2}{2k}t}\sin\alpha\xi t\,\hat{u}_0(\xi))\|_{L^1} \le Ce^{-Ct}\|u_0\|_{L^1}, \tag{2.2.54}$$

$$\|\mathcal{F}^{-1}(e^{-\frac{\beta^2}{2k}t}\xi\cos\alpha\xi t\,\hat{u}_0(\xi))\|_{L^1} \le Ce^{-Ct}\|Du_0\|_{L^1}, \tag{2.2.55}$$

$$\|\mathcal{F}^{-1}(e^{-\frac{\beta^2}{2k}t}\xi\sin\alpha\xi t\,\hat{u}_0(\xi))\|_{L^1} \le Ce^{-Ct}\|Du_0\|_{L^1}, \tag{2.2.56}$$

$$\|\mathcal{F}^{-1}(e^{-\mu\xi^2 t}\cos\gamma\xi t\,\hat{u}_0(\xi))\|_{L^1} \le C\|u_0\|_{L^1}, \tag{2.2.57}$$

$$\|\mathcal{F}^{-1}(\xi e^{-\mu\xi^2 t}\cos\gamma\xi t\,\hat{u}_0(\xi))\|_{L^1} \le C(1+t)^{-\frac{1}{2}}\|u_0\|_{W^{1,1}}, \tag{2.2.58}$$

and for $t \ge 1$,

$$\|\mathcal{F}^{-1}(\xi e^{-\mu\xi^2 t}\cos\gamma\xi t\,\hat{u}_0(\xi))\|_{L^1} \le C(1+t)^{-\frac{1}{2}}\|u_0\|_{L^1} \tag{2.2.59}$$

where C is a positive constant.

Proof. (2.2.53) and (2.2.55) follows from the inverse Fourier transform

$$\mathcal{F}^{-1}(\cos\alpha\xi t\,\hat{u}_0(\xi)) = \frac{1}{2}(u_0(x+\alpha t) + u_0(x-\alpha t)), \tag{2.2.60}$$

$$\mathcal{F}^{-1}(i\xi\cos\alpha\xi t\,\hat{u}_0(\xi)) = \frac{1}{2}D(u_0(x+\alpha t) + u_0(x-\alpha t)). \tag{2.2.61}$$

The proof of (2.2.54) and (2.2.56) can be done in the same way. (2.2.57)–(2.2.59) can be easily derived from the inverse Fourier transform

$$\mathcal{F}^{-1}(e^{-\mu\xi^2 t}\cos\gamma\xi t\,\hat{u}_0(\xi))$$
$$= \frac{1}{4\sqrt{\pi\mu t}}\int_{\mathrm{IR}}\left(e^{-\frac{(x-\xi+\gamma t)^2}{4\mu t}} + e^{-\frac{(x-\xi-\gamma t)^2}{4\mu t}}\right)u_0(\xi)d\xi, \tag{2.2.62}$$

$$\mathcal{F}^{-1}(i\xi e^{-\mu\xi^2 t}\cos\gamma\xi t\,\hat{u}_0(\xi))$$
$$= \frac{\partial}{\partial x}\frac{1}{4\sqrt{\pi\mu t}}\int_{\mathrm{IR}}\left(e^{-\frac{(x-\xi+\gamma t)^2}{4\mu t}} + e^{-\frac{(x-\xi-\gamma t)^2}{4\mu t}}\right)u_0(\xi)d\xi. \tag{2.2.63}$$

Thus the proof of Lemma 2.2.2 is completed.   □

We now continue to prove Theorem 2.2.1. It turns out from (2.2.37) that we only need to discuss the properties of $\hat{G}_{ij}$. Let

$$\hat{G}_{ij} = a_{ij}e^{\lambda_1 t} + b_{ij}e^{\lambda_2 t} + c_{ij}e^{\lambda_3 t}. \tag{2.2.64}$$

A straightforward calculation shows that

as $\xi \to 0$,

$$
\begin{cases}
a_{11} - \dfrac{\beta^2}{\alpha^2 + \beta^2} = O(\xi^2), \quad b_{11} - \dfrac{\alpha^2}{2(\alpha^2 + \beta^2)} = O(\xi), \\[3mm]
c_{11} - \dfrac{\alpha^2}{2(\alpha^2 + \beta^2)} = O(\xi).
\end{cases}
\tag{2.2.65}
$$

as $\xi \to \infty$,

$$
a_{11} = O(\xi^{-4}), \; b_{11} - \frac{1}{2} = O(\xi^{-2}), \; c_{11} - \frac{1}{2} = O(\xi^{-2}).
\tag{2.2.66}
$$

Let

$$
\hat{I}_{11} = e^{-\frac{\beta^2}{2k}t} \cos \alpha\xi t,
\tag{2.2.67}
$$

$$
\hat{Z}_{11} = \frac{\beta^2}{\alpha^2 + \beta^2} e^{-\frac{k\alpha^2}{\alpha^2+\beta^2}\xi^2 t} + \frac{\alpha^2}{\alpha^2 + \beta^2} e^{-\frac{k\beta^2}{2(\alpha^2+\beta^2)}\xi^2 t} \cos \sqrt{\alpha^2 + \beta^2}\xi t
\tag{2.2.68}
$$

and

$$
\hat{h}_{11} = \hat{I}_{11} + \hat{Z}_{11}.
\tag{2.2.69}
$$

It is easy to see from (2.2.30)–(2.2.33) and (2.2.65)–(2.2.66) that when $\xi \to 0$, the major part of $\hat{G}_{11}$ is $\hat{Z}_{11}$ and when $\xi \to \infty$, the major part of $\hat{G}_{11}$ is $\hat{I}_{11}$. Let $r_1 > 0$ be a small number and $r_2 > 0$ be a large number. It follows from (2.2.30), (2.2.65) that

$$
\int_{|\xi| \leq r_1} |\hat{G}_{11} - \hat{h}_{11}|^2 d\xi
$$
$$
\leq C \left(\int_{|\xi| \leq r_1} \xi^2 e^{-C\xi^2 t} d\xi + \int_{|\xi| \leq r_1} \xi^6 t^2 e^{-C\xi^2 t} d\xi + \int_{|\xi| \leq r_1} |\hat{I}_{11}|^2 d\xi \right).
\tag{2.2.70}
$$

Thus for $t \geq 1$, we have

$$
\int_{|\xi| \leq r_1} |\hat{G}_{11} - \hat{h}_{11}|^2 d\xi \leq C(1 + t)^{-\frac{3}{2}} + C e^{-Ct}.
\tag{2.2.71}
$$

It can be easily seen from (2.2.38), (2.2.67)–(2.2.69) that

$$
\int_{r_1 \leq |\xi| \leq r_2} |\hat{G}_{11} - \hat{h}_{11}|^2 d\xi \leq C e^{-Ct}.
\tag{2.2.72}
$$

Similarly, by (2.2.32) and (2.2.66) we have

$$
\int_{|\xi| \geq r_2} |\hat{G}_{11} - \hat{h}_{11}|^2 d\xi \leq C \int_{|\xi| \geq r_2} \xi^{-2} e^{-Ct} d\xi + \int_{|\xi| \geq r_2} |\hat{Z}_{11}|^2 d\xi \leq C e^{-Ct}.
\tag{2.2.73}
$$

Thus, finally we arrive at

$$\|\widehat{G}_{11} - \widehat{h}_{11}\| \leq C(1+t)^{-\frac{3}{4}}, \quad \forall t \geq 1. \tag{2.2.74}$$

It is easy to see that the same estimate (2.2.74) holds for $0 \leq t \leq 1$. The following estimate can be carried out in the same way:

$$\|\frac{\partial}{\partial \xi}(\widehat{G}_{11} - \widehat{h}_{11})\| \leq C(1+t)^{\frac{1}{4}}. \tag{2.2.75}$$

Applying Corollary 2.2.1 with $T = (1+t)$, we obtain

$$\|G_{11} * u_0\|_{L^1} \leq C \left(\sqrt{2T}(1+t)^{-\frac{3}{4}} + \sqrt{\frac{2}{T}}(1+t)^{\frac{1}{4}} \right) \|u_0\|_{L^1} + \|\mathcal{F}^{-1}(\widehat{h}_{11}\widehat{u}_0)\|_{L^1}$$

$$\leq C(1+t)^{-\frac{1}{4}}\|u_0\|_{L^1} + \|\mathcal{F}^{-1}(\widehat{h}_{11}\widehat{u}_0)\|_{L^1}, \quad t \geq 1. \tag{2.2.76}$$

For $0 \leq t \leq 1$, it follows from (2.2.30)–(2.2.32), (2.2.65)–(2.2.67) that

$$\int_{\mathbb{R}} |\widehat{G}_{11} - \widehat{I}_{11}|^2 d\xi \leq C, \quad \int_{\mathbb{R}} |\frac{\partial}{\partial \xi}|\widehat{G}_{11} - \widehat{I}_{11}|^2 d\xi \leq C. \tag{2.2.77}$$

Applying Lemma 2.2.1 with $T = 1$ yields

$$\|G_{11} * u_0\|_{L^1} \leq C\|u_0\|_{L^1} + \|\mathcal{F}^{-1}(\widehat{I}_{11}\widehat{u}_0)\|_{L^1}. \tag{2.2.78}$$

Combining (2.2.76) with (2.2.78) and applying Lemma 2.2.2, we get

$$\|G_{11} * u_0\|_{L^1} \leq C\|u_0\|_{L^1}, \quad t \geq 0. \tag{2.2.79}$$

To prove (2.2.15) , we introduce

$$\widehat{M}_{11} = \xi(\widehat{I}_{11} + \widehat{Z}_{11}) - \frac{\alpha_2}{k^2}te^{-\frac{\beta^2}{2k}t}\sin\alpha\xi t. \tag{2.2.80}$$

In the same way as above we can obtain for $t \geq 1$

$$\int_{|\xi|\leq r_1} |\xi\widehat{G}_{11} - \widehat{M}_{11}|^2 d\xi \leq C(1+t)^{-\frac{5}{2}}, \tag{2.2.81}$$

$$\int_{r_1\leq|\xi|\leq r_2} |\xi\widehat{G}_{11} - \widehat{M}_{11}|^2 d\xi \leq Ce^{-Ct}, \tag{2.2.82}$$

$$\int_{|\xi|\geq r_2} |\xi\widehat{G}_{11} - \widehat{M}_{11}|^2 d\xi \leq Ce^{-Ct}, \tag{2.2.83}$$

$$\int_{|\xi|\leq r_1} |\frac{\partial}{\partial\xi}(\xi\widehat{G}_{11} - \widehat{M}_{11})|^2 d\xi \leq C(1+t)^{-\frac{1}{2}}, \tag{2.2.84}$$

$$\int_{r_1\leq|\xi|\leq r_2} |\frac{\partial}{\partial\xi}(\xi\widehat{G}_{11} - \widehat{M}_{11})|^2 d\xi \leq Ce^{-Ct}, \tag{2.2.85}$$

$$\int_{|\xi|\geq r_2} |\frac{\partial}{\partial\xi}(\xi\widehat{G}_{11} - \widehat{M}_{11})|^2 d\xi \leq Ce^{-Ct}. \tag{2.2.86}$$

Applying Corollary 2.2.1 again with $T = 1+t$ and Lemma 2.2.2, we obtain for $t \geq 1$

$$\|\frac{\partial}{\partial x}(G_{11} * u_0)\|_{L^1} \leq C(1+t)^{-\frac{3}{4}}\|u_0\|_{L^1} + \|\mathcal{F}^{-1}(\widehat{M}_{11}u_0)\|_{L^1}$$
$$\leq C(1+t)^{-\frac{1}{2}}\|u_0\|_{W^{1,1}}. \tag{2.2.87}$$

For $t \leq 1$, we define

$$\widehat{N}_{11} = \xi\widehat{I}_{11} - \frac{\alpha_2}{k^2}te^{-\frac{\beta^2}{2k}t}\sin\alpha\xi t \tag{2.2.88}$$

and in the same way as before we can obtain

$$\|\xi\widehat{G}_{11} - \widehat{N}_{11}\|_{L^2} \leq C, \quad \|\frac{\partial}{\partial\xi}(\xi\widehat{G}_{11} - \widehat{N}_{11})\|_{L^2} \leq C, \quad 0 \leq t \leq 1. \tag{2.2.89}$$

Applying Corollary 2.2.1 again with $T = 1$ and Lemma 2.2.2 yields

$$\|\frac{\partial}{\partial x}(G_{11} * u_0)\|_{L^1} \leq C\left(\|u_0\|_{L^1} + \|\mathcal{F}^{-1}(\widehat{N}_{11}\widehat{u}_0)\|_{L^1}\right) \leq C\|u_0\|_{W^{1,1}}, \quad t \leq 1. \tag{2.2.90}$$

Thus we can deduce from (2.2.87), (2.2.90) that

$$\|\frac{\partial}{\partial x}(G_{11} * u_0)\|_{L^1} \leq C(1+t)^{-\frac{1}{2}}\|u_0\|_{W^{1,1}}, \quad for\ t \geq 0. \tag{2.2.91}$$

For other $G_{ij} * u_0$, similar estimates can be obtained in the same manner.

Thus (2.2.14)–(2.2.15) have been proved. Estimates (2.2.16)–(2.2.17) can be easily obtained from (2.2.14)–(2.2.15). (2.2.18) can be estimated in the same manner as (2.2.15), while (2.2.19) is deduced from (2.2.18). Thus the proof of Theorem 2.2.1 is complete. □

Let

$$V_{ij}(u_0) = \mathcal{F}^{-1}(\widehat{Z}_{ij}\widehat{u}_0). \tag{2.2.92}$$

A typical expression $\hat{Z}_{11}$ has been given by (2.2.68). Straightforward calculation of the inverse Fourier transform shows that

$$V_{11}(u_0) = \mathcal{F}^{-1}(\hat{Z}_{11}\hat{u}_0) = Z_{11} * u_0 \qquad (2.2.93)$$

where $*$ denotes convolution and

$$Z_{11} = \frac{\beta^2}{2\alpha\sqrt{k\pi(\alpha^2+\beta^2)t}}e^{-\frac{(\alpha^2+\beta^2)x^2}{4k\alpha^2 t}} + \frac{\alpha^2}{2\beta\sqrt{2(\alpha^2+\beta^2)k\pi t}}$$
$$\times \left(e^{-\frac{(\alpha^2+\beta^2)(x-\sqrt{\alpha^2+\beta^2}t)^2}{2k\beta^2 t}} + e^{-\frac{(\alpha^2+\beta^2)(x+\sqrt{\alpha^2+\beta^2}t)^2}{2k\beta^2 t}} \right). \qquad (2.2.94)$$

Thus $V_{ij}(u_0)$ have the same property as the heat potential operator. We can easily deduce from the proof of Theorem 2.2.1

Corollary 2.2.2 *Suppose $u_0 \in W^{1,1}(\mathbb{R})$. Then for $t \geq 1$, the following estimates hold:*

$$\|G_{ij} * u_0 - V_{ij}(u_0)\|_{L^1} \leq C(1+t)^{-\frac{1}{4}}\|u_0\|_{L^1}, \qquad (2.2.95)$$

$$\|\frac{\partial}{\partial x}G_{ij} * u_0 - \frac{\partial}{\partial x}V_{ij}(u_0)\|_{L^1} \leq C(1+t)^{-\frac{3}{4}}\|u_0\|_{W^{1,1}} \qquad (2.2.96)$$

Theorem 2.2.2 *Suppose $u_0 \in W^{2,1}(\mathbb{R})$. Then the following estimates hold ($i, j = 1, 2, 3$) :*

$$\|D^2 G_{ij} * u_0\|_{L^1} \leq C(1+t)^{-1}\|u_0\|_{W^{2,1}}, \ \forall t \geq 0, \qquad (2.2.97)$$

$$\|D_t G_{ij} * u_0\|_{L^1} \leq C(1+t)^{-1}\|u_0\|_{W^{2,1}}, \ \forall t \geq 0, \qquad (2.2.98)$$

$$\|D^2(G_{ij} * u_0 - V_{ij}(u_0))\|_{L^1} \leq C(1+t)^{-\frac{5}{4}}\|u_0\|_{W^{2,1}}, \ \forall t \geq 1, \qquad (2.2.99)$$

$$\|D_t(G_{ij} * u_0 - V_{ij}(u_0))\|_{L^1} \leq C(1+t)^{-\frac{5}{4}}\|u_0\|_{W^{2,1}}, \ \forall t \geq 1. \qquad (2.2.100)$$

Since the proof is similar to that of Theorem 2.2.1, we omit the details here. □

In what follows we derive some more delicate $L^p(p = 2, \infty)$ norm estimates of the solution to the initial value problem (2.2.11)–(2.2.12). By the properties of the Fourier transform, we can easily prove the following

Lemma 2.2.3 *Suppose $u_0 \in W^{1,1} \cap H^2$. Then for $t \geq 1$ the following estimates hold:*

$$\begin{cases} \|\mathcal{F}^{-1}(e^{-\frac{\beta^2}{2k}t}\xi \cos \alpha\xi t \,\hat{u}_0(\xi))\| \leq Ce^{-Ct}\|Du_0\|, \\ \|\mathcal{F}^{-1}(e^{-\frac{\beta^2}{2k}t}\xi \sin \alpha\xi t \,\hat{u}_0(\xi))\| \leq Ce^{-Ct}\|Du_0\|, \end{cases} \qquad (2.2.101)$$

$$\begin{cases} \|\mathcal{F}^{-1}(te^{-\frac{\beta^2}{2k}t} \cos \alpha\xi t \,\hat{u}_0(\xi))\| \leq Ce^{-Ct}\|u_0\|, \\ \|\mathcal{F}^{-1}(te^{-\frac{\beta^2}{2k}t} \sin \alpha\xi t \,\hat{u}_0(\xi))\| \leq Ce^{-Ct}\|u_0\|, \end{cases} \qquad (2.2.102)$$

$$\|\mathcal{F}^{-1}(e^{-\mu\xi^2 t}\xi \cos \gamma\xi t \,\hat{u}_0(\xi))\| \leq C(1+t)^{-\frac{3}{4}}\|u_0\|_{L^1}, \qquad (2.2.103)$$

$$\begin{cases} \|\mathcal{F}^{-1}(e^{-\frac{\beta^2}{2k}t}\xi \cos \alpha\xi t \,\hat{u}_0(\xi))\|_{L^\infty} \leq Ce^{-Ct}\|Du_0\|_{H^1}, \\ \|\mathcal{F}^{-1}(e^{-\frac{\beta^2}{2k}t}\xi \sin \alpha\xi t \,\hat{u}_0(\xi))\|_{L^\infty} \leq Ce^{-Ct}\|Du_0\|_{H^1}, \end{cases} \qquad (2.2.104)$$

$$\begin{cases} \|\mathcal{F}^{-1}(te^{-\frac{\beta^2}{2k}t} \cos \alpha\xi t \,\hat{u}_0(\xi))\|_{L^\infty} \leq Ce^{-Ct}\|Du_0\|_{L^1}, \\ \|\mathcal{F}^{-1}(te^{-\frac{\beta^2}{2k}t} \sin \alpha\xi t \,\hat{u}_0(\xi))\|_{L^\infty} \leq Ce^{-Ct}\|Du_0\|_{L^1}, \end{cases} \qquad (2.2.105)$$

$$\begin{cases} \|\mathcal{F}^{-1}(\xi e^{-\mu\xi^2 t} \cos \gamma\xi t \,\hat{u}_0(\xi))\|_{L^\infty} \leq C(1+t)^{-1}\|u_0\|_{L^1}, \\ \|\mathcal{F}^{-1}(\xi e^{-\mu\xi^2 t} \sin \gamma\xi t \,\hat{u}_0(\xi))\|_{L^\infty} \leq C(1+t)^{-1}\|u_0\|_{L^1}. \end{cases} \qquad (2.2.106)$$

Proof. Estimates (2.2.101)–(2.2.102), (2.2.104)–(2.2.105) and (2.2.106) can be directly obtained by the expression of the inverse Fourier transform. We use (2.2.106), (2.2.59) and

$$\|u\|^2 \leq \|u\|_{L^\infty}\|u\|_{L^1} \qquad (2.2.107)$$

to obtain (2.2.103). $\square$

Theorem 2.2.3 *Let U be a solution to the initial value problem (2.2.11)–(2.2.12). Then the following estimates hold:*

$$\|DU(t)\| \leq C(1+t)^{-\frac{3}{4}}\|U_0\|_{L^1} + Ce^{-Ct}\|U_0\|_{H^1}, \ \forall t \geq 1, \qquad (2.2.108)$$

$$\|DU(t)\| \leq C(1+t)^{-\frac{1}{2}}\|U_0\|_{H^2}, \ \forall t \geq 0, \qquad (2.2.109)$$

$$\|DU(t)\|_{L^\infty} \leq C(1+t)^{-1}\|U_0\|_{W^{1,1}}$$

$$+Ce^{-Ct}(\|DU_0\|_{H^1} + \|DU_0\|_{L^1}), \ \forall t \geq 1, \qquad (2.2.110)$$

$$\|DU(t)\|_{L^\infty} \leq C(1+t)^{-\frac{3}{4}}\|U_0\|_{H^2}, \ \forall t \geq 0. \qquad (2.2.111)$$

Proof. The solution to problem (2.2.11)–(2.2.12) can be expressed as

$$U(x,t) = \mathcal{F}^{-1}(\widehat{G}(\xi,t)\widehat{U}_0(\xi)), \tag{2.2.112}$$

$$DU = \mathcal{F}^{-1}(i\xi\widehat{G}(\xi,t)\widehat{U}_0(\xi)). \tag{2.2.113}$$

Thus it follows from (2.2.81)–(2.2.83) that

$$
\begin{aligned}
\|DU\| &\le \|\mathcal{F}^{-1}(i(\xi\widehat{G}-\widehat{M})\widehat{U}_0)\| + \|\mathcal{F}^{-1}(i\widehat{M}\widehat{U}_0)\| \\
&= \|(\xi\widehat{G}-\widehat{M})\widehat{U}_0\| + \|\mathcal{F}^{-1}(i\widehat{M}\widehat{U}_0)\| \\
&\le \|\widehat{U}_0\|_{L^\infty}\|\xi\widehat{G}-\widehat{M}\| + \|\mathcal{F}^{-1}(i\widehat{M}\widehat{U}_0)\| \\
&\le C(1+t)^{-\frac{5}{4}}\|U_0\|_{L^1} + \|\mathcal{F}(i\widehat{M}\widehat{U}_0)\|_{L^2},
\end{aligned}
\tag{2.2.114}
$$

where $\widehat{M}(\xi)$ is a 3×3 matrix function whose typical element $\widehat{M}_{11}(\xi)$ is given by (2.2.80).

By Lemma 2.2.3, we have

$$\|\mathcal{F}^{-1}(\widehat{M}\widehat{U}_0)\| \le C(1+t)^{-\frac{3}{4}}\|U_0\|_{L^1} + Ce^{-Ct}\|U_0\|_{H^1}. \tag{2.2.115}$$

Combining (2.2.114) with (2.2.115) yields (2.2.108).

To prove (2.2.110), we first notice that

$$
\begin{aligned}
\|DU\|_{L^\infty} &\le \|\mathcal{F}^{-1}(i(\xi\widehat{G}-\widehat{M})\widehat{U}_0)\|_{L^\infty} + \|\mathcal{F}^{-1}(i\widehat{M}\widehat{U}_0)\|_{L^\infty} \\
&\le C\|(\xi\widehat{G}-\widehat{M})\widehat{U}_0\|_{L^1} + \|\mathcal{F}^{-1}(i\widehat{M}\widehat{U}_0)\|_{L^\infty}
\end{aligned}
\tag{2.2.116}
$$

and

$$\|(\xi\widehat{G}-\widehat{M})\widehat{U}_0\|_{L^1} \le \left\|\frac{\xi\widehat{G}-\widehat{M}}{(1+|\xi|^2)^{\frac{1}{2}}}\right\|_{L^1} \|(1+|\xi|^2)^{\frac{1}{2}}\widehat{U}_0\|_{L^\infty}. \tag{2.2.117}$$

In the same manner as we derived (2.2.81)–(2.2.83), we can obtain

$$\left\|\frac{\xi\widehat{G}-\widehat{M}}{(1+|\xi|^2)^{\frac{1}{2}}}\right\|_{L^1} \le C(1+t)^{-1}. \tag{2.2.118}$$

Using the properties of the Fourier transform, we have

$$\|(1+|\xi|^2)^{\frac{1}{2}}\widehat{U}_0\|_{L^\infty} \le C\|U_0\|_{W^{1,1}}. \tag{2.2.119}$$

By Lemma 2.2.3, we have

$$\|\mathcal{F}^{-1}(i\widehat{M}\widehat{U}_0)\|_{L^\infty} \le C(1+t)^{-1}\|U_0\|_{L^1} + Ce^{-Ct}(\|DU_0\|_{L^1} + \|DU_0\|_{H^1}). \tag{2.2.120}$$

Thus combining (2.2.116)–(2.2.120) yields (2.2.110).

In what follows we prove (2.2.109) and (2.2.111). By the well-known Young–Hausdorff inequality (see Reed & Simon [1]), we have

$$\|DU\|_{L^p} \leq C \left\| i\xi \frac{\hat{G}(\xi,t)}{(1+|\xi|^2)^{\frac{l}{2}}} \right\|_{L^r} \|U_0\|_{W^{l,q}} \tag{2.2.121}$$

where $p \geq 2, \frac{1}{p} = \frac{1}{q} - \frac{1}{r}$ and l is an integer with $l > 1 + \frac{1}{r}$.

For $t \geq 1$, by the means which we used to derive (2.2.118), we can obtain

$$\left\| i\xi \frac{\hat{G}(\xi,t)}{(1+|\xi|^2)^{\frac{l}{2}}} \right\|_{L^r} \leq Ct^{-(\frac{1}{2r}+\frac{1}{2})}. \tag{2.2.122}$$

For $t \leq 1$, it follows from the expression for $\hat{G}$ that

$$\left\| i\xi \frac{\hat{G}(\xi,t)}{(1+|\xi|^2)^{\frac{l}{2}}} \right\|_{L^r}^r \leq C \int_0^\infty \frac{1}{(1+\xi^2)^{\frac{r(l-1)}{2}}} d\xi \leq C. \tag{2.2.123}$$

Thus combining (2.2.121)–(2.2.123) yields

$$\|DU\|_{L^p} \leq C(1+t)^{-(\frac{1}{2r}+\frac{1}{2})}\|U_0\|_{W^{l,q}}. \tag{2.2.124}$$

We obtain (2.2.109)–(2.2.111) by letting $p = 2, q = 2, r = \infty, l = 2$ and $p = \infty, q = 2, r = 2, l = 2$, respectively. The proof is completed. $\qquad\square$

2.2.2 Initial Boundary Value Problems

In this subsection, without loss of generality, we are concerned with the following one-dimensional thermoelastic system for $(x,t) \in (0,l) \times (0,\infty)$,

$$u_{tt} - u_{xx} + \gamma\theta_x = 0, \tag{2.2.125}$$

$$\theta_t + \gamma u_{xt} - k\theta_{xx} = 0 \tag{2.2.126}$$

with initial conditions

$$u|_{t=0} = \varphi(x), \ u_t|_{t=0} = \psi(x), \ \theta|_{t=0} = \theta_0(x), \ 0 \leq x \leq l, \tag{2.2.127}$$

and boundary conditions which very often are one of the following four pairs at $x = 0$, or $x = l$:

$$(i) \begin{cases} u = 0, \\ \theta = 0. \end{cases} \quad (ii) \begin{cases} u = 0, \\ \theta_x = 0. \end{cases} \quad (iii) \begin{cases} \sigma = 0, \\ \theta = 0. \end{cases} \quad (iv) \begin{cases} u = 0, \\ \theta_x = 0 \end{cases} \tag{2.2.128}$$

where u is the displacement, θ is the temperature, $\sigma = u_x - \gamma\theta$ is the stress and $k > 0, \gamma \neq 0$ are given constants depending on the material properties.

The energy function of the system is defined by

$$E(t) = \|u_x\|^2 + \|u_t\|^2 + \|\theta\|^2 \qquad (2.2.129)$$

where $\|\cdot\|$ denotes, as before, the L^2 norm in x. In this subsection we will show that if $\varphi \in H^1, \psi \in L^2, \theta_0 \in L^2$, then the energy function defined in (2.2.129) exponentially decays to zero as $t \to \infty$, namely,

$$E(t) \leq Me^{-Ct}E(0), \quad \forall t > 0 \qquad (2.2.130)$$

with M, C being two positive constants. The material of this subsection is based on Burns, Liu & Zheng [1].

We would like to mention that exponential stability, such as (2.2.130), is not only important for the global existence of solutions to the corresponding nonlinear system with small initial data, but is also closely related to the feedback control of the system, namely linear Gaussian quadratic optimal control. We refer to Gibson, Rosen & Tao [1] and the references cited there for more details.

Before going into detail, we first recall some related results. Dafermos [1] probably was the first to investigate the asymptotic behaviour of solutions to the initial boundary value problem for the linear thermoelastic system. It was shown in his paper that the energy function converges to zero. However, no decay rate was given. In 1981, among other things, Slemrod [1] used the energy method to prove that if u and θ satisfy (ii) (or (iii)) of (2.2.128) at both ends and if $\varphi \in H^2, \psi \in H^1, \theta_0 \in H^2$ and satisfy the compatibility conditions, then there are positive constants M and C such that

$$\|u_t(t)\|^2 + \|u_x(t)\|^2 + \|u_{tt}(t)\|^2 + \|u_{xt}(t)\|^2 + \|u_{xx}(t)\|^2$$
$$+\|\theta(t)\|^2 + \|\theta_t(t)\|^2 + \|\theta_x(t)\|^2 + \|\theta_{xx}(t)\|^2$$
$$\leq Me^{-Ct}\left(\|\varphi\|_{H^2}^2 + \|\psi\|_{H^1}^2 + \|\theta_0\|_{H^2}^2\right), \quad \forall t > 0. \qquad (2.2.131)$$

In 1990, Revira [1] used the energy method to prove that the estimate (2.2.131) still holds if both u and θ satisfy the Dirichlet boundary condition, i.e., (i) of (2.2.128) at both ends. We refer the reader to Jiang [1] for the related results when u and θ satisfy the boundary condition (iv) at both ends and when initial data are in a more regular space. As far as (2.2.130) is concerned, Hansen [1] in 1990 succeeded in establishing (2.2.130) using the Fourier series expansion method and a decoupling technique. We refer to Gibson, Rosen & Tao [1] for another approach, i.e., a combination of semigroup theory and the energy method. When u and θ both satisfy the Dirichlet boundary conditions at both ends, Kim [3] and Liu & Zheng [1] independently proved the assertion (2.2.130) using completely different methods. Kim's method is based on a control theory approach and a unique continuation theorem by J.L.Lions while in Liu & Zheng [1], the authors used the following theorem by Huang [1] (for the proof, see also Liu & Zheng [2]).

Lemma 2.2.4 *(Huang [1]) A C_o-semigroup $S(t) = e^{tA}$ on a Hilbert space H is exponentially stable if and only if*

$$\sup\{Re\,\lambda;\ \lambda \in \sigma(A)\} < 0 \tag{2.2.132}$$

and

$$\sup_{Re\lambda\geq 0} \|(\lambda - A)^{-1}\| < \infty \tag{2.2.133}$$

hold.

In the above, $\sigma(A)$ stands for the spectrum of A. Recall that a C_o-semigroup $S(t)$ on a Hilbert space H is exponentially stable if there exist positive constants M and C such that

$$\|S(t)\|_{\mathcal{L}(H,H)} \leq Me^{-Ct}, \quad \forall t \geq 0. \tag{2.2.134}$$

In the following we will show that (2.2.130) is equivalent to the exponential stability of an appropriate C_o-semigoup $S(t)$.

Based on Lemma 2.2.4, Burns, Liu & Zheng [1] completely solved the problem of exponential stability for one-dimensional thermoelastic systems with all possible combinations of boundary conditions (2.2.128) at both ends.

In what follows we give another approach, a variant of the method used in Burns, Liu & Zheng [1] and in Liu & Zheng [1], which was also used to treat the thermoelastic Kirchhoff plate problem (see Liu & Zheng [5]). To illustrate the method, we only state and prove the related results here for the boundary conditions

$$u|_{x=0} = u|_{x=l} = \theta|_{x=0} = \theta|_{x=l} = 0, \tag{2.2.135}$$

and

$$u|_{x=0} = 0, \quad \sigma|_{x=l} = 0, \quad \theta|_{x=0} = 0, \quad \theta_x|_{x=l} = 0, \tag{2.2.136}$$

respectively.

We first consider the case of boundary conditions (2.2.135). If we introduce new dependent variables

$$v_1 = u, \quad v_2 = u_t, \quad v_3 = \theta, \tag{2.2.137}$$

then the system (2.2.125)–(2.2.127) can be rewritten as the following first-order evolution system:

$$\begin{cases} V_t = AV, \\ V|_{t=0} = (\varphi, \psi, \theta_0)^T \end{cases} \tag{2.2.138}$$

with $V = (v_1, v_2, v_3)^T$ and

$$A = \begin{pmatrix} 0 & I & 0 \\ D^2 & 0 & -\gamma D \\ 0 & -\gamma D & kD^2 \end{pmatrix}. \tag{2.2.139}$$

Let $\mathcal{H} = H_o^1(0, l) \times L^2(0, l) \times L^2(0, l)$ with the inner product

$$\langle U, V \rangle = \int_0^l (Du_1 Dv_1 + u_2 v_2 + u_3 v_3) dx \tag{2.2.140}$$

and

$$D(A) = \left\{ V \mid v_1 \in H^2 \cap H_o^1, \ v_2 \in H_o^1, \ v_3 \in H^2 \cap H_o^1 \right\}. \tag{2.2.141}$$

Then $D(A)$ is dense in $\mathcal{H}$ and the dissipativeness of A follows from the following calculation

$$\langle AV, V \rangle = \int_0^l (Dv_1 Dv_2 + v_2 D^2 v_1 - \gamma v_2 Dv_3 - \gamma v_3 Dv_2 + kv_3 D^2 v_3) dx$$
$$= -k\|Dv_3\|^2 \leq 0. \tag{2.2.142}$$

It is easy to see that A is a closed linear operator and its adjoint operator A^* is given by

$$A^* = \begin{pmatrix} 0 & -I & 0 \\ -D^2 & 0 & \gamma D \\ 0 & \gamma D & kD^2 \end{pmatrix} \tag{2.2.143}$$

with $D(A^*) = D(A)$.

A straightforward calculation like (2.2.142) also shows that A^* is dissipative. Thus by Theorem 1.3.5 we conclude that A is the infinitesimal generator of a C_o-semigroup $S(t)$ of contractions on $\mathcal{H}$. Furthermore, we have

Theorem 2.2.4 *The C_o-semigroup $S(t)$ is exponentially stable, i.e., there exist positive constants M, C such that*

$$\|S(t)\|_{\mathcal{L}(\mathcal{H}, \mathcal{H})} \leq Me^{-Ct}, \ \forall t \geq 0 \tag{2.2.144}$$

which is equivalent to (2.2.130).

Proof. We use Lemma 2.2.4 and a contradiction argument to prove the theorem. Suppose that (2.2.144) is not true. Then by Lemma 2.2.4, (2.2.132) or (2.2.133) must fail to hold. If (2.2.133) fails to hold, then there must exist a sequence of $\lambda_n \in \mathcal{C}$ and a sequence of $h_n = (h_n^{(1)}, h_n^{(2)}, h_n^{(3)})^T \in D(A)$ with $Re\,\lambda_n \geq 0$, $\|h_n\|_{\mathcal{H}} = 1$ such that as $n \to \infty$,

$$(\lambda_n I - A)h_n \to 0, \quad in \ \mathcal{H}. \tag{2.2.145}$$

As a result, we have

$$Re\,\langle(\lambda_n I - A)h_n, h_n\rangle = Re\,\lambda_n \|h_n\|_{\mathcal{H}}^2 + k\|Dh_n^{(3)}\|^2 \to 0. \tag{2.2.146}$$

Therefore, it follows that

$$Re\,\lambda_n \to 0, \ \ \|Dh_n^{(3)}\| \to 0. \tag{2.2.147}$$

Since $h_n^{(3)} \in H^2 \cap H_o^1$, by the Poincaré inequality, we obtain

$$h_n^{(3)} \to 0 \ \ in \ L^2. \tag{2.2.148}$$

It turns out that

$$\|h_n^{(1)}\|_{H^1}^2 + \|h_n^{(2)}\|^2 \to 1. \tag{2.2.149}$$

In what follows we show that this is a contradiction.

We now first prove that there exists a constant $\delta > 0$ such that $|\lambda_n| \geq \delta$ for n large enough. Otherwise, it follows from (2.2.145) and (2.2.147) that

$$
\begin{cases}
h_n^{(2)} \to 0, & in \; H_o^1, \\[2mm]
D^2 h_n^{(1)} \to 0, & in \; L^2, \\[2mm]
-\gamma D h_n^{(2)} + k D^2 h_n^{(3)} \to 0, & in \; L^2.
\end{cases}
\tag{2.2.150}
$$

Taking the inner product in L^2 with $h_n^{(1)}$ for the second one of (2.2.150) and using the fact that $h_n^{(1)} \in H_o^1$, we obtain

$$
h_n^{(1)} \to 0 \quad in \; H_o^1.
\tag{2.2.151}
$$

Thus combining the first one of (2.2.150) with (2.2.151) contradicts (2.2.149).

Now we can divide (2.2.145) by λ_n to obtain

$$
\begin{cases}
h_n^{(1)} - \dfrac{1}{\lambda_n} h_n^{(2)} \to 0, & in \; H_o^1, \\[4mm]
h_n^{(2)} - \dfrac{1}{\lambda_n} D^2 h_n^{(1)} + \dfrac{\gamma}{\lambda_n} D h_n^{(3)} \to 0, & in \; L^2, \\[4mm]
h_n^{(3)} + \dfrac{\gamma}{\lambda_n} D h_n^{(2)} - \dfrac{k}{\lambda_n} D^2 h_n^{(3)} \to 0, & in \; L^2.
\end{cases}
\tag{2.2.152}
$$

Combining the first and the third one of (2.2.152) with (2.2.148) yields

$$
\gamma D h_n^{(1)} - \frac{k}{\lambda_n} D^2 h_n^{(3)} \to 0, \quad in \; L^2
\tag{2.2.153}
$$

which also implies that $\dfrac{1}{\lambda_n} D^2 h_n^{(3)}$ is bounded in L^2.

It follows from the second one of (2.2.152) that

$$
h_n^{(2)} - \frac{1}{\lambda_n} D^2 h_n^{(1)} \to 0, \quad in \; L^2
\tag{2.2.154}
$$

and $\dfrac{1}{\lambda_n} D^2 h_n^{(1)}$ is bounded in L^2.

Taking the inner product with $D h_n^{(1)}$ in (2.2.153) yields

$$
\gamma \| D h_n^{(1)} \|^2 - \frac{k}{\lambda_n} (D^2 h_n^{(3)}, D h_n^{(1)}) \to 0,
\tag{2.2.155}
$$

$$
\gamma \| D h_n^{(1)} \|^2 + \frac{k}{\lambda_n} (D h_n^{(3)}, D^2 h_n^{(1)}) - \frac{k}{\lambda_n} D h_n^{(3)} \overline{D h_n^{(1)}} \Big|_0^l \to 0.
\tag{2.2.156}
$$

By the Nirenberg inequality, we have

$$|Dh_n^{(i)}|_{L^\infty} \leq C \|Dh_n^{(i)}\|^{\frac{1}{2}} \|D^2 h_n^{(i)}\|^{\frac{1}{2}}, \quad (i = 1, 3). \tag{2.2.157}$$

Thus it follows that

$$\left| \frac{k}{\lambda_n} Dh_n^{(3)} \overline{Dh_n^{(1)}} \right|_0^l \right|$$
$$\leq C \left\| \frac{1}{\lambda_n} D^2 h_n^{(1)} \right\|^{\frac{1}{2}} \left\| \frac{1}{\lambda_n} D^2 h_n^{(3)} \right\|^{\frac{1}{2}} \|Dh_n^{(1)}\|^{\frac{1}{2}} \|Dh_n^{(3)}\|^{\frac{1}{2}}$$
$$\rightarrow 0. \tag{2.2.158}$$

The second term in (2.2.156) can also be estimated as follows:

$$\left| \frac{k}{\lambda_n} (Dh_n^{(3)}, D^2 h_n^{(1)}) \right| \leq C \left\| \frac{1}{\lambda_n} D^2 h_n^{(1)} \right\| \|Dh_n^{(3)}\| \rightarrow 0. \tag{2.2.159}$$

Combining (2.2.156) with (2.2.158), (2.2.159) yields

$$h_n^{(1)} \rightarrow 0, \quad in \ H_o^1. \tag{2.2.160}$$

Taking the inner product of (2.2.154) with $h_n^{(2)}$ yields

$$\|h_n^{(2)}\|^2 + \frac{1}{\lambda_n} (Dh_n^{(1)}, Dh_n^{(2)}) \rightarrow 0. \tag{2.2.161}$$

Thus it follows from the first one of (2.2.152), (2.2.160) and (2.2.161) that

$$h_n^{(2)} \rightarrow 0, \quad in \ L^2, \tag{2.2.162}$$

a contradiction.

If (2.2.132) fails to hold, then there must exist a sequence of $\lambda_n \in C$ and a sequence of $h_n \in D(A) \subset \mathcal{H}$ with $\lambda_n \in \sigma(A)$, $Re \ \lambda_n \rightarrow 0$, $\|h_n\|_{\mathcal{H}} = 1$, such that

$$(\lambda_n I - A) h_n = 0. \tag{2.2.163}$$

Thus using the argument along the same lines as above yields a contradiction. The proof of the theorem is complete. □

We now turn to the case of boundary conditions (2.2.136). Let $H_l^1 = \{f(x) \,|\, f(x) \in H^1, \ f(0) = 0\}$ and

$$\mathcal{H} = H_l^1 \times L^2 \times L^2, \tag{2.2.164}$$

equipped with the norm

$$\|V\|_{\mathcal{H}} = \left(\|Dv_1\|^2 + \|v_2\|^2 + \|v_3\|^2\right)^{\frac{1}{2}}. \tag{2.2.165}$$

Again we have the first-order evolution system (2.2.138). However, $D(A)$ has to be modified as

$$D(A) = \left\{ V \left| \begin{array}{lll} v_1 \in H^2 \cap H^1_l, & v_2 \in H^1_l, & v_3 \in H^2, \\ v_3|_{x=0} = 0, & Dv_3|_{x=l} = 0, & (Dv_1 - \gamma v_3)|_{x=l} = 0 \end{array} \right. \right\}. \tag{2.2.166}$$

For any $V \in D(A)$, we have

$$\begin{aligned}
\langle AV, V \rangle &= \int_0^l (Dv_1 Dv_2 + v_2 D^2 v_1 - \gamma v_2 Dv_3 - \gamma v_3 Dv_2 + kv_3 D^2 v_3) dx \\
&= \int_0^l \left(-D(Dv_1 - \gamma v_3)v_2 + v_2 D^2 v_1 - \gamma v_2 Dv_3 - k(Dv_3)^2 \right) dx \\
&= -k\|Dv_3\|^2 \le 0.
\end{aligned} \tag{2.2.167}$$

Moreover, the adjoint operator A^* is again given by (2.2.143) with $D(A) = D(A^*)$ and therefore A is the infinitesimal generator of a C_o-semigroup $S_1(t)$ of contractions on $\mathcal{H}$.

Similarly, we have

Theorem 2.2.5 *The C_o-semigroup $S_1(t)$ is exponentially stable, i.e, there exist positive constants M, C such that*

$$\|S_1(t)\|_{\mathcal{L}(\mathcal{H}, \mathcal{H})} \le Me^{-Ct}, \ \forall\, t \ge 0. \tag{2.2.168}$$

Proof. The proof is essentially the same as that in Theorem 2.2.4. except for certain slight modifications due to the changes in the boundary conditions. In order to avoid redundant work, we only point out the differences from the proof of Theorem 2.2.4.
(a) Since $h_n^{(3)}|_{x=0} = 0$, it still follows from the Poincaré inequality and (2.2.147) that

$$h_n^{(3)} \to 0, \ in\ L^2. \tag{2.2.169}$$

(b) Since $h_n^{(2)}|_{x=0} = 0$, $(Dh_n^{(1)} - \gamma h_n^{(3)})|_{x=l} = 0$, instead of the second one of (2.2.150) we get

$$D^2 h_n^{(1)} - \gamma Dh_n^{(3)} \to 0, \tag{2.2.170}$$

and

$$\|Dh_n^{(1)}\|^2 - \gamma(Dh_n^{(3)}, Dh_n^{(1)}) \to 0. \tag{2.2.171}$$

Therefore, we again obtain (2.2.151).

(c) Instead of (2.2.161), we get

$$\|h_n^{(2)}\|^2 + \frac{1}{\lambda_n}(Dh_n^{(1)} - \gamma h_n^{(3)},\, Dh_n^{(2)}) \to 0, \tag{2.2.172}$$

taking the inner product of the second one of (2.2.152) with $h_n^{(2)}$. Thus it follows from (2.2.160), (2.2.169) and the uniform boundedness of $\frac{1}{\lambda_n}Dh_n^{(2)}$ that

$$h_n^{(2)} \to 0, \quad in\ L^2. \tag{2.2.173}$$

Thus the proof is complete. $\qquad\qquad\qquad\qquad\qquad\qquad\qquad\qquad\qquad\square$

2.3   Linear Thermoviscoelastic Systems

In this section we consider the following linear hyperbolic–parabolic coupled system

$$\begin{cases} u_{1t} - \alpha u_{2x} = 0, \\ u_{2t} - \alpha u_{1x} + \beta u_{3x} - \mu u_{2xx} = 0, \\ u_{3t} + \beta u_{2x} - k u_{3xx} = 0 \end{cases} \tag{2.3.1}$$

with constants $\alpha > 0, \beta \neq 0, \mu > 0, k > 0$. Here the subscripts t and x again denote the partial derivatives with respect to t and x, respectively. This system is a model of linear one-dimensional thermoviscoelasticity with u_1 being the scaled deformation gradient, u_2 the velocity and u_3 the temperature. (2.3.1) is also a linearized system of motion of compressible viscous and heat-conductive fluids in Lagrange coordinates. In this section we will derive similar decay estimates to those in Section 2.2.

2.3.1   Initial Value Problems

Consider the initial value problem for system (2.3.1):

$$u_1|_{t=0} = u_1^0(x), \quad u_2|_{t=0} = u_2^0(x), \quad u_3|_{t=0} = u_3^0(x). \tag{2.3.2}$$

As in the previous section, if we denote $U = (u_1, u_2, u_3)^T$ and

$$A = \begin{pmatrix} 0 & \alpha D & 0 \\ \alpha D & \mu D^2 & -\beta D \\ 0 & -\beta D & kD^2 \end{pmatrix}, \qquad (2.3.3)$$

then (2.3.1), (2.3.2) can be rewritten as the initial value problem for a first-order evolution system:

$$\begin{cases} U_t = AU, \\ U|_{t=0} = U_0(x) = (u_1^0, u_2^0, u_3^0)^T. \end{cases} \qquad (2.3.4)$$

We apply the Fourier transform method again to problem (2.3.4) and obtain the same decay estimates as in the previous section (see Zheng & Shen [3]).

In particular, we have

Theorem 2.3.1 *Suppose $U_0(x) \in H^2(\mathrm{IR}) \cap W^{2,1}(\mathrm{IR})$. Then problem (2.3.4) admits a unique solution U such that $U \in C([0,\infty); H^2 \cap W^{2,1})$, $U_t \in C(\mathrm{IR}^+; L^2 \cap L^1)$, Du_2, $Du_3 \in L^2(\mathrm{IR}^+; H^2)$. Moreover, the following decay estimates hold:*

$$\|U(t)\|_{L^1} \leq C\|U_0\|_{L^1}, \quad t \geq 0, \qquad (2.3.5)$$

$$\|DU(t)\|_{L^1} \leq C(1+t)^{-\frac{1}{2}}\|U_0\|_{W^{1,1}}, \quad t \geq 0, \qquad (2.3.6)$$

$$\|D^2U(t)\|_{L^1} \leq C(1+t)^{-1}\|U_0\|_{W^{2,1}}, \quad t \geq 0, \qquad (2.3.7)$$

$$\|\frac{\partial}{\partial t}U(t)\|_{L^1} \leq C(1+t)^{-1}\|U_0\|_{W^{2,1}}, \quad t \geq 0, \qquad (2.3.8)$$

$$\|U(t)\|_{L^\infty} \leq C(1+t)^{-\frac{1}{2}}\|U_0\|_{W^{1,1}}, \quad t \geq 0, \qquad (2.3.9)$$

$$\|DU(t)\|_{L^\infty} \leq C(1+t)^{-1}(\|U_0\|_{W^{1,1}} + \|U_0\|_{H^2}), \quad t \geq 0, \qquad (2.3.10)$$

$$\|U(t)\| \leq C(1+t)^{-\frac{1}{4}}\|U_0\|_{W^{1,1}}, \quad t \geq 0, \qquad (2.3.11)$$

$$\|DU(t)\| \leq C(1+t)^{-\frac{3}{4}}(\|U_0\|_{H^2} + \|U_0\|_{L^1}), \quad t \geq 0. \qquad (2.3.12)$$

Proof. Since the proof is essentially the same as in the previous section, we only point out the differences. Taking the Fourier transform in (2.3.4), we obtain

$$\begin{cases} \hat{U}_t = \mathcal{A}\hat{U} \\ \hat{U}|_{t=0} = \hat{U}_0 \end{cases} \qquad (2.3.13)$$

with

$$A = \begin{pmatrix} 0 & i\alpha\xi & 0 \\ i\alpha\xi & -\mu\xi^2 & -i\beta\xi \\ 0 & -i\beta\xi & -k\xi^2 \end{pmatrix}. \tag{2.3.14}$$

The corresponding characteristic equation becomes

$$\lambda^3 + (\mu + k)\xi^2\lambda^2 + ((\alpha^2 + \beta^2)\xi^2 + \mu k\xi^4)\lambda + k\alpha^2\xi^4 = 0 \tag{2.3.15}$$

with the following asymptotic behaviour (see Zheng & Shen [3] and also Matsumura [2]).

As $\xi \to 0$,

$$\begin{cases} \lambda_1 = -\dfrac{k\alpha^2}{\alpha^2 + \beta^2}\xi^2 + O(\xi^3), \\[3mm] \lambda_2 = -\dfrac{1}{2}(\mu + \dfrac{k\beta^2}{\alpha^2 + \beta^2})\xi^2 + i\sqrt{\alpha^2 + \beta^2}\,\xi + O(\xi^3), \\[3mm] \lambda_3 = \bar{\lambda}_2 = -\dfrac{1}{2}(\mu + \dfrac{k\beta^2}{\alpha^2 + \beta^2})\xi^2 - i\sqrt{\alpha^2 + \beta^2}\,\xi + O(\xi^3). \end{cases} \tag{2.3.16}$$

As $|\xi| \to \infty$, if $\mu \neq k$, then

$$\begin{cases} \lambda_1 = -\dfrac{\alpha^2}{\mu} + O(\xi^{-2}), \\[3mm] \lambda_2 = -\mu\xi^2 + \dfrac{(\mu - k)\alpha^2 + k\beta^2}{\mu(\mu - k)} + O(\xi^{-2}), \\[3mm] \lambda_3 = -k\xi^2 + \dfrac{\beta^2}{k - \mu} + O(\xi^{-2}) \end{cases} \tag{2.3.17}$$

and if $\mu = k$, then

$$\begin{cases} \lambda_1 = -\dfrac{\alpha^2}{k} + O(\xi^{-2}), \\[3mm] \lambda_2 = -k\xi^2 - \dfrac{\alpha^2}{4k} + i\beta\xi + O(\xi^{-1}), \\[3mm] \lambda_3 = \bar{\lambda}_2 = -k\xi^2 - \dfrac{\alpha^2}{4k} - i\beta\xi + O(\xi^{-1}). \end{cases} \tag{2.3.18}$$

Then we proceed along the same lines of argument as in the previous section, and we omit the details here.   □

Accordingly, we have a similar corollary to Corollary 2.2.2 and similar estimates to (2.2.95)–(2.2.96). We omit the details here.

2.3.2   Initial Boundary Value Problems

Consider the system (2.3.1) in $(0, l) \times (0, \infty)$ with various kinds of boundary conditions as described in the previous section and with the initial condition (2.3.2). The method used in Section 2.2.2 still works. But we will use the energy method in this section, instead. However, the usual energy method (see e.g., Slemrod [1], Revira [1] for the linear thermoelastic system) gives the exponential decay estimates in $D(A)$ as in (2.2.131) which is a weaker statement than the exponential stability of the C_o-semigroup associated with

$$\begin{cases} U_t = AU, \\ U|_{t=0} = U_0(x). \end{cases} \tag{2.3.19}$$

Using an idea in Kim [3], we can prove that under certain conditions both statements are equivalent:

Theorem 2.3.2 *Suppose A is the infinitesimal generator of a C_o-semigroup $S(t)$ of contractions on a Hilbert space H. Suppose that A is invertible, i.e., A is one to one and onto H. Then the exponential stability of the semigroup $S(t)$ is equivalent to*

$$\|U(t)\|_{D(A)} \leq M e^{-Ct} \|U_0\|_{D(A)}, \quad \forall t \geq 0. \tag{2.3.20}$$

Proof. First we prove that the exponential stability of the semigroup implies (2.3.20). Indeed, if $U_0 \in D(A)$, then it follows from the well-known result in semigroup theory (see Pazy [1]) that for $t \geq 0$, $S(t)U_0 \in D(A)$ and

$$\frac{d}{dt} S(t)U_0 = AS(t)U_0 = S(t)AU_0. \tag{2.3.21}$$

Since A is invertible, A^{-1} is a bounded operator and $\|AU\|_H$ is an equivalent norm in $D(A)$ equipped with the graph norm.

Combining (2.3.21) with the exponential stability yields

$$\|S(t)U_0\|_{D(A)} = \|AS(t)U_0\|_H = \|S(t)AU_0\|_H$$
$$\leq M e^{-Ct} \|AU_0\|_H = M e^{-Ct} \|U_0\|_{D(A)}. \tag{2.3.22}$$

It remains to prove that A being invertible and (2.3.20) imply

$$\|U(t)\|_H \le M e^{-Ct} \|U_0\|_H, \quad for \ t \ge 0. \tag{2.3.23}$$

Let $U_0 \in H$ and

$$V(t) = \int_0^t U(\tau) d\tau + A^{-1} U_0. \tag{2.3.24}$$

Then it follows (see Pazy [1]) that

$$AV(t) = A \int_0^t U(\tau) d\tau + U_0 = A \int_0^t S(\tau) U_0 d\tau + U_0 = S(t) U_0. \tag{2.3.25}$$

On the other hand,

$$\frac{dV}{dt} = U(t) = S(t) U_0. \tag{2.3.26}$$

Thus $V(t)$ satisfies

$$\begin{cases} V_t = AV, \\ V|_{t=0} = A^{-1} U_0 \in D(A). \end{cases} \tag{2.3.27}$$

By (2.3.20) and (2.3.27), we obtain

$$\|V(t)\|_{D(A)} \le M e^{-Ct} \|V(0)\|_{D(A)} = M e^{-Ct} \|U_0\|_H. \tag{2.3.28}$$

$$\|U(t)\|_H = \|\frac{dV}{dt}(t)\|_H = \|AV(t)\|_H = \|V(t)\|_{D(A)} \le M e^{-Ct} \|U_0\|_H. \tag{2.3.29}$$

Thus the proof is complete. $\qquad\square$

To illustrate the energy method, let us consider the following boundary condition for system (2.3.1)

$$u_2|_{x=0} = u_2|_{x=l} = u_3|_{x=0} = u_3|_{x=l} = 0. \tag{2.3.30}$$

It follows from the first equation of (2.3.1) and the boundary conditions $u_2|_{x=0,l} = 0$ that $\int_0^l u_{1t} dx = 0$, i.e., $\int_0^l u_1 dx$ is conserved for $t \ge 0$. Without loss of generality, we assume that $\int_0^l u_1 dx = 0$. Otherwise, we can make the substitution $\tilde{u}_1 = u_1 - \frac{1}{l} \int_0^l u_1^0(x) dx$ which does not change the system (2.3.1). Now for problem (2.3.1), (2.3.2) and (2.3.30), we can rewrite it as a first-order evolution system on the Hilbert space $\mathcal{H} = \{V | v_1 \in L^2, \ v_2 \in L^2, \ v_3 \in L^2, \ \int_0^l v_1 dx = 0\}$:

$$\begin{cases} U_t = AU, \\ U|_{t=0} = U_0(x) = (u_1^0, u_2^0, u_3^0)^T \end{cases} \tag{2.3.31}$$

with $U = (u_1, u_2, u_3)^T$,

$$A = \begin{pmatrix} 0 & \alpha D & 0 \\ \alpha D & \mu D^2 & -\beta D \\ 0 & -\beta D & k D^2 \end{pmatrix} \tag{2.3.32}$$

and

$$D(A) = \left\{ U \in \mathcal{H} \,\middle|\, \alpha u_1 + \mu D u_2 \in H^1, \int_0^l u_1 dx = 0, \, u_2 \in H_o^1, u_3 \in H^2 \cap H_o^1 \right\}. \tag{2.3.33}$$

For $U \in D(A)$, we have

$$\langle AU, U \rangle$$
$$= \int_0^l (\alpha u_1 D u_2 + u_2 D(\alpha u_1 + \mu D u_2) - \beta u_2 D u_3 - \beta u_3 D u_2 + k u_3 D^2 u_3) dx$$
$$= -\mu \|D u_2\|^2 - k \|D u_3\|^2 \le 0. \tag{2.3.34}$$

In the same manner as in the previous section, we can prove that the dissipative operator A is the infinitesimal generator of a C_o-semigroup $S(t)$ of contractions on $\mathcal{H}$. Furthermore, we have

Theorem 2.3.3 *The C_o-semigroup $S(t)$ associated with (2.3.32) is exponentially stable.*

Proof. We use the energy method to prove the theorem. The spirit of the method is to show that a suitable energy function $E(t)$, which is equivalent to $\|U(t)\|_{D(A)}$, satisfies a differential inequality of the following form:

$$\frac{dE}{dt} + CE(t) \le 0 \tag{2.3.35}$$

with a positive constant C.

Suppose $U_0 \in D(A)$. Then by the standard energy method, we have

$$\frac{1}{2}\frac{d}{dt}(\|u_1\|^2 + \|u_2\|^2 + \|u_3\|^2) + \mu\|u_{2x}\|^2 + k\|u_{3x}\|^2 = 0 \tag{2.3.36}$$

and

$$\frac{1}{2}\frac{d}{dt}(\|u_{1t}\|^2 + \|u_{2t}\|^2 + \|u_{3t}\|^2) + \mu\|u_{2xt}\|^2 + k\|u_{3xt}\|^2 = 0. \tag{2.3.37}$$

It follows from the first equation of (2.3.1) that

$$\frac{\alpha^2}{4}\frac{d}{dt}\|u_{2x}\|^2 = \frac{1}{4}\frac{d}{dt}\|u_{1t}\|^2. \tag{2.3.38}$$

Multiplying the third equation of (2.3.1) by εu_{3t} and integrating with respect to x yields

$$\frac{\varepsilon k}{2}\frac{d}{dt}\|u_{3x}\|^2 + \varepsilon\|u_{3t}\|^2 + \varepsilon\beta\int_0^l u_{2x}u_{3t}dx = 0 \tag{2.3.39}$$

with $\varepsilon > 0$ being a small constant.

It follows from the second equation of (2.3.1) that

$$\varepsilon\|D(\alpha u_1 + \mu Du_2)\|^2 \le 2\varepsilon(\|u_{2t}\|^2 + \beta^2\|u_{3x}\|^2). \tag{2.3.40}$$

Adding (2.3.36)–(2.3.39), (2.3.40) together yields

$$\frac{dE(t)}{dt} + \mu(\|u_{2x}\|^2 + \|u_{2xt}\|^2) + k(\|u_{3x}\|^2 + \|u_{3xt}\|^2)$$
$$+\varepsilon\|u_{3t}\|^2 + \varepsilon\beta\int_0^l u_{2x}u_{3t}dx + \varepsilon\|D(\alpha u_1 + \mu Du_2)\|^2$$
$$\le 2\varepsilon(\|u_{2t}\|^2 + \|u_{3x}\|^2) \tag{2.3.41}$$

with $E(t)$ being the energy function defined by

$$E(t) = \frac{1}{2}\left(\|u_1\|^2 + \|u_2\|^2 + \|u_3\|^2 + \|u_{2t}\|^2 + \|u_{3t}\|^2\right) + \frac{1}{4}\|u_{1t}\|^2$$
$$+\frac{\alpha^2}{4}\|u_{2x}\|^2 + \frac{\varepsilon k}{2}\|u_{3x}\|^2. \tag{2.3.42}$$

Since $\int_0^l u_1 dx = 0, u_2|_{x=0,l} = u_3|_{x=0,l} = 0$, by the Poincaré inequality, we have

$$\|\alpha u_1 + \mu Du_2\| \le C\|D(\alpha u_1 + \mu Du_2)\|, \quad \|u_2\|^2 \le C\|u_{2x}\|^2, \tag{2.3.43}$$

$$\|u_3\|^2 \le C\|u_{3x}\|^2, \quad \|u_{2t}\|^2 \le C\|u_{2xt}\|^2, \quad \|u_{3t}\|^2 \le C\|u_{3xt}\|^2. \tag{2.3.44}$$

with positive constant C.

It turns out from (2.3.43) that

$$\varepsilon\|u_1\|^2 \le C\varepsilon(\|Du_2\|^2 + \|D(\alpha u_1 + \mu Du_2)\|^2). \tag{2.3.45}$$

Thus taking ε appropriately small, we derive from system (2.3.1), (2.3.41)–(2.3.45) and the Cauchy inequality that (2.3.35) is satisfied.

It follows from (2.3.35) and the Gronwall inequality that

$$E(t) \le e^{-Ct}E(0). \tag{2.3.46}$$

On the other hand, it is easy to see from system (2.3.1) and (2.3.43), (2.3.45) that $E(t)$ is equivalent to

$$\|U(t)\|^2_{D(A)} = \|u_1\|^2 + \|u_2\|^2 + \|u_{2x}\|^2 + \|D(\alpha u_1 + \mu D u_2)\|^2$$
$$+\|u_3\|^2 + \|u_{3x}\|^2 + \|u_{3xx}\|^2 \tag{2.3.47}$$

i.e., there exist positive constants C_1, C_2 such that

$$C_1 E(t) \leq \|U(t)\|^2_{D(A)} \leq C_2 E(t). \tag{2.3.48}$$

Thus it turns out from (2.3.46) that

$$\|U(t)\|_{D(A)} \leq M e^{-Ct} \|U_0\|_{D(A)}. \tag{2.3.49}$$

In order to apply Theorem 2.3.2, it remains to prove that A is invertible. For any $F = (f_1, f_2, f_3)^T \in \mathcal{H}$, we consider

$$AU = F. \tag{2.3.50}$$

It turns out from the first equation of the vector form (2.3.50) that u_2 is uniquely given by

$$u_2 = \frac{1}{\alpha} \int_0^x f_1 dx \in H_o^1. \tag{2.3.51}$$

Substituting this expression of u_2 into the third equation of (2.3.50), we get

$$kD^2 u_3 = f_3 + \beta D u_2 \in L^2. \tag{2.3.52}$$

It follows from (2.3.52) that there is a unique solution $u_3 \in H^2 \cap H_o^1$. Once u_2, u_3 have been obtained, it turns out from the second equation of (2.3.50) that u_1 is uniquely given by

$$u_1 = \frac{1}{\alpha}(\int_0^x f_2(\xi)\, d\xi + C - \mu D u_2 + \beta u_3) \tag{2.3.53}$$

with

$$C = -\frac{1}{l}\left(\beta \int_0^l u_3 dx + \int_0^l \int_0^x f_2\, d\xi\, dx\right). \tag{2.3.54}$$

Thus A is invertible and the proof is complete. $\square$

Chapter 3

Global Existence for Small Initial Data

Based on the decay results obtained in Chapter 2, in this chapter we establish the global existence and uniqueness of *small* smooth solutions to both initial value problems and initial boundary value problems for fully nonlinear parabolic equations and nonlinear one-dimensional thermoelastic and thermoviscoelastic systems. In the final section of this chapter we give some blow-up results for nonlinear evolution equations with *small* initial data.

In Section 3.1 we consider the initial value problems for fully nonlinear parabolic equations and establish the global existence and uniqueness of small smooth solutions using the weighted norms and the globally iterative method. It turns out that the space dimension n and the order of nonlinearity near the origin are two crucial factors for global existence or nonexistence of classical solutions. The results presented in this section are not only sharp with respect to the space dimension as obtained by Zheng & Chen [1], Zheng [6] and Ponce [1], independently, but also are a refinement of the previous results in the following sense: less requirement is put on initial data; sharp decay rates for lower-order derivatives of solutions are obtained; the results also give decay rates of higher-order derivatives of solutions.

In Section 3.2, the global existence and uniqueness of small smooth solutions to the initial boundary value problems for fully nonlinear parabolic equations is also obtained by introducing the weighted norms and using the globally iterative method. In contrast to the initial value problems, the results do not depend on the space dimension n.

In Section 3.3 and Section 3.4 we are concerned with two classes of one-dimensional quasilinear hyperbolic–parabolic coupled systems. The nonlinear thermoelastic system, the equations of motion of compressible viscous and heat-conductive fluids, the nonlinear thermoviscoelastic system and the equations of radiation hydrodynamics are important examples of these two classes of systems. Based on the results obtained in Chapter 2, we prove the global existence and uniqueness of small smooth solutions using the weighted norms and the continuation argument.

3.1   Fully Nonlinear Parabolic Equations: IVP

In this section we consider the following initial value problem for fully nonlinear parabolic equations:

$$\begin{cases} u_t - \Delta u = F(u, Du, D^2 u), \\ u|_{t=0} = \varphi(x) \end{cases} \tag{3.1.1}$$

where $x = (x_1, \cdots, x_n) \in {I\!\!R}^n, t \in {I\!\!R}^+, Du = (u_{x_1}, \cdots, u_{x_n}), D^2 u = (u_{x_i x_j}), (i, j = 1, \cdots, n)$ and F, φ are smooth functions. The basic assumption on F is:

$$F(\omega) = F(\omega_0, \omega_i, \omega_{ij}) = O(|\omega|^{\alpha+1}), \quad near \ \ \omega = 0 \tag{3.1.2}$$

with $\alpha \geq 1$ being an integer.

In Fujita [1], [2], the author considered the following initial value problem

$$\begin{cases} u_t - \Delta u = u^{\alpha+1}, \\ u|_{t=0} = \varphi(x) \end{cases} \tag{3.1.3}$$

and proved that if $\alpha < \frac{2}{n}$, then for any smooth initial data $\varphi \geq 0, \varphi \not\equiv 0$, the solution u must blow up in finite time no matter how small φ is. On the other hand, he also proved global existence if $\alpha > \frac{2}{n}$ and initial data are small. Later, the blow-up results for the case $\alpha = \frac{2}{n}$ were obtained by Kobayashi et al.[1] and Weissler [1], respectively. In 1982, Klainerman [1] considered the problem (3.1.1) and proved the global existence of small smooth solutions when $\frac{1}{\alpha}(1 + \frac{1}{\alpha}) < \frac{n}{2}$, using the decay rates of the L^∞ norm. Zheng & Chen [1] and Ponce [1] independently obtained sharp results on the relation between α and n: if $\alpha > \frac{2}{n}$, then for any φ sufficiently small in $H^S \cap W^{S,1}$ with an integer $s \geq n + 4$, the Cauchy problem (3.1.1) admits a unique global small solution. The main ingredient in these two papers is that the decay rates of the L^2 norm as well as the L^∞ norm of the solution to the heat equation have been used. In this section we use a globally iterative method and combine the decay estimates of the solution to the heat equation with the technique used in Zheng [6] to obtain a refined result.

Now let us begin by introducing some results concerning estimates of composite functions, the so-called 'calculus of inequalities' (see Theorems 4.1–4.4 in Li & Chen [1], also Klainerman [1] and Ponce [1]).

Lemma 3.1.1 *Suppose that*

$$\frac{1}{r} = \frac{1}{p} + \frac{1}{q}, \quad 1 \leq r, p, q \leq \infty \tag{3.1.4}$$

and suppose that all the norms appearing below are bounded. Then for any given integer $s \geq 0$ we have

$$\|D^S(fg)\|_{L^r} \leq C_S \left(\|f\|_{L^p} \|D^S g\|_{L^q} + \|D^S f\|_{L^p} \|g\|_{L^q} \right), \tag{3.1.5}$$

$$\|D^S(fg)\|_{L^r} \leq C_S \left(\|f\|_{L^p} \|D^S g\|_{L^q} + \|D^S f\|_{L^q} \|g\|_{L^p} \right). \tag{3.1.6}$$

For $s \geq 1$, we have

$$\|D^S(fg) - f(D^S g)\|_{L^r} \leq C_S \left(\|Df\|_{L^p} \|D^{S-1} g\|_{L^q} + \|D^S f\|_{L^p} \|g\|_{L^q} \right), \tag{3.1.7}$$

$$\|D^S(fg) - f(D^S g)\|_{L^r} \leq C_S \left(\|Df\|_{L^p} \|D^{S-1} g\|_{L^q} + \|D^S f\|_{L^q} \|g\|_{L^p} \right). \tag{3.1.8}$$

Lemma 3.1.2 *Suppose that $F(w) : \mathbb{R}^m \mapsto \mathbb{R}$ is a smooth function satisfying*

$$F(w) = O(|w|^{1+\alpha}), \quad near \ w = 0 \tag{3.1.9}$$

where $\alpha \geq 1$ is an integer. Suppose that for a given vector function $w(x)$,

$$\|w\|_{L^\infty} \leq \nu_0 \leq 1 \tag{3.1.10}$$

and all the norms appearing below are bounded. Then for any given integer $s \geq 0$,

$$\|D^S F(w)\|_{L^p} \leq C_S \|w\|_{L^\infty}^\alpha \|D^S w\|_{L^p}, \tag{3.1.11}$$

$$\|F(w)\|_{W^{s,r}} \leq C_S \|w\|_{W^{s,q}} \prod_{i=1}^{\alpha} \|w\|_{L^{p_i}}, \tag{3.1.12}$$

where $p \geq 1$ and r, p_i and q satisfy

$$\frac{1}{r} = \sum_{i=1}^{\alpha} \frac{1}{p_i} + \frac{1}{q}, \quad 1 \leq p_i, q, r \leq \infty \tag{3.1.13}$$

and C_S is a positive constant depending on s and ν_0. In particular, we have

$$\|F(w)\|_{W^{s,r}} \leq C_S \|w\|_{W^{s,q}} \|w\|_{L^p} \|w\|_{L^\infty}^{\alpha-1} \tag{3.1.14}$$

with r, p, q satisfying (3.1.4).

For proof, see Theorems 4.1–4.2 in Li & Chen [1]. □

In the following, we often need to estimate the integral

$$J = \int_0^t (1 + t - \tau)^{-\beta} (1 + \tau)^{-\gamma} d\tau \tag{3.1.15}$$

with positive constants β, γ.

Notice that if we make change of variables: $\sigma = t - \tau$, then

$$J = \int_0^t (1 + \sigma)^{-\beta} (1 + t - \sigma)^{-\gamma} d\sigma. \tag{3.1.16}$$

Therefore, without loss of generality, we can assume that $\beta \geq \gamma > 0$.

We now have

Lemma 3.1.3 Suppose that $\beta > 1$, $\beta \geq \gamma > 0$. Then there is a positive constant C such that for all $t \geq 0$,

$$J \leq C(1 + t)^{-\gamma}. \tag{3.1.17}$$

Proof. A straightforward calculation gives

$$J = \int_0^{\frac{t}{2}} (1 + t - \tau)^{-\beta} (1 + \tau)^{-\gamma} d\tau + \int_{\frac{t}{2}}^t (1 + t - \tau)^{-\beta} (1 + \tau)^{-\gamma} d\tau$$

$$\leq C(1 + t)^{-\beta} \int_0^{\frac{t}{2}} (1 + \tau)^{-\gamma} d\tau + C(1 + t)^{-\gamma} \int_{\frac{t}{2}}^t (1 + t - \tau)^{-\beta} d\tau. \tag{3.1.18}$$

Thus (3.1.17) easily follows from (3.1.18) and the assumption on β and γ. □

For the results previously obtained by Zheng & Chen [1], Zheng [6] and Ponce [1], we now state and prove the following refined version:

Theorem 3.1.1 *Let α and s be integers such that $s \geq n+4$, $\alpha > \frac{2}{n}$. Suppose that $F \in C^{S+1}$ and F satisfies (3.1.2). Then the Cauchy problem (3.1.1) admits a unique global smooth solution u such that $u \in C([0,\infty); H^S \cap L^1) \cap C^1([0,\infty); H^{S-2}) \cap C((0,+\infty); W^{2,1})$, $Du \in L^2(\mathbb{R}^+; H^S)$, provided that $\|\varphi\|_{H^S \cap L^1} = \|\varphi\|_{H^S} + \|\varphi\|_{L^1}$ is sufficiently small.*

Moreover, for $t > 0$,

$$\|D^k u(t)\| \leq C(1+t)^{-\frac{n+2k}{4}}, \quad k = 0,1,2, \tag{3.1.19}$$

$$\|D^k u(t)\|_{L^\infty} \leq C(1+t)^{-\frac{n+k}{2}}, \quad k = 0,1,2, \tag{3.1.20}$$

$$\|D^k u(t)\|_{L^1} \leq Ct^{-\frac{k}{2}}, \quad k = 0,1,2, \tag{3.1.21}$$

$$\|D^j u(t)\| \leq C(1+t)^{-\frac{1}{2}}, \quad j = 3, \cdots, s. \tag{3.1.22}$$

Proof. We would like to mention that the refinement of Theorem 3.1.1 is reflected by the following facts: we only assume $\varphi \in H^S \cap L^1$ instead of $H^S \cap W^{S,1}$; the results (3.1.19)–(3.1.21) are exactly the same as the decay rates of the derivatives of the solution to the heat equation (see Lemma 2.1.2); the theorem also gives the decay rates of higher-order derivatives (3.1.22).

Instead of the method of combining a local existence theorem with uniform a priori estimates, we use a globally iterative method which was used by the author in Zheng [9–10] and later on by Li Tatsien and his students in a series of papers (see Li & Chen [1] and the references cited there).

Let $T > 0$ be an arbitrary positive number and let s be an integer such that $s \geq n + 4$. We introduce:

$$X_T = \left\{ v(x,t) \,\middle|\, \begin{array}{l} v \in C([0,T]; H^S \cap L^1) \cap L^2([0,T]; H^{S+1}) \cap C((0,T]; W^{2,1}) \\[4pt] \|v\|_{X_T} < \infty \end{array} \right\} \tag{3.1.23}$$

equipped with the weighted norm

$$\|v\|_{X_T} = \sup_{0<t\leq T} \sum_{k=0}^{2} t^{\frac{k}{2}} \|D^k v(t)\|_{L^1}$$

$$+ \max_{0\leq t\leq T} \left(\sum_{k=0}^{2} \left((1+t)^{\frac{n+2k}{4}} \|D^k v(t)\| + (1+t)^{\frac{n+k}{2}} \|D^k v(t)\|_{L^\infty} \right) \right.$$

$$\left. + \sum_{k=3}^{S} (1+t)^{\frac{k}{2}} \|D^k v(t)\| + \left(\int_0^t \sum_{k=1}^{S+1} (1+\tau)^{k-1} \|D^k v(\tau)\|^2 d\tau \right)^{\frac{1}{2}} \right). \tag{3.1.24}$$

Then it is easy to see that X_T is a Banach space. For $0 < \varepsilon < 1$, we denote by $\mathcal{S}_{T,\varepsilon}$ a convex closed subset of X_T:

$$\mathcal{S}_{T,\varepsilon} = \left\{ v(x,t) \,|\, v(x,t) \in X_T, \ v|_{t=0} = \varphi(x), \ \|v\|_{X_T} \leq \varepsilon \right\}. \tag{3.1.25}$$

Since the function $v = \int_{\mathbb{R}^n} K(x - y, t)\varphi(y)dy \in X_T$ and $v|_{t=0} = \varphi$, it is easy to see that for any given T and ε, $\mathcal{S}_{T,\varepsilon}$ is nonempty provided that φ is sufficiently small in $H^s \cap L^1$.

For any $v \in \mathcal{S}_{T,\varepsilon}$, we consider the following auxiliary linear problem:

$$\begin{cases} u_t - \Delta u = F(\Lambda v), \\ u|_{t=0} = \varphi(x) \end{cases} \tag{3.1.26}$$

where we simply denote $\Lambda v = (v, Dv, D^2 v)$.

The unique classical solution u of problem (3.1.26) is given by

$$u = \int_{\mathbb{R}^n} K(x - y, t)\varphi(y)dy + \int_0^t \int_{\mathbb{R}^n} K(x - y, t - \tau)F(\Lambda v(y, \tau))dy d\tau \tag{3.1.27}$$

where $K(x - y, t)$ is the fundamental solution to the heat equation given in (2.1.2).

We now prove that u also belongs to X_T. Indeed, it is easy to see from Lemma 2.1.2 that the function u_I given by

$$u_I = \int_{\mathbb{R}^n} K(x - y, t)\varphi(y)dy \tag{3.1.28}$$

satisfies the homogeneous heat equation and belongs to X_T. Moreover, by the usual energy method, we have for $0 \leq k \leq s$

$$\frac{1}{2}\frac{d}{dt}\|D^k u_I\|^2 + \|D^{k+1}u_I\|^2 = 0. \tag{3.1.29}$$

For $1 \leq k \leq s$, we multiply (3.1.29) by $\delta^k t^k$ to obtain

$$\frac{\delta^k}{2}\frac{d}{dt}\left(t^k\|D^k u_I\|^2\right) - \frac{\delta^k k}{2}t^{k-1}\|D^k u_I\|^2 + \delta^k t^k\|D^{k+1}u_I\|^2 = 0. \tag{3.1.30}$$

Adding up with k from 0 to s in (3.1.29) and from 1 to s in (3.1.30), taking $\delta < \frac{1}{s}$, then integrating with respect to t, we obtain

$$\sum_{k=0}^{s}(1 + t^k)\|D^k u_I(t)\|^2 + \sum_{k=0}^{s}\int_0^t(1 + \tau^k)\|D^{k+1}u_I\|^2 d\tau \leq C\|\varphi\|_{H^s}^2. \tag{3.1.31}$$

Thus combining (3.1.31) with (2.1.16), (2.1.12) and (2.1.19) yields

$$\|u_I\|_{X_T} \leq C\|\varphi\|_{H^s \cap L^1}. \tag{3.1.32}$$

¿From now on we denote by C a positive constant independent of T, u and v, which may vary in different places.

By the Duhamel principle, the function u_{II} given by

$$u_{II} = \int_0^t \int_{\mathbb{R}^n} K(x - y, t - \tau) F(\Lambda v(y, \tau)) \, dy \, d\tau \tag{3.1.33}$$

satisfies the nonhomogeneous heat equation with homogeneous initial data. By $v \in X_T$ and the assumption on F we can conclude that

$$F(\Lambda v) \in C([0, T]; H^{S-2} \cap W^{S-2,1}) \cap L^2([0, T]; H^{S-1}) \cap L^1([0, T]; W^{S-1,1}).$$

It turns out that we have the following estimates on u_{II} :

(i) Energy norm estimates

For $0 \leq k \leq 2$ and $t > 0$, it follows from (2.1.17), (2.1.18) and Lemma 3.1.2 that

$$\|D^k u_{II}(t)\| \leq C \int_0^{\frac{t}{2}} (1 + t - \tau)^{-\frac{n+2k}{4}} \|F\|_{W^{k+n,1}} \, d\tau + C \int_{\frac{t}{2}}^t (1 + t - \tau)^{-\frac{k}{2}} \|F\|_{H^k} \, d\tau$$

$$\leq C \int_0^{\frac{t}{2}} (1 + t - \tau)^{-\frac{n+2k}{4}} \|\Lambda v\|_{L^\infty}^{\alpha-1} \|\Lambda v\| \, \|\Lambda v\|_{H^{k+n}} d\tau$$

$$+ C \int_{\frac{t}{2}}^t (1 + t - \tau)^{-\frac{k}{2}} \|\Lambda v\|_{L^\infty}^\alpha \|\Lambda v\|_{H^k} d\tau. \tag{3.1.34}$$

Noticing that

$$\|\Lambda v(t)\|_{H^{k+n}} \leq C \left(\sum_{j=0}^{2} \|D^j v(t)\| + \sum_{j=3}^{k+n+2} \|D^j v(t)\| \right)$$

$$\leq C\varepsilon \left((1 + t)^{-\frac{n}{4}} + (1 + t)^{-\frac{3}{2}} \right), \tag{3.1.35}$$

we deduce from (3.1.34) that

$$\|D^k u_{II}(t)\| \leq C\varepsilon^{\alpha+1} \int_0^{\frac{t}{2}} (1 + t - \tau)^{-\frac{n+2k}{4}} \left((1 + \tau)^{-\frac{n\alpha}{2}} + (1 + \tau)^{-\left(\frac{n(\alpha-1)}{2} + \frac{n}{4} + \frac{3}{2}\right)} \right) d\tau$$

$$+ C\varepsilon^{\alpha+1} \int_{\frac{t}{2}}^t (1 + t - \tau)^{-\frac{k}{2}} \left((1 + \tau)^{-\left(\frac{n\alpha}{2} + \frac{n}{4}\right)} + (1 + \tau)^{-\left(\frac{n\alpha}{2} + \frac{3}{2}\right)} \right) d\tau$$

$$\leq C\varepsilon^{\alpha+1}(1+t)^{-\frac{n+2k}{4}}\int_0^{\frac{t}{2}}\left((1+\tau)^{-\frac{n\alpha}{2}}+(1+\tau)^{-\left(\frac{n(\alpha-1)}{2}+\frac{n}{4}+\frac{3}{2}\right)}\right)d\tau$$

$$+C\varepsilon^{\alpha+1}\left((1+t)^{-\left(\frac{n\alpha}{2}+\frac{n}{4}\right)}+(1+t)^{-\left(\frac{n\alpha}{2}+\frac{3}{2}\right)}\right)\int_{\frac{t}{2}}^t(1+t-\tau)^{-\frac{k}{2}}d\tau. \tag{3.1.36}$$

Since $0\leq k\leq 2$, $n\alpha>2$, we can deduce from (3.1.36) that

$$\|D^k u_{II}(t)\|\leq C\varepsilon^{\alpha+1}(1+t)^{-\frac{n+2k}{4}}. \tag{3.1.37}$$

In the same manner as we derived the energy estimates for u_I, we get

$$\sum_{k=0}^S(1+t^k)\|D^k u_{II}(t)\|^2+\sum_{k=0}^S\int_0^t(1+\tau^k)\|D^{k+1}u_{II}\|^2d\tau$$

$$\leq C\left(\int_0^t\|u_{II}\|\,\|F\|d\tau+\sum_{k=0}^{S-1}\int_0^t(1+\tau^{k+1})\|D^k F\|^2d\tau\right)$$

$$\leq C\left(\int_0^t\|u_{II}\|\,\|\Lambda v\|_{L^\infty}^\alpha\|\Lambda v\|d\tau+\sum_{k=0}^{S-1}\int_0^t(1+\tau^{k+1})\|\Lambda v\|_{L^\infty}^{2\alpha}\|D^k\Lambda v\|^2d\tau\right)$$

$$\leq C\left(\varepsilon^{\alpha+1}\int_0^t\|u_{II}\|(1+\tau)^{-\left(\frac{n\alpha}{2}+\frac{n}{4}\right)}d\tau+\varepsilon^{2(\alpha+1)}\sum_{k=0}^{S-2}\int_0^t(1+\tau^{k+1})(1+\tau)^{-n\alpha-k-\frac{n}{2}}d\tau\right.$$

$$\left.+\varepsilon^{2\alpha}\int_0^t(1+\tau^S)(1+\tau)^{-n\alpha}\|D^{S+1}v\|^2d\tau\right)$$

$$\leq C\left(\varepsilon^{\alpha+1}\sup_{0\leq\tau\leq t}(1+\tau)^{\frac{n}{4}}\|u_{II}\|+\varepsilon^{2(\alpha+1)}\right)$$

$$\leq C\varepsilon^{2(\alpha+1)}. \tag{3.1.38}$$

(ii) L^∞ norm estimates

For $0\leq k\leq 2$ and $t>0$, it follows from (2.1.20), (2.1.21) and Lemma 3.1.2 that

$$\|D^k u_{II}\|_{L^\infty}$$

$$\leq C\int_0^{\frac{t}{2}}(1+t-\tau)^{-\frac{n+k}{2}}\|F\|_{W^{n+k,1}}d\tau+C\int_{\frac{t}{2}}^t(1+t-\tau)^{-\frac{n}{4}}\|D^k F\|_{H^n}d\tau$$

$$\leq C \left(\int_0^{\frac{t}{2}} (1+t-\tau)^{-\frac{n+k}{2}} \|\Lambda v\|_{L^\infty}^{\alpha-1} \|\Lambda v\| \, \|\Lambda v\|_{H^{n+k}} d\tau \right.$$

$$\left. + \int_{\frac{t}{2}}^t (1+t-\tau)^{-\frac{n}{4}} \|\Lambda v\|_{L^\infty}^{\alpha} \|D^k \Lambda v\|_{H^n} d\tau \right)$$

$$\leq C\varepsilon^{\alpha+1} \left(\int_0^{\frac{t}{2}} (1+t-\tau)^{-\frac{n+k}{2}} \left((1+\tau)^{-\frac{n\alpha}{2}} + (1+\tau)^{-\left(\frac{n(\alpha-1)}{2}+\frac{n}{4}+\frac{3}{2}\right)} \right) d\tau \right.$$

$$\left. + \int_{\frac{t}{2}}^t (1+t-\tau)^{-\frac{n}{4}} \left((1+\tau)^{-\left(\frac{n\alpha}{2}+\frac{n+2k}{4}\right)} + (1+\tau)^{-\left(\frac{n\alpha}{2}+\frac{3}{2}\right)} \right) d\tau \right)$$

$$\leq C\varepsilon^{\alpha+1}(1+t)^{-\frac{n+k}{2}}. \tag{3.1.39}$$

Therefore,

$$\sup_{0 \leq t \leq T} (1+t)^{\frac{n+k}{2}} \|D^k u_{II}\|_{L^\infty} \leq C\varepsilon^{\alpha+1}, \quad 0 \leq k \leq 2. \tag{3.1.40}$$

(iii) L^1 norm estimates

By (2.1.15) and Lemma 3.1.2, we have for $0 \leq k \leq 2$:

$$\|D^k u_{II}(t)\|_{L^1}$$

$$\leq \int_0^t (1+t-\tau)^{-\frac{k}{2}} \|F\|_{W^{k,1}} d\tau \leq C \int_0^t (1+t-\tau)^{-\frac{k}{2}} \|\Lambda v\|_{H^k} \|\Lambda v\| \, \|\Lambda v\|_{L^\infty}^{\alpha-1} d\tau$$

$$\leq C\varepsilon^{\alpha+1} \int_0^t (1+t-\tau)^{-\frac{k}{2}} \left((1+\tau)^{-\frac{n\alpha}{2}} + (1+\tau)^{-\left(\frac{n(\alpha-1)}{2}+\frac{n}{4}+\frac{3}{2}\right)} \right) d\tau. \tag{3.1.41}$$

Applying Lemma 3.1.3 to (3.1.41) yields

$$\|D^k u_{II}(t)\|_{L^1} \leq C(1+t)^{-\frac{k}{2}} \varepsilon^{\alpha+1}, \quad 0 \leq k \leq 2, \ t \geq 0. \tag{3.1.42}$$

Combining the results obtained in the previous three steps with (3.1.32) yields

$$\|u\|_{X_T} \leq C_1 \left(\|\varphi\|_{H^s \cap L^1} + \varepsilon^{\alpha+1} \right). \tag{3.1.43}$$

Thus if we take

$$\varepsilon_0 = \min\left(1, \left(\frac{1}{2C_1}\right)^{\frac{1}{\alpha}}\right), \tag{3.1.44}$$

then when $\varepsilon \le \varepsilon_0$ and

$$\|\varphi\|_{H^s \cap L^1} \le \frac{\varepsilon}{2C_1}, \tag{3.1.45}$$

we have

$$\|u\|_{X_T} \le \varepsilon. \tag{3.1.46}$$

For any given $T > 0$, the auxiliary linear problem (3.1.26) defines a nonlinear operator $\mathcal{N} : v \in \mathcal{S}_{T,\varepsilon} \mapsto u \in \mathcal{S}_{T,\varepsilon}$. In what follows we prove that $\mathcal{N}$ is a contraction provided that ε is suitably small.

Indeed, for any v_1, $v_2 \in \mathcal{S}_{T,\varepsilon}$, we denote $u_1 = \mathcal{N}v_1$, $u_2 = \mathcal{N}v_2$, $v = v_1 - v_2$ and $u = u_1 - u_2$. Then u satisfies

$$\begin{cases} u_t - \Delta u = F(\Lambda v_1) - F(\Lambda v_2), \\ u|_{t=0} = 0. \end{cases} \tag{3.1.47}$$

We can write

$$F(\Lambda v_1) - F(\Lambda v_2) = \int_0^1 F'(\sigma \Lambda v_1 + (1 - \sigma)\Lambda v_2)\Lambda v \, d\sigma \tag{3.1.48}$$

with

$$F'\Lambda v = F_\omega v + \sum_{i=1}^n F_{\omega_i} v_{x_i} + \sum_{i,j=1}^n F_{\omega_{ij}} v_{x_i x_j}. \tag{3.1.49}$$

Now we apply Lemma 2.1.2, Lemma 3.1.1 and Lemma 3.1.2 to obtain the energy norm estimates, L^∞ norm estimates and L^1 norm estimates again.

(i) Energy norm estimates

For $0 \le k \le 2$, $t \ge 0$,

$$\|D^k u(t)\| \le C \int_0^{\frac{t}{2}} (1 + t - \tau)^{-\frac{n+2k}{4}} \|F(\Lambda v_1) - F(\Lambda v_2)\|_{W^{k+n,1}} \, d\tau$$

$$+ C \int_{\frac{t}{2}}^t (1 + t - \tau)^{-\frac{k}{2}} \|F(\Lambda v_1) - F(\Lambda v_2)\|_{H^k} d\tau$$

$$\le C \int_0^{\frac{t}{2}} (1 + t - \tau)^{-\frac{n+2k}{4}} \left(\|F'\| \|\Lambda v\|_{H^{k+n}} + \|\Lambda v\| \|F'\|_{H^{k+n}} \right) d\tau$$

$$+ C \int_{\frac{t}{2}}^t (1 + t - \tau)^{-\frac{k}{2}} \left(\|F'\|_{L^\infty} \|\Lambda v\|_{H^k} + \|\Lambda v\|_{L^\infty} \|F'\|_{H^k} \right) d\tau$$

$$\le C\varepsilon^\alpha \int_0^{\frac{t}{2}} (1+t-\tau)^{-\frac{n+2k}{4}} \left((1+\tau)^{-\left(\frac{n(\alpha-1)}{2}+\frac{n}{4}\right)} \|\Lambda v\|_{H^{k+n}} \right.$$

$$\left. +(1+\tau)^{-\left(\frac{n(\alpha-1)}{2}+\frac{n}{4}\right)} \|\Lambda v\| + (1+\tau)^{-\left(\frac{n(\alpha-1)}{2}+\frac{3}{2}\right)} \|\Lambda v\| \right) d\tau$$

$$+ C\varepsilon^\alpha \int_{\frac{t}{2}}^t (1+t-\tau)^{-\frac{k}{2}} \left((1+\tau)^{-\frac{n\alpha}{2}} \left(\|\Lambda v\| + \|D^3 v\| + \|D^4 v\| \right) \right.$$

$$\left. + \left((1+\tau)^{-\left(\frac{n(\alpha-1)}{2}+\frac{3}{2}\right)} + (1+\tau)^{-\left(\frac{n(\alpha-1)}{2}+\frac{n}{4}\right)} \right) \|\Lambda v\|_{L^\infty} \right) d\tau$$

$$\le C\varepsilon^\alpha (1+t)^{-\frac{n+2k}{4}} \sup_{0 \le \tau \le t} \left((1+\tau)^{\frac{n}{4}} \|\Lambda v\| + (1+\tau)^{\frac{3}{2}} \sum_{j=3}^{k+n+2} \|D^j v\| \right)$$

$$\times \left(\int_0^{\frac{t}{2}} (1+\tau)^{-\frac{n\alpha}{2}} d\tau + \int_0^{\frac{t}{2}} (1+\tau)^{-\left(\frac{n(\alpha-1)}{2}+\frac{n}{4}+\frac{3}{2}\right)} d\tau \right)$$

$$+ C\varepsilon^\alpha \left((1+t)^{-\left(\frac{n\alpha}{2}+\frac{n}{4}\right)} + (1+t)^{-\left(\frac{n\alpha}{2}+\frac{3}{2}\right)} \right) \left(\sup_{0 \le \tau \le t} (1+\tau)^{\frac{n}{4}} \|\Lambda v(\tau)\| \right.$$

$$\left. + \sup_{0 \le \tau \le t} (1+\tau)^{\frac{3}{2}} \sum_{j=3}^4 \|D^j v(\tau)\| + \sup_{0 \le \tau \le t} (1+\tau)^{\frac{n}{2}} \|\Lambda v(\tau)\|_{L^\infty} \right) \cdot \int_{\frac{t}{2}}^t (1+t-\tau)^{-\frac{k}{2}} d\tau$$

$$\le C\varepsilon^\alpha (1+t)^{-\frac{n+2k}{4}} \|v\|_{X_T}. \tag{3.1.50}$$

Therefore,

$$\sup_{0 \le t \le T} \sum_{k=0}^2 (1+t)^{\frac{n+2k}{4}} \|D^k u(t)\| \le C\varepsilon^\alpha \|v\|_{X_T}, \tag{3.1.51}$$

In the same manner as before, we get

$$\sum_{k=0}^S (1+t^k) \|D^k u(t)\|^2 + \sum_{k=0}^S \int_0^t (1+\tau^k) \|D^{k+1} u\|^2 \, d\tau$$

$$\le C \left(\int_0^t \|u\| \, \|F(\Lambda v_1) - F(\Lambda v_2)\| \, d\tau \right.$$

$$\left. + \sum_{k=0}^{S-1} \int_0^t (1+\tau^{k+1}) \|D^k (F(\Lambda v_1) - F(\Lambda v_2))\|^2 \, d\tau \right)$$

$$\leq C \left(\int_0^t \|u\| \, \|F'\|_{L^\infty} \|\Lambda v\| \, d\tau \right.$$

$$\left. + \sum_{k=0}^{S-1} \int_0^t (1+\tau^{k+1}) \left(\|D^k \Lambda v\|^2 \|F'\|_{L^\infty}^2 + \|D^k F'\|^2 \|\Lambda v\|_{L^\infty}^2 \right) d\tau \right)$$

$$\leq C\varepsilon^\alpha \sup_{0\leq\tau\leq t} (1+\tau)^{\frac{n}{4}} \|u(\tau)\| \sup_{0\leq\tau\leq t} (1+\tau)^{\frac{n}{4}} \|\Lambda v(\tau)\| \int_0^t (1+\tau)^{-\frac{n(\alpha+1)}{2}} d\tau$$

$$+ C\varepsilon^{2\alpha} \sum_{k=0}^{S-1} \int_0^t (1+\tau^{k+1})(1+\tau)^{-n\alpha} \|D^k v\|_{H^2}^2 \, d\tau$$

$$+ C \sup_{0\leq\tau\leq t} (1+\tau)^n \|\Lambda v(\tau)\|_{L^\infty}^2 \sum_{k=0}^{S-1} \int_0^t (1+\tau^{k+1})(1+\tau)^{-n}$$

$$\left(\|D^k \Lambda v_1\|^2 + \|D^k \Lambda v_2\|^2 \right) \left(\|\Lambda v_1\|_{L^\infty}^{2(\alpha-1)} + \|\Lambda v_2\|_{L^\infty}^{2(\alpha-1)} \right) d\tau$$

$$\leq C\varepsilon^{2\alpha} \|v\|_{X_T}^2 + C\varepsilon^{2\alpha} \left(\int_0^t \|v\|_{H^2}^2 \, d\tau + \sum_{k=1}^{S-1} \int_0^t (1+\tau)^{k-1} \|D^k v\|_{H^2}^2 \, d\tau \right)$$

$$+ C\varepsilon^{2\alpha} \sup_{0\leq\tau\leq t} (1+\tau)^n \|\Lambda v\|_{L^\infty}^2$$

$$\leq C\varepsilon^{2\alpha} \|v\|_{X_T}^2. \tag{3.1.52}$$

(ii) L^∞ norm estimates

For $0 \leq k \leq 2$, we have

$$\|D^k u(t)\|_{L^\infty} \leq C \int_0^{\frac{t}{2}} (1+t-\tau)^{-\frac{n+k}{2}} \|F(\Lambda v_1) - F(\Lambda v_2)\|_{W^{n+k,1}} \, d\tau$$

$$+ C \int_{\frac{t}{2}}^t (1+t-\tau)^{-\frac{n}{4}} \|D^k(F(\Lambda v_1) - F(\Lambda v_2))\|_{H^n} \, d\tau$$

$$\leq C \left(\int_0^{\frac{t}{2}} (1+t-\tau)^{-\frac{n+k}{2}} \left(\|F'\| \, \|\Lambda v\|_{H^{n+k}} + \|\Lambda v\| \, \|F'\|_{H^{n+k}} \right) d\tau \right.$$

$$+ \int_{\frac{t}{2}}^{t} (1 + t - \tau)^{-\frac{n}{4}} \left(\|F'\|_{L^\infty} \|D^k \Lambda v\|_{H^n} + \|\Lambda v\|_{L^\infty} \|D^k F'\|_{H^n} \right) d\tau \Bigg)$$

$$\leq C(1+t)^{-\frac{n+k}{2}} \varepsilon^\alpha \left(\int_0^{\frac{t}{2}} (1+\tau)^{-\left(\frac{n(\alpha-1)}{2} + \frac{n}{4} \right)} \|\Lambda v\|_{H^{n+k}} \, d\tau \right.$$

$$+ \int_0^{\frac{t}{2}} \left((1+\tau)^{-\left(\frac{n(\alpha-1)}{2} + \frac{n}{4} \right)} + (1+\tau)^{-\left(\frac{n(\alpha-1)}{2} + \frac{3}{2} \right)} \right) \|\Lambda v\| \, d\tau \Bigg)$$

$$+ C\varepsilon^\alpha \left((1+t)^{-\left(\frac{n\alpha}{2} + \frac{n+2k}{4} \right)} \sup_{0 \leq \tau \leq t} \left((1+\tau)^{\frac{n+2k}{4}} \|D^k v(\tau)\|_{H^{2-k}} \right. \right.$$

$$+ (1+\tau)^{\frac{n}{2}} \|\Lambda v(\tau)\|_{L^\infty} \Big) + (1+t)^{-\left(\frac{n\alpha}{2} + \frac{3}{2} \right)} \sup_{0 \leq \tau \leq t} \left((1+\tau)^{\frac{3}{2}} \|D^3 v(\tau)\|_{H^{n+k-1}} \right.$$

$$+ (1+\tau)^{\frac{n}{2}} \|\Lambda v(\tau)\|_{L^\infty} \Big) \Big) \int_{\frac{t}{2}}^{t} (1+t-\tau)^{-\frac{n}{4}} \, d\tau$$

$$\leq C(1+t)^{-\frac{n+k}{2}} \varepsilon^\alpha \|v\|_{X_T}. \tag{3.1.53}$$

(iii) L^1 norm estimates

For $0 \leq k \leq 2$, $t > 0$,

$$\|D^k u(t)\|_{L^1} \leq C \int_0^t (1+t-\tau)^{-\frac{k}{2}} \|F(\Lambda v_1) - F(\Lambda v_2)\|_{W^{k,1}} \, d\tau$$

$$\leq C \int_0^t (1+t-\tau)^{-\frac{k}{2}} \left(\|F'\|_{H^k} \|\Lambda v\| + \|\Lambda v\|_{H^k} \|F'\| \right) d\tau$$

$$\leq C\varepsilon^\alpha \int_0^t (1+t-\tau)^{-\frac{k}{2}} \left((1+\tau)^{-\left(\frac{n(\alpha-1)}{2} + \frac{n+2k}{4} \right)} + (1+\tau)^{-\left(\frac{n(\alpha-1)}{2} + \frac{3}{2} \right)} \right) \|\Lambda v\| \, d\tau$$

$$+ C\varepsilon^\alpha \int_0^t (1+t-\tau)^{-\frac{k}{2}} (1+\tau)^{-\left(\frac{n(\alpha-1)}{2} + \frac{n}{4} \right)} \|\Lambda v\|_{H^k} \, d\tau$$

$$\leq C\varepsilon^\alpha (1+t)^{-\frac{k}{2}} \left(\sup_{0 \leq \tau \leq t} (1+\tau)^{\frac{n}{4}} \|\Lambda v(\tau)\| + \sup_{0 \leq \tau \leq t} (1+\tau)^{\frac{3}{2}} \|D^3 v(\tau)\|_{H^1} \right)$$

$$\leq C\varepsilon^\alpha (1+t)^{-\frac{k}{2}} \|v\|_{X_T}. \tag{3.1.54}$$

Combining (3.1.51)–(3.1.54) yields

$$\|u\|_{X_T} \le C\varepsilon^{\alpha}\|v\|_{X_T}. \tag{3.1.55}$$

Since the constant C appearing in (3.1.55) is independent of ε, T, u and v, we can take ε small enough, if necessary, so that

$$C\varepsilon^{\alpha} \le \frac{1}{2}. \tag{3.1.56}$$

Thus the nonlinear operator $\mathcal{N}$ is a contraction from $\mathcal{S}_{T,\varepsilon}$ to $\mathcal{S}_{T,\varepsilon}$. It turns out that by the contraction mapping theorem, it has a unique fixed point $u \in \mathcal{S}_{T,\varepsilon}$. Since T is arbitrary, it is the unique global solution to the Cauchy problem (3.1.1). The decay rates (3.1.19)–(3.1.22) are deduced from the definition of $\mathcal{S}_{T,\varepsilon}$. Thus the proof of Theorem 3.1.1 is complete.   □

It is clear that the usual energy method will produce good terms $\int_0^t \|D^k u\|^2 d\tau$ with $1 \le k \le s+1$, and when F depends only on derivatives of u, the terms $\int_0^t \int_{\mathbb{R}} uF dx d\tau$ and $\int_0^t \int_{\mathbb{R}} D^{k+1}uD^{k-1}F dx d\tau$ $(1 \le k \le s)$ can be bounded by $\int_0^t \|D^k u\|^2 d\tau, (1 \le k \le s)$ provided that u is small. Based on this observation, Zheng [6] further proved that if F does not explicitly involve u, then for any $n \ge 1$ and any φ small only in H^S, the Cauchy problem (3.1.1) admits a unique global smooth solution. More precisely, we have

Theorem 3.1.2 *Suppose that (i)* $F = F(Du, D^2u)$ *does not involve* u *explicitly; (ii)* F *satisfies (3.1.2) with* $\alpha \ge 1$; *(iii)* $F \in C^{S+1}$ *with an integer* $s > \frac{n}{2} + 3$, $\varphi(x) \in H^S(\mathbb{R}^n)$. *Then for any* $n \ge 1$, *the Cauchy problem (3.1.1) admits a unique global smooth solution* u *such that* $u \in C([0,\infty); H^S) \cap C^1([0,\infty); H^{S-2})$, $Du \in L^2([0,\infty); H^S)$ *provided that* $\|\varphi\|_{H^S}$ *is sufficiently small.*

Moreover, the solution of (3.1.1) has the following decay rates:

$$\|D^j u(t)\| = O(t^{-\frac{j}{2}}), \quad (1 \le j \le s). \tag{3.1.57}$$

Proof. Indeed, when F does not explicitly involve u, we can use only the energy estimates (see Zheng [6]). In this case we define

$$\|u\|_{\tilde{X}_T} = \sup_{0 \le t \le T} \sum_{k=0}^{S}(1 + t^{\frac{k}{2}})\|D^k u(t)\| + \left(\sum_{k=0}^{S} \int_0^T (1 + \tau^k)\|D^{k+1}u\|^2 \, d\tau\right)^{\frac{1}{2}} \tag{3.1.58}$$

and

$$\tilde{\mathcal{S}}_{T,\varepsilon} = \left\{ u \left| \begin{array}{l} u \in C([0,T]; H^S) \cap L^2([0,T]; H^{S+1}), \\ u|_{t=0} = \varphi(x), \quad \|u\|_{\tilde{X}_T} \le \varepsilon \end{array} \right. \right\}. \tag{3.1.59}$$

For $v \in \tilde{\mathcal{S}}_{T,\varepsilon}$, let u be the solution to the problem

$$\begin{cases} u_t - \Delta u = F(\tilde{\Lambda}v), \\ u|_{t=0} = \varphi(x) \end{cases} \tag{3.1.60}$$

where $\tilde{\Lambda}v = (Dv, D^2v)$.

Then in the same manner as before, we obtain

$$\|u\|_{\tilde{X}_T}^2 \le C \left(\|\varphi\|_{H^S}^2 + \int_0^T \|u\| \|F\| \, d\tau + \sum_{k=1}^S \int_0^T (1+\tau^k) \|D^{k-1}F\|^2 \, d\tau \right)$$

$$\le C \left(\|\varphi\|_{H^S}^2 + \int_0^T \|u\| \|\tilde{\Lambda}v\|_{L^\infty}^\alpha \|\tilde{\Lambda}v\| \, d\tau + \sum_{k=1}^S \int_0^T (1+\tau^k) \|\tilde{\Lambda}v\|_{L^\infty}^{2\alpha} \|D^{k-1}\tilde{\Lambda}v\|^2 \, d\tau \right)$$

$$\le C \left(\|\varphi\|_{H^S}^2 + \sup_{0\le\tau\le T} \|u(\tau)\| \sup_{0\le\tau\le T} \|\tilde{\Lambda}v\|_{H^{[\frac{n}{2}]+1}}^{\alpha-1} \int_0^T \|\tilde{\Lambda}v\|_{H^{[\frac{n}{2}]+1}}^2 \, d\tau \right.$$

$$\left. + \sup_{0\le\tau\le T} \left(\tau \|\tilde{\Lambda}v\|_{H^{[\frac{n}{2}]+1}}^{2\alpha} \right) \sum_{k=1}^S \int_0^T (1+\tau^{k-1}) \|D^{k-1}\tilde{\Lambda}v\|^2 d\tau \right)$$

$$\le C \left(\|\varphi\|_{H^S}^2 + \varepsilon^{\alpha+1} \sup_{0\le\tau\le T} \|u(\tau)\| + \varepsilon^{2(\alpha+1)} \right)$$

$$\le C \left(\|\varphi\|_{H^S}^2 + \frac{\delta}{2} \sup_{0\le\tau\le T} \|u(\tau)\|^2 + \frac{\varepsilon^{2(\alpha+1)}}{2\delta} + \varepsilon^{2(\alpha+1)} \right). \tag{3.1.61}$$

Taking $\delta = \frac{1}{C}$, we deduce from (3.1.61) that

$$\|u\|_{\tilde{X}_T}^2 \le C \left(\|\varphi\|_{H^S}^2 + \varepsilon^{2(\alpha+1)} \right). \tag{3.1.62}$$

Let

$$\varepsilon_0 = \min \left(1, \left(\frac{1}{2C} \right)^{\frac{1}{2\alpha}} \right). \tag{3.1.63}$$

Then it turns out from (3.1.62) that when $\|\varphi\|_{H^S} \le \frac{\varepsilon}{\sqrt{2C}} \le \frac{\varepsilon_0}{\sqrt{2C}}$, the nonlinear operator $\mathcal{N}$ defined by (3.1.60) maps $\tilde{\mathcal{S}}_{T,\varepsilon}$ into itself. Moreover, in the same manner as before if necessary, by taking ε small enough, we deduce that $\mathcal{N}$ is a contraction.

The remaining part of the proof is the same as for the proof of Theorem 3.1.1. and we can omit the details.   □

3.2   Fully Nonlinear Parabolic Equations: IBVP

Consider the following initial boundary value problem for fully nonlinear parabolic equations:

$$
\begin{cases}
\dfrac{\partial u}{\partial t} - \displaystyle\sum_{i,j=1}^{n} \dfrac{\partial}{\partial x_i}\left(a_{ij}(x)\dfrac{\partial u}{\partial x_j}\right) + c(x)u = F(x, u, Du, D^2u), \\[4mm]
u|_\Gamma = 0, \\[2mm]
u|_{t=0} = u_0(x)
\end{cases}
\tag{3.2.1}
$$

where Γ is a smooth boundary of a bounded domain $\Omega \subset \mathbb{R}^n$, $x = (x_1, \cdots, x_n)$, $Du = (u_{x_1}, \cdots, u_{x_n})$, $D^2u = (u_{x_i x_j})$, $(i, j = 1, \cdots, n)$.

We make the following assumptions on problem (3.2.1):

(i) Suppose that for simplicity of exposition, $F \in C^\infty, u_0, a_{ij}, c \in C^\infty(\bar{\Omega})$.

(ii) $c(x) > 0$ and there is a positive constant α such that

$$
\sum_{ij=1}^{n} a_{ij}(x)\xi_i\xi_j \geq \alpha \sum_{i=1}^{n} \xi_i^2, \quad \forall \xi \in \mathbb{R}^n, \ x \in \bar{\Omega}.
\tag{3.2.2}
$$

As can be seen from Section 2.1.2, the assumption (ii) implies that the operator

$$
Au = - \sum_{i,j=1}^{n} \frac{\partial}{\partial x_i}\left(a_{ij}(x)\frac{\partial u}{\partial x_j}\right) + c(x)u
\tag{3.2.3}
$$

with $D(A) = H^2 \cap H_o^1$ is a sectorial operator. Moreover, the first eigenvalue λ_1 of A is strictly positive.

Let s be an integer with $s > \frac{n}{2} + 1$. We can successively derive from (3.2.1) the derivatives of u with respect to t at $t = 0$:

$$
u_1(x) = u_t|_{t=0} = F(x, u_0, Du_0, D^2u_0) - Au_0, \cdots, u_S(x) = \left.\frac{\partial^S u}{\partial t^S}\right|_{t=0}
\tag{3.2.4}
$$

which can be explicitly expressed by u_0, a_{ij}, c, F and their derivatives with respect to x. By assumption (i), it is easy to see that

$$
u_j(x) \in C^\infty(\bar{\Omega}), \quad (1 \leq j \leq s).
\tag{3.2.5}
$$

(iii) The following compatibility conditions are satisfied:

$$u_j(x) \in H_o^1(\Omega), \quad (0 \le j \le s). \tag{3.2.6}$$

(iv) $F(x, \omega_0, \omega_i, \omega_{ij}) = O(|\omega|^2)$ near $\omega = 0$ uniformly with respect to $x \in \bar{\Omega}$ where $\omega = (\omega_0, \omega_1, \omega_{ij}) \in \mathbb{R} \times \mathbb{R}^n \times \mathbb{R}^{n^2}$. More precisely,

$$|D^k F(x, \omega_0, \omega_i, \omega_{ij})| \le Const.|\omega|^2, \quad as \ |\omega| \le 1, \ k \le 2s + 1, \tag{3.2.7}$$

$$|D^k D_\omega F(x, \omega_0, \omega_i, \omega_{ij})| \le Const.|\omega|, \quad as \ |\omega| \le 1, \ k + 1 \le 2s + 1, \tag{3.2.8}$$

$$|D^k D_\omega^l F(x, \omega_0, \omega_i, \omega_{ij})| \le Const., \quad as \ |\omega| \le 1, \ k + l \le 2s + 1, \tag{3.2.9}$$

uniformly with respect to $x \in \bar{\Omega}$.

   Then we have

Theorem 3.2.1 *Suppose that the assumptions (i)–(iv) are satisfied. Then problem (3.2.1) admits a unique global smooth solution* $u \in \bigcap\limits_{k=0}^{s} C^k([0, \infty); H^{2(S-k)+1})$, $u^{(k)} = \frac{\partial^k u}{\partial t^k} \in L^2(\mathbb{R}^+; H^{2(S-k+1)}), (0 \le k \le s)$ *provided that* $\|u_0\|_{H^{2s+1}}$ *is sufficiently small. Moreover,* $\|u^{(k)}\|_{H^{2(S-k)+1}}$ *exponentially decay to zero as* $t \to \infty$.

Remark 3.2.1 *It can be seen from the proof of Theorem 3.2.1 given below that the assumption on positivity of c(x) can be weakened and the assertion of the theorem still holds as long as* $\lambda_1 > 0$. *On the other hand, as will be shown in the final section of this chapter, if* $\lambda_1 \le 0$, *the solution u may blow up in finite time no matter how small the initial data are.*

Remark 3.2.2 *We can also deal with the following second initial boundary value problem*

$$\begin{cases} \dfrac{\partial u}{\partial t} - \sum\limits_{i,j=1}^{n} \dfrac{\partial}{\partial x_i}(a_{ij}(x)\dfrac{\partial u}{\partial x_j}) + c(x)u = F(x, u, Du, D^2 u), \\ Bu|_\Gamma = \sum\limits_{i,j=1}^{n} a_{ij}(x)\dfrac{\partial u}{\partial x_j}\cos(\boldsymbol{n}, x_i)|_\Gamma = 0, \\ u|_{t=0} = u_0(x) \end{cases} \tag{3.2.10}$$

where Γ *is a smooth boundary of a bounded domain* $\Omega \subset \mathbb{R}^n$ *and* $\boldsymbol{n}$ *is the outward normal direction to* Γ. *Then under the same assumptions (i), (ii), (iv) and a compatibility condition similar to (iii), similar assertions to Theorem 3.2.1 still hold (see Zheng [9]).*

Remark 3.2.3 *Theorem 3.2.1 shows that in contrast to the Cauchy problem described in the previous section, the result on global existence of a solution to problem (3.2.1) is independent of the space dimension n. This is essentially due to the fact that the solution of the linearized problem, as shown in Section 2.1.2, has an exponential decay rate as $t \to \infty$.*

The results in this subsection are based on Zheng [9], [10]. Before giving the proof, let us briefly recall some related works. There are a great deal of references in the literature concerning the global existence of solutions to the initial boundary value problem for quasilinear parabolic equations. We refer to Friedman [2], Ladyzhenskaja, Solonnikov & Uraltzeva [1] for the earlier results and refer to a series of papers by Amann [1]–[9] for recent developments. We refer to Krylov [1] for the first initial boundary value problem for other classes of fully nonlinear parabolic equations. What they are all concerned with is the global existence of solutions with arbitrary initial data. The price to pay is to put various restrictions on nonlinear terms such as the sign condition, monotonicity or convexity (concavity) conditions, growth condition, etc. We will devote Chapter 4 to the investigation of global existence with arbitrary initial data and display some techniques for obtaining uniform a priori estimates.

As far as the small solution is concerned, in 1948 Bellman [1] investigated problem (3.2.1) with F only involving u and proved the global existence and uniqueness of classical solutions when the initial data are sufficiently small. Ebihara & Nanbu [1] also proved the global existence of small solutions for quasilinear parabolic equations of special form. We should notice that, as was shown by Fujita [1] (also see Section 3.5 in this book), the smallness assumption on initial data is needed to guarantee the global existence of solutions if one does not put any sign condition or other structure conditions on nonlinear terms.

Proof of Theorem 3.2.1 Let $\lambda_1 > 0$ be the first eigenvalue of the operator A defined in (3.2.3). We now introduce the following Banach space:

For any fixed $T > 0$,

$$X_T = \left\{ v(x,t) \,\bigg|\, \frac{\partial^k v}{\partial t^k} \in C([0,T]; H^{2(S-k)+1} \cap H_o^1) \cap L^2([0,T]; H^{2(S-k+1)}), \, 0 \le k \le s \right\}$$

$$(3.2.11)$$

equipped with the weighted norm

$$\|v\|_{X_T} = \left(\sup_{0 \le t \le T} e^{2\beta t} \sum_{k=0}^{S} \|\frac{\partial^k v}{\partial t^k}(t)\|^2_{H^{2(S-k)+1}} + \int_0^T \sum_{k=0}^{S} \|\frac{\partial^k v}{\partial t^k}\|^2_{H^{2(S-k+1)}} d\tau \right)^{\frac{1}{2}} \quad (3.2.12)$$

where β is a positive number satisfying $\beta < \lambda_1$. Let

$$S_{T,\mu} = \left\{ v \, \Big| \, v \in X_T, \; v|_{t=0} = u_0, \; \frac{\partial^k v}{\partial t^k}\Big|_{t=0} = u_k(x), \; (1 \le k \le s), \|v\|_{X_T} \le \mu < 1 \right\} \tag{3.2.13}$$

where $u_k(x)$ $(1 \le k \le s)$ are defined by (3.2.4) and μ is a small positive constant specified later on. It is easy to see that for fixed μ and T, if $\|u_0\|_{H^{2s}}$ is sufficiently small, then $\sum_{j=0}^{S} \frac{u_j(x)t^j}{j!} \in S_{T,\mu}$ and $S_{T,\mu}$ is nonempty. Clearly, $S_{T,\mu}$ is a closed convex subset of X_T.

For any $v \in S_{T,\mu}$, consider the following linear auxiliary problem

$$\begin{cases} u_t - \sum_{i,j=1}^{n} \frac{\partial}{\partial x_i}(a_{ij}(x)\frac{\partial u}{\partial x_j}) + c(x)u = F(x, v, Dv, D^2v) \stackrel{\text{def}}{=} f(x,t), \\ u|_\Gamma = 0, \\ u|_{t=0} = u_0(x). \end{cases} \tag{3.2.14}$$

We need the following lemma

Lemma 3.2.1 *Suppose* $v \in X_T$. *Then* $f(x, t) = F(x, v, Dv, D^2v)$ *satisfies:*

$$\frac{\partial^k f}{\partial t^k} \in C([0, T]; H^{2(S-k)-1}) \cap L^2([0, T]; H^{2(S-k)}), \quad 0 \le k \le s - 1, \tag{3.2.15}$$

$$\frac{\partial^S f}{\partial t^S} \in L^2([0, T]; L^2(\Omega)). \tag{3.2.16}$$

Moreover, for $v \in S_{T,\mu}$,

$$\max_{0 \le t \le T} \sum_{k=0}^{S-1} e^{2\beta t} \|\frac{\partial^k f}{\partial t^k}\|^2_{H^{2(S-k)-1}} \le C\mu^4, \tag{3.2.17}$$

$$\int_0^T \sum_{k=0}^{S} e^{2\beta t} \|\frac{\partial^k f}{\partial t^k}\|^2_{H^{2(S-k)}} d\tau \le C\mu^4 \tag{3.2.18}$$

where $C > 0$ *is a constant independent of* v, T, t, *and* μ.

Proof. We first prove (3.2.15). For $k = 0$, this follows from $v \in X_T$ and the well-known result that H^l forms a Banach algebra as $l > \frac{n}{2}$ (see Adams [1]).

For $1 \le k \le s - 1$, it can be seen that

$$D_x^{2(S-k)-1} \frac{\partial^k f}{\partial t^k} = \sum_{i,j,m} \left(D_x^m F_\omega^{i+j} \prod_{\sigma=1}^{i} D_x^{\alpha_\sigma} v \prod_{l=1}^{j} D_x^{\beta_l} D_t^{\nu_l} v \right) \tag{3.2.19}$$

where $F_\omega^{(i+j)}$ stands for the $(i+j)$th-order derivatives of F with respect to ω and

$$0 \le i \le 2(s-k) - 1, \quad 1 \le j \le k, \quad 1 \le i + j \le 2(s-k) - 1 + k, \tag{3.2.20}$$

$$\nu_1 + \cdots + \nu_j = k, \quad 0 \le m \le 2(s-k) - 1, \tag{3.2.21}$$

$$0 \le \alpha_\sigma \le 2(s-k) + 1, \quad (1 \le \sigma \le i), \tag{3.2.22}$$

$$0 \le \beta_l \le 2(s-k) + 1 \quad (1 \le l \le j) \tag{3.2.23}$$

$$2(s-k) - 1 \le \sum_{\sigma=1}^{i} \alpha_\sigma + \sum_{l=1}^{j} \beta_l \le 2(s-k) - 1 + 2(i+j). \tag{3.2.24}$$

We use the Nirenberg inequality and the Hölder inequality to estimate the L^2 norm of $D_x^{2(s-k)-1} \frac{\partial^k f}{\partial t^k}$ in (3.2.19). We first choose a_σ, $\tilde{a}_l$ so that they satisfy:

$$\frac{\alpha_\sigma}{2s + 1} \le a_\sigma \le 1, \quad \frac{\beta_l}{2(s - \nu_l) + 1} \le \tilde{a}_l \le 1 \tag{3.2.25}$$

and

$$\sum_{\sigma=1}^{i} \alpha_\sigma + \sum_{l=1}^{j} \beta_l + \frac{(i+j-1)n}{2}$$

$$= \sum_{\sigma=1}^{i} a_\sigma(2s + 1) + \sum_{l=1}^{j} \tilde{a}_l(2s - 2\nu_l + 1). \tag{3.2.26}$$

In what follows we prove that such a choice can always be made. Indeed, for $a_\sigma = 1$, $\tilde{a}_l = 1\,(1 \le \sigma \le i, 1 \le l \le j)$, it follows from (3.2.20)–(3.2.24) that

$$\sum_{\sigma=1}^{i}(2s + 1) + \sum_{l=1}^{j}(2s - 2\nu_l + 1) = (2s + 1)(i + j) - 2k$$

$$= (2s + 1)(i + j - 1) + 2s + 1 - 2k > \left(\frac{n}{2} + 2\right)(i + j - 1) + 2(s - k) + 1$$

$$\ge \sum_{\sigma=1}^{i} \alpha_\sigma + \sum_{l=1}^{j} \beta_l + \frac{(i+j-1)n}{2}. \tag{3.2.27}$$

On the other hand, for $a_\sigma = \frac{\alpha_\sigma}{2s+1}$, $\tilde{a}_l = \frac{\beta_l}{2(s-\nu_l)+1}$, $(1 \le \sigma \le i, 1 \le l \le j)$, we have

$$\sum_{\sigma=1}^{i} a_\sigma(2s + 1) + \sum_{l=1}^{j} \tilde{a}_l(2s - 2\nu_l + 1) = \sum_{\sigma=1}^{i} \alpha_\sigma + \sum_{l=1}^{j} \beta_l$$

$$< \sum_{\sigma=1}^{i} \alpha_\sigma + \sum_{l=1}^{j} \beta_l + \frac{(i+j-1)n}{2}. \tag{3.2.28}$$

Thus, (3.2.27), (3.2.28) imply that the hyperplane (3.2.26) must intersect the set

$$\left\{(a_\sigma, \tilde{a}_l) \left| \frac{\alpha_\sigma}{2s+1} \leq a_\sigma \leq 1, \frac{\beta_l}{2(s-\nu_l)+1} \leq \tilde{a}_l \leq 1, (1 \leq \sigma \leq i, \ 1 \leq l \leq j)\right.\right\}$$

(3.2.29)

which proves the claim. Once we have chosen $a_\sigma, \tilde{a}_l$, then $p_\sigma, \tilde{p}_l$ defined by

$$\frac{1}{p_\sigma} = \frac{\alpha_\sigma - a_\sigma(2s+1)}{n} + \frac{1}{2}, \quad 1 \leq \sigma \leq i,$$

(3.2.30)

$$\frac{1}{\tilde{p}_l} = \frac{\beta_l - \tilde{a}_l(2s - 2\nu_l + 1)}{n} + \frac{1}{2}, \quad 1 \leq l \leq j$$

(3.2.31)

must satisfy

$$p_\sigma \geq 2, \ \tilde{p}_l \geq 2, \ \sum_{\sigma=1}^{i} \frac{1}{p_\sigma} + \sum_{l=1}^{j} \frac{1}{\tilde{p}_l} = \frac{1}{2}.$$

(3.2.32)

Then by the Hölder inequality with $p_\sigma, \tilde{p}_l$ satisfying (3.2.32) and the Nirenberg inequality, we get

$$\left\| D_x^m F_\omega^{(i+j)} \prod_{\sigma=1}^{i} D_x^{\alpha_\sigma} v \prod_{l=1}^{j} D_x^{\beta_l} D_t^{\nu_l} v \right\|$$

$$\leq \max_{\bar{\Omega}} |D_x^m F_\omega^{(i+j)}| \prod_{\sigma=1}^{i} \|D_x^{\alpha_\sigma} v\|_{L^{p_\sigma}} \prod_{l=1}^{j} \|D_x^{\beta_l} D_t^{\nu_l} v\|_{L^{\tilde{p}_l}}$$

$$\leq C \max_{\bar{\Omega}} |D_x^m F_\omega^{(i+j)}| \prod_{\sigma=1}^{i} \|D_x^{2S+1} v\|^{a_\sigma} \|v\|^{1-a_\sigma}$$

$$\times \prod_{l=1}^{j} \|D_x^{2(S-\nu_l)+1} D_t^{\nu_l} v\|^{\tilde{a}_l} \|D_t^{\nu_l} v\|^{1-\tilde{a}_l}$$

$$\leq C \max_{\bar{\Omega}} |D_x^m F_\omega^{(i+j)}| \prod_{\sigma=1}^{i} (\|D_x^{2S+1} v\| + \|v\|)$$

$$\times \prod_{l=1}^{j} (\|D_x^{2(S-\nu_l)+1} D_t^{\nu_l} v\| + \|D_t^{\nu_l} v\|).$$

(3.2.33)

Then we can easily deduce from the definition of $C([0,T]; H^{2(S-k)-1})$, (3.2.19) and (3.2.33) that $\frac{\partial^k f}{\partial t^k} \in C([0,T]; H^{2(S-k)-1})$, $(1 \leq k \leq s-1)$ and (3.2.17) follows from (3.2.33) and assumption (iv). Similarly, we can prove that $\frac{\partial^k f}{\partial t^k} \in L^2([0,T]; H^{2(S-k)})$, $(0 \leq k \leq s)$ and the estimate (3.2.18) holds. The proof of Lemma 3.2.1 is complete.

$\square$

Now we continue our proof of Theorem 3.2.1. For $0 \leq k \leq s$, consider the prolonged system

$$\begin{cases} \dfrac{\partial u^{(k)}}{\partial t} + Au^{(k)} = \dfrac{\partial^k f}{\partial t^k}, \\ u^{(k)}|_\Gamma = 0, \\ u^{(k)}|_{t=0} = u_k(x). \end{cases} \tag{3.2.34}$$

where $u^{(0)} = u$, A is defined in (3.2.3) and $u_k(x)$ are given in (3.2.4). It follows from Theorem 1.3.1 that there is a unique solution $u^{(k)} \in C([0,T]; H_o^1) \cap L^2([0,T]; H^2)$ $(0 \leq k \leq s)$ satisfying

$$\frac{1}{2}\frac{d}{dt}\|u^{(k)}\|_{H^1}^2 + \|Au^{(k)}\|^2 = (Au^{(k)}, \frac{\partial^k f}{\partial t^k}) \tag{3.2.35}$$

where the equivalent norm $\|u\|_{H^1} = (Au, u)^{\frac{1}{2}}$ is used. By the Young inequality, we have

$$\frac{d}{dt}\|u^{(S)}\|_{H^1}^2 + 2\|Au^{(S)}\|^2 \leq 2\|Au^{(S)}\|\|\frac{\partial^S f}{\partial t^S}\| \leq 2\frac{(\lambda_1 - \beta)}{\lambda_1}\|Au^{(S)}\|^2 + C\|\frac{\partial^S f}{\partial t^S}\|^2. \tag{3.2.36}$$

Since $\|Au\|^2 \geq \lambda_1 \|u\|_{H^1}^2$, it turns out from (3.2.36) that

$$e^{2\beta t}\|u^{(S)}(t)\|_{H^1}^2 \leq \|u_S\|_{H^1}^2 + C\int_0^t e^{2\beta\tau}\|\frac{\partial^S f}{\partial t^S}\|^2 \, d\tau \tag{3.2.37}$$

and

$$\|u^{(S)}\|_{H^1}^2 + \int_0^t \|u^{(S)}\|_{H^2}^2 \, d\tau \leq C\left(\|u_S\|_{H^1}^2 + \int_0^t \|\frac{\partial^S f}{\partial t^S}\|^2 \, d\tau\right) \tag{3.2.38}$$

where the fact that $\|Au^{(S)}\|$ is equivalent to $\|u^{(S)}\|_{H^2}$ has been used.

We now prove that for $1 \leq k \leq s$, $u^{(k)} = \dfrac{\partial^k u}{\partial t^k}$. Indeed, let

$$v^{(S-1)} = u_{S-1} + \int_0^t u^{(S)} \, d\tau. \tag{3.2.39}$$

Then

$$\frac{\partial v^{(S-1)}}{\partial t} = u^{(S)}, \tag{3.2.40}$$

$$-Av^{(S-1)} = -Au_{S-1} - A\int_0^t u^{(S)}(\tau) \, d\tau. \tag{3.2.41}$$

Since $-A$ generates a C_o-semigroup, it easily follows from semigroup theory that

$$-A \int_0^t u^{(S)}(\tau) \, d\tau = u^{(S)}(t) - u_S - \int_0^t \frac{\partial^S f}{\partial t^S} \, d\tau$$

$$= u^{(S)}(t) - u_S - \frac{\partial^{S-1} f(t)}{\partial t^{S-1}} + \frac{\partial^{S-1} f}{\partial t^{S-1}}\bigg|_{t=0}. \tag{3.2.42}$$

Since $v \in \mathcal{S}_{T,\mu}$, by the definition of $u_k, (1 \leq k \leq s)$, we have

$$u_{k+1} = -Au_k + \frac{\partial^k f}{\partial t^k}\bigg|_{t=0}, \quad (0 \leq k \leq s-1). \tag{3.2.43}$$

Therefore, it turns out from (3.2.39)–(3.2.43) that $v^{(S-1)}$ also satisfies (3.2.34) with $k = s-1$. By uniqueness, we must have $v^{(S-1)} = u^{(S-1)}, i.e., \frac{\partial u^{(S-1)}}{\partial t} = u^{(S)}$. Applying the same argument to $u^{(k)}, (k = s-1, \cdots, 0)$ successively, we obtain $u^{(k)} = \frac{\partial^k u}{\partial t^k}, (1 \leq k \leq s)$.

For $t \in [0, T]$, consider the elliptic problems induced from (3.2.34)

$$\begin{cases} Au^{(k)} = \frac{\partial^k f}{\partial t^k} - u^{(k+1)}, & (0 \leq k \leq s-1) \\ u^{(k)}|_\Gamma = 0. \end{cases} \tag{3.2.44}$$

Then applying the standard elliptic estimates stated in Chapter 1 yields

$$\|u^{(S-1)}(t)\|_{H^3} \leq C \left(\|u^{(S)}(t)\|_{H^1} + \|\frac{\partial^{S-1} f}{\partial t^{S-1}}\|_{H^1} \right), \tag{3.2.45}$$

$$\|u^{(S-2)}(t)\|_{H^5} \leq C \left(\|u^{(S-1)}(t)\|_{H^3} + \|\frac{\partial^{S-2} f}{\partial t^{S-2}}(t)\|_{H^3} \right)$$

$$\leq C \left(\|u^{(S)}(t)\|_{H^1} + \|\frac{\partial^{S-1} f}{\partial t^{S-1}}\|_{H^1} + \|\frac{\partial^{S-2} f}{\partial t^{S-2}}\|_{H^3} \right), \tag{3.2.46}$$

$$\cdots \cdots$$

$$\|u(t)\|_{H^{2S+1}} \leq C \left(\|u^{(S)}(t)\|_{H^1} + \sum_{k=0}^{S-1} \|\frac{\partial^k f}{\partial t^k}\|_{H^{2(S-k)-1}} \right). \tag{3.2.47}$$

Thus, combining (3.2.45)–(3.2.47) with (3.2.37) yields

$$\max_{0 \leq t \leq T} \sum_{k=0}^{S} e^{2\beta t} \|u^{(k)}(t)\|_{H^{2(S-k)+1}}^2 \leq C \left(\|u_S\|_{H^1}^2 + \max_{0 \leq t \leq T} \sum_{k=0}^{S-1} e^{2\beta t} \|\frac{\partial^k f}{\partial t^k}\|_{H^{2(S-k)-1}}^2 \right.$$

$$\left. + \int_0^T e^{2\beta \tau} \|\frac{\partial^S f}{\partial t^S}\|^2 \, d\tau \right). \tag{3.2.48}$$

Similarly, we have

$$\int_0^t \|u^{(S-1)}\|_{H^4}^2 \, d\tau \le C \left(\int_0^t \|u^{(S)}\|_{H^2}^2 \, d\tau + \int_0^t \|\frac{\partial^{S-1} f}{\partial t^{S-1}}\|_{H^2}^2 \, d\tau \right)$$

$$\le C \left(\|u_S\|_{H^1}^2 + \int_0^t \left(\|\frac{\partial^S f}{\partial t^S}\|^2 + \|\frac{\partial^{S-1} f}{\partial t^{S-1}}\|_{H^2}^2 \right) d\tau \right), \tag{3.2.49}$$

$$\cdots \cdots$$

$$\int_0^t \|u\|_{H^{2S+2}}^2 \, d\tau \le C \left(\|u_S\|_{H^1}^2 + \sum_{k=0}^S \int_0^t \|\frac{\partial^k f}{\partial t^k}\|_{H^{2(S-k)}}^2 \, d\tau \right). \tag{3.2.50}$$

Adding together with (3.2.38) yields

$$\sum_{k=0}^S \int_0^t \|u^{(k)}\|_{H^{2(S-k+1)}}^2 \, d\tau \le C \left(\|u_S\|_{H^1}^2 + \sum_{k=0}^S \int_0^t \|\frac{\partial^k f}{\partial t^k}\|_{H^{2(S-k)}}^2 \, d\tau \right). \tag{3.2.51}$$

Combining (3.2.48) with (3.2.51) yields

$$\|u\|_{X_T}^2 \le C \left(\|u_S\|_{H^1}^2 + \max_{0 \le t \le T} \sum_{k=0}^{S-1} e^{2\beta t} \|\frac{\partial^k f}{\partial t^k}\|_{H^{2(S-k)-1}}^2 \right.$$

$$\left. + \int_0^T e^{2\beta\tau} \|\frac{\partial^S f}{\partial t^S}\|^2 \, d\tau + \sum_{k=0}^S \int_0^T \|\frac{\partial^k f}{\partial t^k}\|_{H^{2(S-k)}}^2 \, d\tau \right). \tag{3.2.52}$$

For $0 \le k \le s-1$, it follows from equation (3.2.1) that (refer to (3.2.19) and (3.2.33))

$$D_x^{2(s-k)-1} u_{k+1}(x) = D_x^{2(s-k)-1} \left(-A u_k(x) + \frac{\partial^k f}{\partial t^k}\Big|_{t=0} \right) = -D_x^{2(s-k)-1} A u_k(x)$$

$$+ \sum D_x^m F_\omega^{(i+j)}(x, u_0, D_x u_0, D_x^2 u_0) \prod_{\sigma=1}^i D_x^{\alpha_\sigma} u_0 \prod_{l=1}^j D_x^{\beta_l} u_{\nu_l}. \tag{3.2.53}$$

By assumption (iv) and (3.2.33) we can successively obtain that when $\|u_0\|_{H^{2S+1}} \le \mu < 1$,

$$\|u_{k+1}\|_{H^{2(s-k)-1}} \le C \|u_0\|_{H^{2S+1}}, \quad (0 \le k \le s-1) \tag{3.2.54}$$

where $C > 0$ is again independent of μ.

Thus it follows from (3.2.52), (3.2.54) and Lemma 3.2.1 that

$$\|u\|_{X_T}^2 \le C_1 \left(\|u_0\|_{H^{2S+1}}^2 + \mu^4 \right) \tag{3.2.55}$$

with $C_1 > 0$ independent of T, μ, v, u.

Let w_1, w_2 be two solutions of (3.2.14) corresponding to $v_1, v_2 \in S_{T,\mu}$. Then in the same way as above we obtain

$$\|w_1 - w_2\|_{X_T}^2 \le C_2 \mu^2 \|v_1 - v_2\|_{X_T}^2 \tag{3.2.56}$$

with $C_2 > 0$ also being a constant independent of T, μ, v_1, v_2. It turns out from (3.2.55) and (3.2.56) that there exists a positive constant μ_0:

$$\mu_0 = \min\left(1, \frac{1}{2\sqrt{C_1}}, \frac{1}{2\sqrt{C_2}}\right) \tag{3.2.57}$$

such that for $\mu \le \mu_0$ and $\|u_0\|_{H^{2s+1}} \le \frac{\mu}{2\sqrt{C_1}}$, $v \in \mathcal{S}_{T,\mu}$, the solution u defined by problem (3.2.14) also belongs to $\mathcal{S}_{T,\mu}$. Moreover, the nonlinear operator $\mathcal{N} : v \in \mathcal{S}_{T,\mu} \mapsto u \in \mathcal{S}_{T,\mu}$ is a contraction. Thus the well-known contraction mapping theorem yields that $\mathcal{N}$ has a unique fixed point. Since T is arbitrary and independent of μ, the global existence and uniqueness of a smooth solution follows. The exponential decay of $\left\| \frac{\partial^k u}{\partial t^k} \right\|_{H^{2(S-k)+1}}$ is directly derived from the weighted norm defined in (3.2.12). Thus the proof of Theorem 3.2.1 is complete. $\qquad\square$

3.3 Nonlinear Hyperbolic–Parabolic Coupled Systems: IVP

In this section we consider the initial value problem for two classes of *quasilinear hyperbolic–parabolic coupled systems* (QHPCS). The first class of QHPCS is the following:

$$\begin{cases} \dfrac{\partial u_1}{\partial t} - \dfrac{\partial u_2}{\partial x} = 0, \\[3mm] \dfrac{\partial u_2}{\partial t} - \dfrac{\partial p(u_1, v)}{\partial x} = 0, \\[3mm] c_0(u_1, v)\dfrac{\partial v}{\partial t} - p_v(u_1, v)\dfrac{\partial u_2}{\partial x} - \dfrac{\partial\left(K(u_1, v)\frac{\partial v}{\partial x}\right)}{\partial x} + d(u_1, v) = 0. \end{cases} \tag{3.3.1}$$

Under the assumptions that $\frac{\partial p}{\partial u_1} > 0, c_0 > 0, K > 0$, the first two equations of the system (3.3.1) consist of a quasilinear first-order hyperbolic system with respect to u_1, u_2 and the third equation is a quasilinear parabolic equation of second order with respect to v. The system (3.3.1) includes the one-dimensional nonlinear thermoelastic system and the equations of radiation hydrodynamics in one space dimension. In this section we also consider another class of QHPCS, namely,

$$
\begin{cases}
\dfrac{\partial u_1}{\partial t} - \dfrac{\partial u_2}{\partial x} = 0, \\[4mm]
\dfrac{\partial u_2}{\partial t} - \dfrac{\partial p(u_1, v, u_{2x})}{\partial x} = 0, \\[4mm]
c_0(u_1, v)\dfrac{\partial v}{\partial t} - \dfrac{\partial p(u_1, v, u_{2x})}{\partial v}\dfrac{\partial u_2}{\partial x} - \dfrac{\partial q(u_1, v, v_x)}{\partial x} + d(u_1, u_2, v_x) = 0.
\end{cases}
\tag{3.3.2}
$$

Under the assumptions that $\dfrac{\partial p}{\partial u_{2x}} > 0, c_0 > 0, \dfrac{\partial q}{\partial v_x} > 0$, the second and third equations

consist of a quasilinear parabolic system with respect to u_2, v and the first equation

can be considered as a first-order hyperbolic equation with respect to u_1. The sys-

tem (3.3.2) includes the one-dimensional nonlinear thermoviscoelastic system and the

equations of motion of compressible viscous and heat-conductive fluids.

We will consider the initial value problems for the above two classes of QHPCS in

this section and the initial boundary value problems in the next section.

We first recall related results on QHPCS. The Russian mathematicians Vol'pert

and Hudjaev [1] motivated by many physical problems established a local existence

theorem for the initial value problem for a class of QHPCS. Li, Yu & Shen in a se-

ries of papers [1–5] established the local existence and uniqueness theorem in Hölder

spaces for the initial value problems and initial boundary value problems for a general

class of QHPCS in one space dimension. Zheng in [2] established a local existence

and uniqueness theorem for the initial boundary value problems for a general class

of QHPCS in higher space dimensions. The treatment in that paper also includes

the initial boundary value problem with the boundary being the characteristic for

part of a nonlinear hyperbolic system. See D. Li [1–3] for the results on other ini-

tial boundary value problems. Kawashima in his doctoral thesis [1] considered some

classes of QHPCS, namely, quasilinear symmetric hyperbolic–parabolic systems and

hyperbolic–parabolic systems of conservation law with a convex extension. He estab-

lished a local existence and uniqueness theorem as well as the global existence of small

solutions to these systems.

As far as the global existence of small smooth solutions is concerned, Zheng [4] con-

sidered other classes of QHPCS in higher dimensions. On the other hand, for concrete

physical systems, around 1980 Matsumura & Nishida [1–4] established the global existence of small solutions to the equations describing the motion of compressible viscous
and heat-conductive fluids in three space dimensions. Ponce [1] extended their results
to the Cauchy problem in the two-dimensional case using more delicate decay estimates. In one space dimension, Kawashima with his collaborators also established a
global existence theorem and obtained some decay rates of the solution to the Cauchy
problem. See Kawashima [1] and the references cited there for the results on elec-
tromagnetodynamic systems. Zheng & Shen [2–3] investigated the Cauchy problem
for the systems (3.3.1) and (3.3.2). In these two papers, on which the present section
is based, the authors obtained the global existence and uniqueness of smooth small
solutions with the same decay rates as the linearized system. For related problems,
also see Liu T-P [1]. We refer to Racke [1–2], Ponce & Racke [1] and the references
cited there especially for the results on the Cauchy problem for three-dimensional
nonlinear thermoelasticity. We also refer to Slemrod [1], Racke, Shibata & Zheng [1]
for the initial boundary value problem for one-dimensional nonlinear thermoelasticity.
In Zheng [3], the global existence of small smooth solutions is established for some
initial boundary value problems for the system (3.3.1). We should also mention that
under certain special constitutive conditions it is also possible to obtain global existence for large initial data. In this respect we refer to Dafermos [1], Dafermos and
Hsiao [1] and some references cited in Kawashima [1].

3.3.1 Global Existence and Uniqueness

In this subsection we first consider the following initial value problem for system
(3.3.1):

$$t = 0: \quad u_1 = u_1^0(x), \quad u_2 = u_2^0(x), \quad v = v^0(x). \tag{3.3.3}$$

For problem (3.3.1), (3.3.3), we make the following assumptions which are also rea-
sonable from a physical point of view:

(H_1) There exist positive numbers γ and a_0, a_1, k_1 depending on γ such that when

$|u_1|, |v| \le \gamma$, the following hold:

$$p_{u_1} = \frac{\partial p}{\partial u_1} \ge a_0 > 0, \quad p_v = \frac{\partial p}{\partial v} \ne 0, \quad c_0 \ge a_1 > 0, K \ge k_1 > 0. \qquad (3.3.4)$$

(H_2) $d(u_1, v)$ is a higher-order term of the following form:

$$d(u_1, v) = d_1(u_1, v)u_{1x}v_x + d_2(u_1, v)v_x^2 \qquad (3.3.5)$$

where d_1, d_2 are smooth functions of u_1, v.

(H_3)

$$p, K \in C^4, \quad c_0, d_1, d_2 \in C^3. \qquad (3.3.6)$$

(H_4)

$$U_0(x) = (u_1^0, u_2^0, v^0)^T \in H^3(\mathrm{I\!R}) \cap W^{1,1}(\mathrm{I\!R}). \qquad (3.3.7)$$

From now on we simply denote by $\| \cdot \|$ the $L^2(\mathrm{I\!R})$ norm and by $U(x, t)$ the vector function $(u_1, u_2, v)^T$. We also denote

$$\|U(t)\|_{L^p} = (\|u_1\|_{L^p}^p + \|u_2\|_{L^p}^p + \|v\|_{L^p}^p)^{\frac{1}{p}}, \quad (p = 1, 2, \infty). \qquad (3.3.8)$$

We now have

Theorem 3.3.1 *Under the above assumptions (H_1)–(H_4) there exists a small positive constant ε such that when $\|U_0\|_{H^3} + \|U_0\|_{W^{1,1}} \le \varepsilon$, problem (3.3.1), (3.3.3) admits a unique global solution (u_1, u_2, v) such that $u_1, u_2 \in C([0, \infty); H^3 \cap W^{1,1}) \cap C^1([0, \infty); H^2), v \in C([0, \infty); H^3 \cap W^{1,1}) \cap C^1([0, \infty); H^1)$. Moreover, the solution has the following decay rates:*

$$\|U(t)\|_{L^\infty} = O((1 + t)^{-\frac{1}{2}}), \quad t \to \infty, \qquad (3.3.9)$$

$$\|U(t)\| = O((1 + t)^{-\frac{1}{4}}), \quad t \to \infty, \qquad (3.3.10)$$

$$\|DU(t)\|_{L^1} = O((1 + t)^{-\frac{1}{2}}), \quad t \to \infty. \qquad (3.3.11)$$

Proof of Theorem 3.3.1 We first prove a local existence and uniqueness result:

Lemma 3.3.1 *Under the assumptions (H_1)–(H_4), there exist $\varepsilon_0 > 0$ and $t_0 > 0$ depending only on $\|U_0\|_{H^3}$ such that when $\|U_0\|_{H^3} \le \varepsilon_0$, problem (3.3.1), (3.3.3) admits a unique local solution U in $\mathrm{I\!R} \times [0, t_0]$ such that $u_1, u_2 \in C([0, t_0]; H^3 \cap W^{1,1}) \cap C^1([0, t_0];$*

H^2), $v \in C([0, t_0]; H^3 \cap W^{1,1}) \cap C^1([0, t_0]; H^1) \cap L^2([0, t_0]; H^4)$. Moreover, for $(x, t) \in \mathbb{R} \times [0, t_0]$,

$$|u_1| \leq \gamma, \quad |v| \leq \gamma. \tag{3.3.12}$$

and

$$\|U(t)\|_{H^3} \leq \tilde{C}_1 \|U_0\|_{H^3}, \quad t \in [0, t_0] \tag{3.3.13}$$

where $\tilde{C}_1 > 1$ depends only on ε_0.

Proof. Since (3.3.1) can be reduced to a symmetric QHPCS by multiplying the first equation of (3.3.1) by $\frac{\partial p}{\partial u_1}(u_1, v)$, it follows from Theorem 1.4.3 with $m' = 2, m'' = 1$ that for $U_0 \in H^3$ in a neighbourhood of the origin, there is a unique local solution such that $u_1, u_2 \in C([0, t_0]; H^3) \cap C^1([0, t_0]; H^2)$, $v \in C([0.t_0]; H^3) \cap C^1([0, t_0]; H^1) \cap L^2([0, t_0]; H^4)$. and (3.3.13), (3.3.12) are satisfied.

To conclude the proof of Lemma 3.3.1, it remains to prove that $U \in C([0, t_0]; W^{1,1})$. To this end, we rewrite (3.3.1), (3.3.3) as

$$\begin{cases} u_{1t} - u_{2x} = 0, \\ u_{2t} - p_0 u_{1x} - p_1 v_x = f_2, \\ v_t - p_2 u_{2x} - k v_{xx} = f_3, \\ U|_{t=0} = U_0(x) \end{cases} \tag{3.3.14}$$

with

$$p_0 = \frac{\partial p}{\partial u_1}(0, 0) > 0, \quad p_1 = \frac{\partial p}{\partial v}(0, 0) \neq 0, \quad p_2 = \frac{p_1}{c_0(0, 0)}. \tag{3.3.15}$$

$$k = \frac{K(0, 0)}{c_0(0, 0)} > 0, \quad f_2 = (p - p_0 u_1 - p_1 v)_x, \tag{3.3.16}$$

$$f_3 = -\frac{\tilde{d}(u_1, v)}{c_0} + \left(\frac{p_v(u_1, v)}{c_0} - \frac{p_v(0, 0)}{c_0(0, 0)} \right) u_{2x} + \frac{\partial}{\partial x} \left(\left(\frac{K}{c_0} - k \right) v_x \right), \tag{3.3.17}$$

$$\tilde{d}(u_1, v) = d(u_1, v) - \frac{K}{c_0} \left(\frac{\partial c_0}{\partial u_1} u_{1x} + \frac{\partial c_0}{\partial v} v_x \right) v_x. \tag{3.3.18}$$

As usual, the subscripts t, x and v denote the corresponding partial derivatives with respect to t, x and v, respectively. Making a linear transform of the unknown functions

$$u_1 = \tilde{u}_1, \quad u_2 = \sqrt{p_0}\tilde{u}_2, \quad v = \sqrt{\frac{p_0}{c_0(0, 0)}}\tilde{v} \tag{3.3.19}$$

and letting

$$\alpha = \sqrt{p_0}, \quad \beta = -\frac{p_1}{\sqrt{c_0(0,0)}}, \tag{3.3.20}$$

and still denoting $(\tilde{u}_1, \tilde{u}_2, \tilde{v})$ by (u_1, u_2, v) and the initial datum by U_0, we can rewrite (3.3.1), (3.3.3) equivalently as

$$\begin{cases} u_{1t} - \alpha u_{2x} = 0, \\ u_{2t} - \alpha u_{1x} + \beta v_x = F_2, \\ v_t + \beta u_{2x} - k v_{xx} = F_3, \\ U|_{t=0} = U_0(x) \end{cases} \tag{3.3.21}$$

where

$$F_2 = \frac{1}{\sqrt{p_0}} f_2, \quad F_3 = \sqrt{\frac{c_0(0,0)}{p_0}} f_3. \tag{3.3.22}$$

It easily follows from $U \in C([0, t_0]; H^3)$ that

$$F = \begin{pmatrix} 0 \\ F_2 \\ F_3 \end{pmatrix} \in C([0, t_0]; W^{1,1}). \tag{3.3.23}$$

Thus by the results in Section 2.2.1, the solution U to (3.3.21) can be expressed as

$$U(x, t) = G(x, t) * U_0 + \int_0^t G(x, t - \tau) * F(x, \tau) d\tau \tag{3.3.24}$$

where G, as seen in Section 2.2.1, is the resolvent matrix of the linearized problem and $*$ denotes convolution. Applying the results in Section 2.2.1 shows that $U \in C([0, t_0]; W^{1,1})$. The proof of Lemma 3.3.1 is completed. $\square$

We now continue our proof of Theorem 3.3.1 by deriving uniform a priori estimates of the solution.

Let $T > 0$ be any fixed positive constant and U be a smooth solution in $\mathbb{R} \times [0, T]$ described in Lemma 3.3.1. For any t, $0 \le t \le T$, define

$$\begin{aligned} N_1(0, t) = &\max_{0 \le \tau \le t} \{ (1 + \tau)^{\frac{1}{4}} \|U(\tau)\| + (1 + \tau)^{\frac{3}{4}} \|DU(\tau)\| \\ &+ (1 + \tau) \|D^2 U(\tau)\|_{H^1} \}, \end{aligned} \tag{3.3.25}$$

$$N_2(0,t) = \max_{0\leq\tau\leq t}\{\|U(\tau)\|_{L^1} + (1+\tau)^{\frac{1}{2}}\|DU(\tau)\|_{L^1} + (1+\tau)^{\frac{1}{2}}\|U(\tau)\|_{L^\infty}$$
$$+ (1+\tau)\|DU(\tau)\|_{L^\infty}\}, \tag{3.3.26}$$

$$N_3(0,t) = \left(\int_0^t (\|DU\|_{H^2}^2 + \tau\|D^2U\|^2 + \tau^2\|D^3U\|^2\right.$$
$$+ (1+\tau)^2\|D^4v\|^2)d\tau\Big)^{\frac{1}{2}}. \tag{3.3.27}$$

where, as usual , $D^i = \dfrac{\partial^i}{\partial x^i}$. Let

$$N(0,t) = N_1(0,t) + N_2(0,t) + N_3(0,t). \tag{3.3.28}$$

Then we have the following results concerning uniform a priori estimates:

Lemma 3.3.2 *Suppose that the assumptions (H_1)–(H_4) are satisfied and U is a solution to (3.3.21) in $\mathbb{R}\times[0,T]$ satisfying $|u_1|\leq\gamma$, $|v|\leq\gamma$. Then there exists a positive constant $\varepsilon_1, \varepsilon_1 \leq \varepsilon_0$, such that when $\|U_0\|_{H^3} + \|U_0\|_{W^{1,1}} \leq \varepsilon_1$, the following estimate holds.*

$$N(0,t) \leq \tilde{C}_2(\|U_0\|_{H^3} + \|U_0\|_{W^{1,1}}), \quad t\in[0,T] \tag{3.3.29}$$

where $\tilde{C}_2$ is a positive constant independent of T, U and ε_1.

Before proving Lemma 3.3.2 we would like to point out that once Lemma 3.3.2 is proved, it follows from Lemma 3.3.1 and Lemma 3.3.2 that when $\|U_0\|_{H^3} + \|U_0\|_{W^{1,1}} \leq \varepsilon_1$, problem (3.3.1), (3.3.3) admits a unique local solution in $\mathbb{R}\times[0,t_0]$ satisfying (3.3.12) and (3.3.29). Let

$$\varepsilon = \min(\varepsilon_1, \frac{\varepsilon_0}{\tilde{C}_2}). \tag{3.3.30}$$

Thus, when $\|U_0\|_{H^3} + \|U_0\|_{W^{1,1}} \leq \varepsilon$, by Lemma 3.3.1, the solution can be extended to $\mathbb{R}\times[0,2t_0]$. Moreover, (3.3.12) is satisfied. Applying Lemma 3.3.2 again yields that (3.3.29) is satisfied in $\mathbb{R}\times[0,2t_0]$. Thus we can extend the local solution step by step to get a unique global smooth solution. Moreover, by the definition of the weighted norm $N(0,t)$, (3.3.29) also gives the desired decay estimates. Thus, in order to complete the proof of Theorem 3.3.1, it remains to prove Lemma 3.3.2. Before going into the details of the proof of Lemma 3.3.2, let us explain the idea first. We will derive the following inequality for $N(0,t)$:

$$N^2(0,t) \leq C(\|U_0\|_{H^3}^2 + \|U_0\|_{W^{1,1}}^2 + N^3(0,t) + N^{10}(0,t)) \tag{3.3.31}$$

where $C > 0$ is a constant independent of the solution and of t. Then we can deduce from this inequality and the positivity of $N(0,t)$ that when $\|U_0\|_{H^3}^2 + \|U_0\|_{W^{1,1}}^2$ is sufficiently small, $N(0,t)$ also remains small for all time, i.e., uniform a priori estimate (3.3.29) holds.

Proof of Lemma 3.3.2 Our proof consists of three major steps.

(i) *Energy norm estimates*

Letting D^l $(l = 0, 1, 2)$ act on the system (3.3.21), then multiplying it by $D^l u_1$, $D^l u_2$, $D^l v$ respectively, adding up and integrating with respect to x, we obtain

$$\frac{1}{2}\frac{d}{dt}\|D^l U\|^2 + k\|D^{l+1}v\|^2 = \int_{\rm I\!R}(D^l u_2 D^l F_2 + D^l v D^l F_3)dx, \quad (0 \le l \le 2). \quad (3.3.32)$$

Acting with D^3 on the system (3.3.1), which is equivalent to (3.3.21), we obtain

$$
\begin{cases}
\dfrac{\partial D^3 u_1}{\partial t} - \dfrac{\partial D^3 u_2}{\partial x} = 0, \\[2ex]
\dfrac{\partial D^3 u_2}{\partial t} - p_{u_1}\dfrac{\partial D^3 u_1}{\partial x} - p_v\dfrac{\partial D^3 v}{\partial x} = D^3(p_{u_1}\dfrac{\partial u_1}{\partial x}) - p_{u_1}\dfrac{\partial D^3 u_1}{\partial x} + D^3(p_v\dfrac{\partial v}{\partial x}) - p_v\dfrac{\partial D^3 v}{\partial x}, \\[2ex]
c_0\dfrac{\partial D^3 v}{\partial t} - p_v\dfrac{\partial D^3 u_2}{\partial x} - \dfrac{\partial}{\partial x}(K\dfrac{\partial D^3 v}{\partial x}) = c_0 D^3(\dfrac{p_v}{c_0}\dfrac{\partial u_2}{\partial x}) - p_v\dfrac{\partial D^3 u_2}{\partial x} \\[2ex]
+ c_0 D^3(\dfrac{1}{c_0}\dfrac{\partial}{\partial x}(K\dfrac{\partial v}{\partial x})) - \dfrac{\partial}{\partial x}(K\dfrac{\partial D^3 v}{\partial x}) - c_0 D^3(\dfrac{d}{c_0}).
\end{cases}
$$
$$\quad (3.3.33)$$

Notice that the solution of (3.3.1), (3.3.3) is linked with the solution of (3.3.21) by (3.3.19). Since the norms of the two solutions are equivalent, in what follows we still denote the solution of (3.3.1), (3.3.3) by U.

Multiplying the equations in (3.3.33) by $p_{u_1}D^3 u_1$, $D^3 u_2$, $D^3 v$, respectively, adding up and integrating with respect to x, we get

$$\frac{1}{2}\frac{d}{dt}\int_{\rm I\!R}\left(p_{u_1}(D^3 u_1)^2 + (D^3 u_2)^2 + c_0(D^3 v)^2\right)dx + \int_{\rm I\!R}K(D^4 v)^2 dx = \int_{\rm I\!R}I dx \quad (3.3.34)$$

with

$$
\begin{aligned}
I &= \frac{1}{2}\frac{\partial p_{u_1}}{\partial t}(D^3 u_1)^2 + \frac{1}{2}\frac{\partial c_0}{\partial t}(D^3 v)^2 - \frac{\partial p_{u_1}}{\partial x}D^3 u_1 D^3 u_2 \\[2ex]
&+ D^3 u_2\left(D^3(p_{u_1}\frac{\partial u_1}{\partial x}) - p_{u_1}D^4 u_1 + D^3(p_v\frac{\partial v}{\partial x}) - p_v D^4 v\right)
\end{aligned}
$$

$$- \frac{\partial p_v}{\partial x} D^3 u_2 D^3 v + D^3 v \left(c_0 D^3 (\frac{p_v}{c_0} \frac{\partial u_2}{\partial x}) - p_v D^4 u_2 \right)$$

$$- D^4 v \left(c_0 D^2 (\frac{1}{c_0} \frac{\partial}{\partial x} (K \frac{\partial v}{\partial x})) - K D^4 v - c_0 D^2 (\frac{d}{c_0}) \right)$$

$$- D^3 v \left(D c_0 D^2 (\frac{1}{c_0} \frac{\partial}{\partial x} (K \frac{\partial v}{\partial x})) - D c_0 D^2 (\frac{d}{c_0}) \right). \tag{3.3.35}$$

Integrating in (3.3.32), (3.3.34) with respect to t yields

$$\frac{1}{2} \| D^l U(t) \|^2 + k \int_0^t \| D^{l+1} v \|^2 \, d\tau$$

$$= \frac{1}{2} \| D^l U_0 \|^2 + \int_0^t \int_{\mathrm{IR}} (D^l u_2 D^l F_2 + D^l v D^l F_3) dx \, d\tau, \quad 0 \le l \le 2 \tag{3.3.36}$$

and

$$\frac{1}{2} \int_{\mathrm{IR}} \left(p_{u_1} (D^3 u_1)^2 + (D^3 u_2)^2 + c_0 (D^3 v)^2 \right) dx + \int_0^t \int_{\mathrm{IR}} K (D^4 v)^2 dx \, d\tau$$

$$= \frac{1}{2} \int_{\mathrm{IR}} \left(p_{u_1} (u_1^0, v^0)(D^3 u_1^0)^2 + (D^3 u_2^0)^2 + c_0 (u_1^0, v^0)(D^3 v^0)^2 \right) dx$$

$$+ \int_0^t \int_{\mathrm{IR}} I dx \, d\tau. \tag{3.3.37}$$

For $l = 1, 2$, multiplying (3.3.32) by t and t^2, respectively, then integrating with respect to t, we obtain

$$t \| DU(t) \|^2 + 2k \int_0^t \tau \| D^2 v \|^2 \, d\tau$$

$$= \int_0^t \| DU \|^2 \, d\tau + 2 \int_0^t \tau \int_{\mathrm{IR}} (Du_2 DF_2 + Dv DF_3) dx \, d\tau \tag{3.3.38}$$

and

$$t^2 \| D^2 U(t) \|^2 + 2k \int_0^t \tau^2 \| D^3 v \|^2 \, d\tau$$

$$= 2 \int_0^t \tau \| D^2 U \|^2 \, d\tau + 2 \int_0^t \tau^2 (D^2 u_2 D^2 F_2 + D^2 v D^2 F_3) dx \, d\tau. \tag{3.3.39}$$

Similarly, multilplying (3.3.34) by t^2, then integrating with respect to t yields

$$t^2 \int_{\mathrm{IR}} (p_{u_1} (D^3 u_1)^2 + (D^3 u_2)^2 + c_0 (D^3 v)^2) dx + 2 \int_0^t \tau^2 \int_{\mathrm{IR}} K (D^4 v)^2 dx \, d\tau$$

$$= 2 \int_0^t \tau \int_{\mathrm{IR}} \left(p_{u_1} (D^3 u_1)^2 + (D^3 u_2)^2 + c_0 (D^3 v)^2 \right) dx \, d\tau$$

$$+ 2 \int_0^t \tau^2 \int_{\mathrm{IR}} I dx \, d\tau. \tag{3.3.40}$$

We use system (3.3.21) to estimate $\int_0^t \|D^l u_i\|^2 \, d\tau$ ($i = 1, 2$, $1 \le l \le 3$) appearing on the right-hand side of (3.3.38)–(3.3.40).

In what follows we first get the estimates of $\int_0^t \|D^l u_2\|^2 d\tau$. Differentiating the second equation of (3.3.21) with respect to x, then multiplying it by βv and multiplying the third equation of (3.3.21) by βu_{2x}, adding up and integrating with respect to x, we obtain

$$\frac{d}{dt} \left(\int_{\rm I\!R} \beta v \, Du_2 dx \right) + \beta^2 \|Du_2\|^2 + \alpha\beta \int_{\rm I\!R} Du_1 Dv dx - \beta^2 \|Dv\|^2 - k\beta \int_{\rm I\!R} Du_2 D^2 v dx$$
$$= \beta \int_{\rm I\!R} (v DF_2 + F_3 Du_2) dx. \tag{3.3.41}$$

Integrating with respect to t and using the inequality $ab \le \frac{\eta a^2}{2} + \frac{b^2}{2\eta}$, we get

$$\beta \int_{\rm I\!R} Du_2 v dx + \frac{\beta^2}{2} \int_0^t \|Du_2\|^2 \, d\tau$$
$$\le C \left(\|U_0\|_{H^1}^2 + \int_0^t (\eta \|Du_1\|^2 + \frac{1}{\eta} \|Dv\|_{H^1}^2) \, d\tau \right.$$
$$\left. + \left| \int_0^t \int_{\rm I\!R} (v DF_2 + F_3 Du_2) dx \, d\tau \right| \right) \tag{3.3.42}$$

where η is an arbitrarily small positive constant.

In the same way as before, we have

$$\frac{d}{dt} \left(\beta \int_{\rm I\!R} Dv D^2 u_2 dx \right) + \beta^2 \|D^2 u_2\|^2 + \alpha\beta \int_{\rm I\!R} D^2 u_1 D^2 v dx - \beta^2 \|D^2 v\|^2$$
$$- k\beta \int_{\rm I\!R} D^2 u_2 D^3 v dx = \beta \int_{\rm I\!R} (Dv D^2 F_2 + DF_3 D^2 u_2) dx \tag{3.3.43}$$

and

$$\frac{d}{dt} \left(\beta \int_{\rm I\!R} D^3 u_2 D^2 v dx \right) + \beta^2 \|D^3 u_2\|^2 + \alpha\beta \int_{\rm I\!R} D^3 u_1 D^3 v dx - \beta^2 \|D^3 v\|^2$$
$$- k\beta \int_{\rm I\!R} D^3 u_2 D^4 v dx = \beta \int_{\rm I\!R} (D^2 v D^3 F_2 + D^3 u_2 D^2 F_3) dx. \tag{3.3.44}$$

Integrating with respect to t, we get

$$\beta \int_{\rm I\!R} D^2 u_2 Dv dx + \frac{\beta^2}{2} \int_0^t \|D^2 u_2\|^2 \, d\tau$$
$$\le C \left(\|U_0\|_{H^2}^2 + \int_0^t (\eta \|D^2 u_1\|^2 + \frac{1}{\eta} \|D^2 v\|_{H^1}^2) \, d\tau \right.$$
$$\left. + \left| \int_0^t \int_{\rm I\!R} (Dv D^2 F_2 + DF_3 D^2 u_2) dx \, d\tau \right| \right) \tag{3.3.45}$$

and

$$\beta \int_{\mathbb{R}} D^3 u_2 D^2 v dx + \frac{\beta^2}{2} \int_0^t \|D^3 u_2\|^2 d\tau$$
$$\leq C \left(\|U_0\|_{H^3}^2 + \int_0^t (\eta \|D^3 u_1\|^2 + \frac{1}{\eta} \|D^3 v\|_{H^1}^2) d\tau \right.$$
$$\left. + \left| \int_0^t \int_{\mathbb{R}} (-D^3 v D^2 F_2 + D^3 u_2 D^2 F_3) dx \, d\tau \right| \right). \tag{3.3.46}$$

Multiplying (3.3.43)–(3.3.44) by t, t^2, respectively, then integrating with respect to t, we obtain

$$\beta t \int_{\mathbb{R}} D^2 u_2 Dv dx + \frac{\beta^2}{2} \int_0^t \tau \|D^2 u_2\|^2 d\tau$$
$$\leq C \left(\int_0^t (\|D^2 u_2\|^2 + \|Dv\|^2 + \eta \tau \|D^2 u_1\|^2 + \frac{\tau}{\eta} \|D^2 v\|_{H^1}^2) d\tau \right.$$
$$\left. + \left| \int_0^t \tau \int_{\mathbb{R}} (Dv D^2 F_2 + D^2 u_2 DF_3) dx \, d\tau \right| \right) \tag{3.3.47}$$

and

$$\beta t^2 \int_{\mathbb{R}} D^3 u_2 D^2 v dx + \frac{\beta^2}{2} \int_0^t \tau^2 \|D^3 u_2\|^2 d\tau$$
$$\leq C \left(\int_0^t (\eta \tau^2 \|D^3 u_1\|^2 + \frac{\tau^2}{\eta} \|D^3 v\|_{H^1}^2 + \|D^2 v\|^2) d\tau \right.$$
$$\left. + \left| \int_0^t \tau^2 \int_{\mathbb{R}} (-D^3 v D^2 F_2 + D^3 u_2 D^2 F_3) dx \, d\tau \right| \right). \tag{3.3.48}$$

Estimates (3.3.38), (3.3.41)–(3.3.44) show that $\int_0^t \|D^l u_2\|^2 d\tau, (1 \leq l \leq 3)$ can be bounded by $\eta \int_0^t \|D^l u_1\|^2 d\tau$, with η being small, and other terms.

To get the estimates of $\int_0^t \|D^l u_1\|^2 d\tau$, $(1 \leq l \leq 3)$, we differentiate the first equation of (3.3.21) with respect to x, then we multiply it by $-\alpha u_2$ and multiply the second equation of (3.3.21) by $-\alpha Du_1$, add up and integrate with respect to x to obtain

$$\frac{d}{dt} \left(-\alpha \int_{\mathbb{R}} u_2 Du_1 dx \right) + \alpha^2 \|Du_1\|^2 - \alpha^2 \|Du_2\|^2 - \alpha\beta \int_{\mathbb{R}} Du_1 Dv dx$$
$$= -\alpha \int_{\mathbb{R}} Du_1 F_2 dx. \tag{3.3.49}$$

Similarly, we have

$$\frac{d}{dt} \left(-\alpha \int_{\mathbb{R}} Du_2 D^2 u_1 dx \right) + \alpha^2 \|D^2 u_1\|^2 - \alpha^2 \|D^2 u_2\|^2 - \alpha\beta \int_{\mathbb{R}} D^2 u_1 D^2 v dx$$
$$= -\alpha \int_{\mathbb{R}} D^2 u_1 DF_2 dx \tag{3.3.50}$$

and

$$\frac{d}{dt}\left(-\alpha \int_{\rm IR} D^2 u_2 D^3 u_1 dx\right) + \alpha^2 \|D^3 u_1\|^2 - \alpha^2 \|D^3 u_2\|^2 - \alpha\beta \int_{\rm IR} D^3 u_1 D^3 v dx$$
$$= -\alpha \int_{\rm IR} D^3 u_1 D^2 F_2 dx. \tag{3.3.51}$$

Integrating with respect to t, we obtain

$$-\alpha \int_{\rm IR} u_2 D u_1 dx + \frac{\alpha^2}{2} \int_0^t \|D u_1\|^2 \, d\tau$$
$$\le C\left(\|U_0\|_{H^1}^2 + \int_0^t (\|D u_2\|^2 + \|D v\|^2) \, d\tau + \left|\int_0^t \int_{\rm IR} D u_1 F_2 dx \, d\tau\right|\right) \tag{3.3.52}$$

and

$$-\alpha \int_{\rm IR} D u_2 D^2 u_1 dx + \frac{\alpha^2}{2} \int_0^t \|D^2 u_1\|^2 \, d\tau$$
$$\le C\left(\|U_0\|_{H^2}^2 + \int_0^t (\|D^2 u_2\|^2 + \|D^2 v\|^2) \, d\tau\right.$$
$$\left. + \left|\int_0^t \int_{\rm IR} D^2 u_1 D F_2 dx \, d\tau\right|\right), \tag{3.3.53}$$

$$-\alpha \int_{\rm IR} D^2 u_2 D^3 u_1 dx + \frac{\alpha^2}{2} \int_0^t \|D^3 u_1\|^2 \, d\tau$$
$$\le C\left(\|U_0\|_{H^3}^2 + \int_0^t (\|D^3 u_2\|^2 + \|D^3 v\|^2) d\tau\right.$$
$$\left. + \left|\int_0^t \int_{\rm IR} D^3 u_1 D^2 F_2 dx \, d\tau\right|\right). \tag{3.3.54}$$

Multiplying (3.3.50), (3.3.51) by t, t^2, respectively, then integrating with respect to t yields

$$-\alpha t \int_{\rm IR} D u_2 D^2 u_1 dx + \frac{\alpha^2}{2} \int_0^t \tau \|D^2 u_1\|^2 \, d\tau$$
$$\le C\left(\int_0^t \tau (\|D^2 u_2\|^2 + \|D^2 v\|^2) \, d\tau + \left|\int_0^t \tau \int_{\rm IR} D^2 u_1 D F_2 dx \, d\tau\right|\right.$$
$$\left. + \int_0^t (\|D u_2\|^2 + \|D^2 u_1\|^2) \, d\tau\right) \tag{3.3.55}$$

and

$$-\alpha t^2 \int_{\rm IR} D^2 u_2 D^3 u_1 dx + \frac{\alpha^2}{4} \int_0^t \tau^2 \|D^3 u_1\|^2 \, d\tau$$
$$\le C\left(\int_0^t (\tau^2 (\|D^3 u_2\|^2 + \|D^3 v\|^2) + \|D^2 u_2\|^2) \, d\tau\right.$$
$$\left. + \left|\int_0^t \tau^2 \int_{\rm IR} D^3 u_1 D^2 F_2 dx \, d\tau\right|\right). \tag{3.3.56}$$

Estimates (3.3.52)–(3.3.56) show that $\int_0^t \|D^l u_1\|^2 \, d\tau, (1 \le l \le 3)$ can be bounded by $\int_0^t \|D^l u_2\|^2 \, d\tau, (1 \le l \le 3)$ and other terms. Now we appropriately make linear combination of (3.3.36)–(3.3.40), with (3.3.42), (3.3.45)–(3.3.48), (3.3.52)–(3.3.56) so as to get the energy estimates.

Let μ, η be small positive constants and set

$$(3.3.36) + (3.3.37) + \mu((3.3.42) + (3.3.45) + (3.3.46))$$
$$+\eta^{\frac{1}{2}}((3.3.52) + (3.3.53) + (3.3.54)))$$
$$+ \mu^2(3.3.38) + \mu^3((3.3.47) + \sqrt{\eta}(3.3.55)) + \mu^4(3.3.39)$$
$$+\mu^5((3.3.48) + \sqrt{\eta}(3.3.56)) + \mu^6(3.3.40). \tag{3.3.57}$$

First taking η appropriately small, then taking μ sufficiently small, by the inequality $ab \le \frac{\varepsilon a^2}{2} + \frac{b^2}{2\varepsilon}$, we can deduce from (3.3.57) that

$$\|U(t)\|_{H^3}^2 + t\|DU(t)\|^2 + t^2\|D^2 U(t)\|_{H^1}^2 + N_3^2(0,t)$$
$$\le C \left(\|U_0\|_{H^3}^2 + \sum_{l=0}^2 \left| \int_0^t \int_{\rm I\!R} (D^l u_2 D^l F_2 + D^l v D^l F_3) dx \, d\tau \right| + \left| \int_0^t \int_{\rm I\!R} I dx \, d\tau \right| \right.$$
$$+ \sum_{l=1}^2 \left| \int_0^t \tau^l \int_{\rm I\!R} (D^l u_2 D^l F_2 + D^l F_3 D^l v) dx \, d\tau \right|$$
$$+ \sum_{l=0}^2 \left| \int_0^t \int_{\rm I\!R} (-D^{l+1} v D^l F_2 + D^{l+1} u_2 D^l F_3) dx \, d\tau \right| + \sum_{l=0}^2 \left| \int_0^t \int_{\rm I\!R} D^{l+1} u_1 D^l F_2 dx \, d\tau \right|$$
$$+ \sum_{l=1}^2 \left| \int_0^t \tau^l \int_{\rm I\!R} (-D^{l+1} v D^l F_2 + D^{l+1} u_2 D^l F_3) dx \, d\tau \right|$$
$$+ \sum_{l=1}^2 \left| \int_0^t \tau^l \int_{\rm I\!R} D^{l+1} u_1 D^l F_2 dx \, d\tau \right| + \left| \int_0^t \tau^2 \int_{\rm I\!R} I dx \, d\tau \right| \right). \tag{3.3.58}$$

In what follows we prove that the integral terms on the right-hand side of (3.3.58) can be bounded by $CN^3(0,t)$ and the higher-order terms of $N(0,t)$. Indeed, it can be seen from the expressions for F_2, F_3 that

$$\left| \int_0^t \int_{\rm I\!R} (F_2 u_2 + F_3 v) dx \, d\tau \right|$$
$$\le C \int_0^t \int_{\rm I\!R} \left(|u_1 u_2 v_x| + |v| \left(|u_{1x}||v_x| + |v_x|^2 \right. \right.$$
$$\left. \left. + |v||u_{2x}| + |u_1||u_{2x}| + |u_1||v_{xx}| + |v||v_{xx}| \right) \right) dx \, d\tau \tag{3.3.59}$$

and

$$\int_0^t \int_{\mathrm{IR}} |u_1 u_2 v_x| dx\, d\tau$$

$$\leq \int_0^t \|u_1\|_{L^\infty} \|u_2\|_{L^\infty} \|v_x\|_{L^1}\, d\tau$$

$$\leq N_2^3(0,t) \int_0^t (1+\tau)^{-\frac{3}{2}}\, d\tau \leq C N^3(0,t). \tag{3.3.60}$$

Notice that the usual L^2 norm energy method is not enough to get uniform bounded-ness of the above term. The appearance and estimate of such terms explains why we need to combine the L^2 norm energy method with L^∞-L^1 norm decay estimates (we also refer to Section 3.1).

Similar estimates lead to

$$\left| \int_0^t \int_{\mathrm{IR}} (F_2 u_2 + F_3 v) dx\, d\tau \right| \leq C N^3(0,t). \tag{3.3.61}$$

The other terms on the right-hand side of (3.3.58) can be bounded in the same manner as before. Since it is quite tedious, in what follows we only pick up some terms and show the method of estimates.

$$\left| \int_0^t \int_{\mathrm{IR}} D^2 u_2 D^2 F_2 dx\, d\tau \right|$$

$$\leq C \int_0^t \int_{\mathrm{IR}} |D^2 u_2| \Big((|u_1| + |v|)(|D^3 u_1| + |D^3 v|)$$

$$+ |Du_1||Dv|^2 + |Du_1|^2|Dv| + |Du_1|^3 + |Dv|^3 \Big) dx\, d\tau$$

$$\leq C \left(\int_0^t (\|u_1\|_{L^\infty} + \|v\|_{L^\infty}) \|D^2 u_2\| (\|D^3 u_1\| + \|D^3 v\|)\, d\tau \right.$$

$$\left. + \int_0^t (\|Du_1\|_{L^\infty} + \|Dv\|_{L^\infty})^2 \|D^2 u_2\| (\|Du_1\| + \|Dv\|)\, d\tau \right)$$

$$\leq C \left(N^3(0,t) \int_0^t (1+\tau)^{-\frac{1}{2}-2} d\tau + N^4(0,t) \int_0^t (1+\tau)^{-2-1-\frac{3}{4}} d\tau \right)$$

$$\leq C \left(N^3(0,t) + N^4(0,t) \right). \tag{3.3.62}$$

$$\left| \int_0^t \tau^2 \int_{\mathrm{IR}} D^2 u_2 D^2 F_2 dx\, d\tau \right|$$

$$\leq C \left(N_2(0,t) \int_0^t \tau^{\frac{3}{2}} \|D^2 u_2\| (\|D^3 u_1\| + \|D^3 v\|)\, d\tau + N^4(0,t) \int_0^t (1+\tau)^{-(1+\frac{3}{4})}\, d\tau \right)$$

$$\leq C \left(N_2(0,t) \left(\int_0^t \tau \|D^2 u_2\|^2\, d\tau \right)^{\frac{1}{2}} \left(\int_0^t \tau^2 (\|D^3 u_1\|^2 + \|D^3 v\|^2)\, d\tau \right)^{\frac{1}{2}} + N^4(0,t) \right)$$

$$\leq C \left(N^3(0,t) + N^4(0,t) \right). \tag{3.3.63}$$

Similarly, we have

$$
\int_0^t \int_{\rm I\!R} |I| dx \, d\tau
$$
$$
\leq C \int_0^t \int_{\rm I\!R} \Big((|Du_2| + |D^2v| + |Du_1||Dv| + |Dv|^2)(|D^3u_1|^2 + |D^3v|^2)
$$
$$
+ |D^3u_2|((|Du_1| + |Dv|)^4 + |D^2u_1|^2 + |D^2v|^2 + (|Du_1| + |Dv|)(|D^3u_1| + |D^3v|))
$$
$$
(|D^3v| + |D^4v|)(|DU|^4 + |DU||D^3U| + |D^2U|^2)
$$
$$
+|D^3v||DU|(|DU|^4 + |D^2U|^2 + |DU||D^3U| + |D^4v|) \Big) dx \, d\tau
$$
$$
\leq C \left(N^3(0,t) + N^4(0,t) + N^5(0,t) + N^6(0,t) \right). \tag{3.3.64}
$$

To estimate $\int_0^t \tau^2 \int_{\rm I\!R} |I| dx \, d\tau$, the crucial term is

$$
\int_0^t \tau^2 \int_{\rm I\!R} |DU||D^3v||D^4v| dx \, d\tau \leq C N_2(0,t) \int_0^t (1+\tau)\|D^3v\|\|D^4v\| \, d\tau
$$
$$
\leq C N_2(0,t) \left(\int_0^t \|D^3v\|^2 \, d\tau \right)^{\frac{1}{2}} \left(\int_0^t (1+\tau)^2 \|D^4v\|^2 \, d\tau \right)^{\frac{1}{2}}
$$
$$
\leq C N^3(0,t). \tag{3.3.65}
$$

Therefore, we get

$$
\int_0^t \tau^2 \int_{\rm I\!R} |I| dx \, d\tau \leq C \left(N^3(0,t) + N^4(0,t) + N^5(0,t) + N^6(0,t) \right). \tag{3.3.66}
$$

The other terms can be bounded in the same way. Finally, we arrive at

$$
\|U(t)\|_{H^3}^2 + t\|DU(t)\|^2 + t^2\|D^2U(t)\|_{H^1}^2 + N_3^2(0,t)
$$
$$
\leq C \left(\|U_0\|_{H^3}^2 + N^3(0,t) + N^4(0,t) + N^5(0,t) + N^6(0,t) \right). \tag{3.3.67}
$$

(ii) $W^{1,1}$ *norm estimates*

It follows from (3.3.21) and the results in Section 2.2.1 that

$$
\|U(t)\|_{L^1} \leq \|G * U_0\|_{L^1} + \| \int_0^t G(t-\tau) * F(\tau) \, d\tau \|_{L^1} \tag{3.3.68}
$$

where

$$
F = \begin{pmatrix} 0 \\ F_2 \\ F_3 \end{pmatrix}. \tag{3.3.69}
$$

By Theorem 2.2.1, we have

$$\|G * U_0\|_{L^1} \leq C\|U_0\|_{L^1}. \tag{3.3.70}$$

Noticing that $F_2 = Dg_2$ with $g_2 = \frac{1}{\sqrt{p_0}}(p - p_0 u_1 - p_1 v)$, by integration by parts and (2.2.15), we obtain for $i = 1, 2, 3$

$$\|\int_0^t G_{i2}(t - \tau) * F_2(\tau)\, d\tau\|_{L^1} = \|\int_0^t \frac{\partial G_{i2}}{\partial x}(t - \tau) * g_2\, d\tau\|_{L^1}$$

$$\leq C \int_0^t \frac{1}{\sqrt{1 + t - \tau}}\left(\|g_2\|_{L^1} + \|Dg_2\|_{L^1}\right) d\tau$$

$$\leq C \int_0^t \frac{1}{\sqrt{1 + t - \tau}}\left(\|U\|^2 + \|U\|_{L^\infty}\|DU\|_{L^1}\right) d\tau$$

$$\leq CN^2(0, t) \int_0^t \frac{1}{\sqrt{1 + t - \tau}} \frac{d\tau}{\sqrt{1 + \tau}} \leq CN^2(0, t). \tag{3.3.71}$$

To estimate $\|\int_0^t G_{i3} * F_3\, d\tau\|_{L^1}$, we rewrite F_3 as

$$F_3 = (h_1 + h_2 + h_3)\sqrt{\frac{c_0(0,0)}{p_0}} \tag{3.3.72}$$

where

$$\begin{cases} h_1 = -\frac{\tilde{d}}{c_0}, \quad h_2 = \left(\frac{p_v}{c_0} - \frac{p_v(0,0)}{c_0(0,0)}\right) u_{2x} = (g_{31}u_1 + g_{32}v)u_{2x}, \\[2mm] h_3 = \left(\left(\frac{K}{c_0} - \frac{K(0,0)}{c_0(0,0)}\right) v_x\right)_x \end{cases} \tag{3.3.73}$$

and g_{31}, g_{32} can be expressed by the integrals of $\frac{p_v}{c_0}$.

Thus in the same manner as before we get

$$\|\int_0^t G_{i3} * h_3\, d\tau\|_{L^1} \leq C \int_0^t \|h_3\|_{L^1}\, d\tau$$

$$\leq C \int_0^t \left(\|DU\|^2 + \|U\|\|v_{xx}\|\right) d\tau \leq CN^2(0, t), \tag{3.3.74}$$

and

$$\|\int_0^t G_{i3} * h_1\, d\tau\|_{L^1} \leq C \int_0^t \|h_1\|_{L^1}\, d\tau \leq C \int_0^t \|DU\|^2\, d\tau \leq CN^2(0, t). \tag{3.3.75}$$

The estimate of $\|\int_0^t G_{i3} * h_2\, d\tau\|_{L^1}$ is more delicate. By the system (3.3.21), we have

$$(g_{31}u_1 + g_{32}v)u_{2x} = \frac{1}{\alpha}g_{31}u_1 u_{1t} + \frac{g_{32}}{\beta}v(-v_t + kv_{xx} + F_3)$$

$$= \frac{1}{\alpha}g_{31}u_1 u_{1t} - \frac{g_{32}}{\beta}vv_t + \frac{kg_{32}}{\beta}vv_{xx} + \frac{g_{32}}{\beta}vF_3. \tag{3.3.76}$$

It is easy to get

$$\left\| \int_0^t G_{i3} * \left(\frac{k g_{32}}{\beta} v v_{xx} + \frac{g_{32}}{\beta} v F_3 \right) d\tau \right\|_{L^1}$$

$$\leq C \int_0^t \left(\|v\| \, \|v_{xx}\| + \|v\|_{L^\infty} \|F_3\|_{L^1} \right) d\tau \leq C \left(N^2(0,t) + N^3(0,t) \right). \quad (3.3.77)$$

Now it remains to estimate $\left\| \int_0^t G_{i3} * \left(\frac{1}{\alpha} g_{31} u_1 u_{1\tau} - \frac{1}{\beta} g_{32} v v_\tau \right) d\tau \right\|_{L^1}$. By integration by parts and Theorem 2.2.2 we have

$$\left\| \int_0^t \frac{1}{\alpha} G_{i3} * g_{31} u_1 u_{1\tau} \, d\tau \right\|_{L^1}$$

$$\leq C \left(\left\| \int_0^t G_{i3} * \frac{\partial(g_{31} u_1^2)}{\partial \tau} d\tau \right\|_{L^1} + \left\| \int_0^t G_{i3} * \left(\frac{\partial g_{31}}{\partial u_1} u_{1\tau} + \frac{\partial g_{31}}{\partial v} v_\tau \right) u_1^2 \, d\tau \right\|_{L^1} \right)$$

$$\leq C \left(\|G_{i3}(t - \tau) * g_{31} u_1^2 \, (\tau)|_0^t \|_{L^1} + \left\| \int_0^t \frac{\partial G_{i3}(t - \tau)}{\partial t} * (g_{31} u_1^2) \, d\tau \right\|_{L^1} \right.$$

$$\left. + \int_0^t \left(\|u_{2x} u_1^2\|_{L^1} + \|(u_{1x} v_x + v_x^2 + v_x x) u_1^2\|_{L^1} \right) d\tau \right). \quad (3.3.78)$$

Since

$$G_{i3}(0) * (g_{31} u_1^2(t)) = \begin{cases} g_{31} u_1^2(t), & i = 3, \\ 0 & i = 1, 2, \end{cases} \quad (3.3.79)$$

we have

$$\|G_{i3}(t - \tau) * g_{31} u_1^2(\tau) \, |_0^t \, \|_{L^1} \leq C \left(N^2(0,t) + \|G_{i3}(t) * g_{31} u_1^2(0)\|_{L^1} \right)$$

$$\leq C N^2(0,t). \quad (3.3.80)$$

It follows from Theorem 2.2.2 and Lemma 3.1.3 that

$$\left\| \int_0^t \frac{\partial G_{i3}}{\partial t} * (g_{31} u_1^2) \, d\tau \right\|_{L^1} \leq C \int_0^t (1 + t - \tau)^{-1} \|g_{31} u_1^2\|_{W^{2,1}} \, d\tau$$

$$\leq C \int_0^t (1 + t - \tau)^{-1} \left(\|U\|^2 + \|U\| \, \|DU\| + \|U\|_{L^\infty} \|U\| \, \|DU\| \right.$$

$$\left. + \|U\|_{L^\infty}^2 \|DU\|^2 + \|U\|_{L^\infty} \|U\| \, \|D^2 U\| + \|DU\|^2 \right) d\tau$$

$$\leq C \left(N^2(0,t) \int_0^t (1 + t - \tau)^{-1} (1 + \tau)^{-\frac{1}{2}} \, d\tau \right.$$

$$+ N^3(0,t) \int_0^t (1 + t - \tau)^{-1} (1 + \tau)^{-(\frac{1}{2} + \frac{1}{4} + \frac{3}{4})} \, d\tau$$

$$\left. + N^4(0,t) \int_0^t (1 + t - \tau)^{-1} (1 + \tau)^{-(1 + \frac{3}{2})} \, d\tau \right)$$

$$\leq C(N^2(0,t) + N^3(0,t) + N^4(0,t)). \quad (3.3.81)$$

$\| \int_0^t G_{i3} * \left(-\frac{g_{32}}{\beta} v v_\tau \right) d\tau \|_{L^1}$ can be estimated in the same manner as before. Thus we arrive at

$$\| U(t) \|_{L^1} \le C \left(\| U_0 \|_{L^1} + N^2(0, t) + N^3(0, t) + N^4(0, t) \right). \tag{3.3.82}$$

We now proceed to estimate $\| DU(t) \|_{L^1}$:

$$\| DU(t) \|_{L^1} \le \| DG * U_0 \|_{L^1} + \| \int_0^t DG(t - \tau) * F(\tau) d\tau \|_{L^1}$$

$$\le C(1 + t)^{-\frac{1}{2}} \| U_0 \|_{W^{1,1}} + \| \int_0^t DG(t - \tau) * F(\tau) d\tau \|_{L^1}. \tag{3.3.83}$$

By integration by parts, we get

$$\| \int_0^t DG_{i2} * F_2 \, d\tau \|_{L^1} \le \| \int_{\frac{t}{2}}^t DG_{i2} * F_2 \, d\tau \|_{L^1} + \| \int_0^{\frac{t}{2}} DG_{i2} * F_2 \, d\tau \|_{L^1}$$

$$\le C \int_{\frac{t}{2}}^t \frac{1}{\sqrt{1 + t - \tau}} \| F_2 \|_{W^{1,1}} \, d\tau + \| \int_0^{\frac{t}{2}} D^2 G_{i2} * g_2 \, d\tau \|_{L^1}$$

$$\le C \left((N^2(0, t) + N^3(0, t)) \int_{\frac{t}{2}}^t \frac{1}{\sqrt{1 + t - \tau}} \frac{d\tau}{1 + \tau} + \int_0^{\frac{t}{2}} \frac{1}{1 + t - \tau} \| g_2 \|_{W^{2,1}} \, d\tau \right)$$

$$\le C \left((N^2(0, t) + N^3(0, t))(1 + t)^{-\frac{1}{2}} \right)$$

$$+ C \left(\int_0^{\frac{t}{2}} \frac{1}{1 + t - \tau} \left(\| U \|^2 + \| U \| \| DU \| + \| U \| \| D^2 U \| + \| DU \|^2 \right) d\tau \right)$$

$$\le C \left((N^2(0, t) + N^3(0, t))(1 + t)^{-\frac{1}{2}} + N^2(0, t) \int_0^{\frac{t}{2}} \frac{1}{1 + t - \tau} \frac{1}{\sqrt{1 + \tau}} \, d\tau \right)$$

$$\le C(1 + t)^{-\frac{1}{2}} (N^2(0, t) + N^3(0, t)). \tag{3.3.84}$$

To estimate $\| \int_0^t DG_{i3} * F_3 \, d\tau \|_{L^1}$, it can be seen from the previous argument that the crucial one is the estimate of $\| \int_0^t DG_{i3} * \left(\frac{1}{\alpha} g_{31} u_1 u_{1\tau} - \frac{1}{\beta} g_{32} v v_\tau \right) d\tau \|_{L^1}$. By integration by parts we have

$$\int_0^t DG_{i3} * (g_{31} u_1 u_{1\tau}) \, d\tau = \int_0^t G_{i3} * (D(\frac{1}{2} g_{31} u_1^2))_\tau \, d\tau$$

$$- \int_0^t DG_{i3} * \left(\frac{1}{2} (\frac{\partial g_{31}}{\partial u_1} u_{1\tau} + \frac{\partial g_{31}}{\partial v} v_\tau) u_1^2 \right) d\tau. \tag{3.3.85}$$

By integration by parts and Theorem 2.2.2, we obtain

$$\| \int_0^t G_{i3} * \frac{\partial}{\partial \tau} \left(D(\frac{1}{2} g_{31} u_1^2) \right) d\tau \|_{L^1}$$

$$\le \| \frac{1}{2} \int_0^t \frac{\partial}{\partial t} G_{i3} * (D(g_{31} u_1^2)) \, d\tau \|_{L^1} + \| \frac{1}{2} G_{i3} * D(g_{31} u_1^2) \big|_0^t \|_{L^1}$$

$$\leq C \int_0^t \frac{1}{1+t-\tau} \|D(g_{31}u_1^2)\|_{W^{2,1}} \, d\tau$$

$$+C \left(\|G_{i3}(0) * D(g_{31}u_1^2)(t)\|_{L^1} + \|G_{i3}(t) * D(g_{31}u_1^2)|_{t=0}\|_{L^1} \right)$$

$$\leq C \int_0^t \frac{1}{1+t-\tau} (\|U\|\|DU\| + \|U\|_{L^\infty}\|U\|\|DU\| + \|U\|\|D^2U\| + \|DU\|^2$$

$$+\|U\|_{L^\infty}\|DU\|^2 + \|U\|\|D^3U\| + \|DU\|\|D^2U\|$$

$$+\|U\|_{L^\infty}\|DU\|\|D^2U\| + \|DU\|_{L^\infty}\|DU\|^2 + \|U\|_{L^\infty}\|DU\|_{L^\infty}\|DU\|^2$$

$$+\|U\|_{L^\infty}^2\|DU\|\|D^2U\| + \|U\|_{L^\infty}\|U\|\|D^3U\|) \, d\tau$$

$$+C \left(\|U(t)\|^2 + \|U(t)\|_{L^\infty}\|U(t)\|\|DU(t)\| + \|DG_{i3}(t) * (g_{31}u_1^2|_{t=0})\|_{L^1} \right)$$

$$\leq C \left(N^2(0,t) + N^3(0,t) + N^4(0,t) \right) \int_0^t \frac{1}{1+t-\tau} \frac{d\tau}{1+\tau}$$

$$+C(N^2(0,t) + N^3(0,t))(1+t)^{-\frac{1}{2}} + C(1+t)^{-\frac{1}{2}}(\|U_0\|_{L^\infty} + \|U_0\|_{L^\infty}^2\|U_0\|_{W^{1,1}})$$

$$\leq C(1+t)^{-\frac{1}{2}}(\|U_0\|_{W^{1,1}} + N^2(0,t) + N^3(0,t) + N^4(0,t)). \tag{3.3.86}$$

Using system (3.3.21), applying Theorem 2.2.1 and Lemma 3.1.3, we get

$$\left\| \int_0^t DG_{i3} * \left(\left(\frac{\partial g_{31}}{\partial u_1} u_{1\tau} + \frac{\partial g_{31}}{\partial v} v_\tau \right) u_1^2 \right) d\tau \right\|_{L^1}$$

$$\leq C \int_0^t \frac{1}{\sqrt{1+t-\tau}} \left(\|U\|_{L^\infty}\|U\|\|DU\| + \|U\|_{L^\infty}\|U\|\|D^2U\| \right.$$

$$+\|U\|_{L^\infty}^2\|DU\|^2 + \|U\|_{L^\infty}\|U\|\|D^3U\| + \|U\|_{L^\infty}^2\|DU\|\|D^2U\|$$

$$+\|U\|_{L^\infty}^2\|DU\|_{L^\infty}\|DU\|^2 \,) \, d\tau$$

$$\leq C(N^3(0,t) + N^4(0,t) + N^5(0,t)) \int_0^t \frac{1}{\sqrt{1+t-\tau}} \frac{1}{(1+\tau)^{\frac{3}{2}}} \, d\tau$$

$$\leq C(1+t)^{-\frac{1}{2}}(N^3(0,t) + N^4(0,t) + N^5(0,t)). \tag{3.3.87}$$

$\left\| \int_0^t DG_{i3} * (g_{32}vv_\tau) \, d\tau \right\|_{L^1}$ can be estimated in the same way.

Thus we finally arrive at

$$\|U(t)\|_{L^1} + (1+t)^{\frac{1}{2}}\|DU(t)\|_{L^1}$$

$$\leq C(\|U_0\|_{W^{1,1}} + N^2(0,t) + N^3(0,t) + N^4(0,t) + N^5(0,t)). \tag{3.3.88}$$

(iii) L^∞ norm estimates

It follows from (3.3.88) that

$$(1+t)^{\frac{1}{2}}\|U(t)\|_{L^\infty} \leq (1+t)^{\frac{1}{2}}\|DU(t)\|_{L^1}$$

$$\leq C(\|U_0\|_{W^{1,1}} + N^2(0,t) + N^3(0,t) + N^4(0,t) + N^5(0,t)). \tag{3.3.89}$$

We now proceed to estimate $\|DU(t)\|_{L^\infty}$. In what follows we prove that for $t \geq 0$

$$(1+t)\|DU(t)\|_{L^\infty} \leq C(\|U_0\|_{W^{1,1}} + \|U_0\|_{H^2} + N^2(0,t) + N^3(0,t) + N^4(0,t)). \quad (3.3.90)$$

It can be easily seen from Theorem 2.2.3 and the expression of the solution

$$U(x,t) = G(t) * U_0 + \int_0^t G(t-\tau) * F(\tau)\, d\tau \quad (3.3.91)$$

that (3.3.90) holds for $0 \leq t \leq 2$. Therefore, it suffices to prove (3.3.90) for $t \geq 2$. By Theorem 2.2.3, we have

$$\|DU(t)\|_{L^\infty} \leq \|DG * U_0\|_{L^\infty} + \|\int_0^t DG(t-\tau) * F(\tau)\, d\tau\|_{L^\infty}$$

$$\leq C(1+t)^{-1}(\|U_0\|_{W^{1,1}} + \|U_0\|_{H^2}) + \|\int_0^t D^2 G(t-\tau) * F(\tau)\, d\tau\|_{L^1}$$

$$\leq C(1+t)^{-1}(\|U_0\|_{W^{1,1}} + \|U_0\|_{H^2})$$

$$+\|\int_0^{\frac{t}{2}} D^2 G(t-\tau) * F(\tau)\, d\tau\|_{L^1} + \|\int_{\frac{t}{2}}^t D^2 G(t-\tau) * F(\tau)\, d\tau\|_{L^1}. \quad (3.3.92)$$

For $\tau \in [0, \frac{t}{2}]$, $t - \tau \geq \frac{t}{2} \geq 1$. It follows from (2.2.99) that

$$\|\int_0^{\frac{t}{2}} D^2 G_{i2}(t-\tau) * F_2(\tau)\, d\tau\|_{L^1}$$

$$\leq \|\int_0^{\frac{t}{2}}(D^2 G_{i2} * F_2 - D^2 V_{i2}(F_2))\, d\tau\|_{L^1} + \|\int_0^{\frac{t}{2}} D^2 V_{i2}(F_2)\, d\tau\|_{L^1}$$

$$\leq C\int_0^{\frac{t}{2}}(1+t-\tau)^{-\frac{5}{4}}\|F_2\|_{W^{2,1}}\, d\tau + \|\int_0^{\frac{t}{2}} D^2 V_{i2}(F_2)\, d\tau\|_{L^1}$$

$$\leq C(N^2(0,t) + N^3(0,t) + N^4(0,t))\int_0^{\frac{t}{2}}(1+t-\tau)^{-\frac{5}{4}}(1+\tau)^{-1}\, d\tau$$

$$+\|\int_0^{\frac{t}{2}} D^2 V_{i2}(F_2)\, d\tau\|_{L^1}$$

$$\leq C(1+t)^{-1}(N^2(0,t)+N^3(0,t)+N^4(0,t))+\|\int_0^{\frac{t}{2}} D^2 V_{i2}(F_2)\, d\tau\|_{L^1} \quad (3.3.93)$$

where $V_{ij}(u_0)$ are the integral operators defined in (2.2.92) and possess the same properties as the heat potential operator.

Thus by integration by parts, for $t \geq 2$ we have

$$\|\int_0^{\frac{t}{2}} D^2 V_{i2}(F_2)\, d\tau\|_{L^1} = \|\int_0^{\frac{t}{2}} D^3 V_{i2}(g_2)\, d\tau\|_{L^1}$$

$$\leq C\int_0^{\frac{t}{2}}(t-\tau)^{-\frac{3}{2}}\|g_2\|_{L^1}\, d\tau$$

$$\leq C N^2(0,t)\int_0^{\frac{t}{2}}(t-\tau)^{-\frac{3}{2}}(1+\tau)^{-\frac{1}{2}}\, d\tau$$

$$\leq C(1+t)^{-1}N^2(0,t). \quad (3.3.94)$$

Combining (3.3.94) with (3.3.93) yields

$$\left\| \int_0^{\frac{t}{2}} D^2 G_{i2} * F_2 \, d\tau \right\|_{L^1} \leq C(1+t)^{-1} \left(N^2(0,t) + N^3(0,t) + N^4(0,t) \right). \tag{3.3.95}$$

Similarly, in order to estimate $\| \int_0^{\frac{t}{2}} D^2 G_{i3} * F_3 \, d\tau \|_{L^1}$, it suffices to estimate the terms such as $\| \int_0^{\frac{t}{2}} D^2 V_{i3}(\frac{\partial}{\partial \tau}(g_{31} u_1^2)) \, d\tau \|_{L^1}$.

By integration by parts we have

$$\left\| \int_0^{\frac{t}{2}} D^2 V_{i3} \left(\frac{\partial}{\partial \tau}(g_{31} u_1^2) \right) d\tau \right\|_{L^1}$$

$$= \| D^2 V_{i3}(g_{31} u_1^2(\tfrac{t}{2})) - D^2 V_{i3}(g_{31} u_1^2(t)) - \int_0^{\frac{t}{2}} D^2 D_\tau V_{i3}(g_{31} u_1^2) \, d\tau \|_{L^1}$$

$$\leq C(1+t)^{-1} \left(\|U_0\|^2 + \|u_1(\tfrac{t}{2})\|^2 \right) + C \int_0^{\frac{t}{2}} (1+t-\tau)^{-2} \|g_{31} u_1^2\|_{L^1} \, d\tau$$

$$\leq C(1+t)^{-1} N^2(0,t) + C N^2(0,t) \int_0^{\frac{t}{2}} (1+t-\tau)^{-2}(1+\tau)^{-\frac{1}{2}} \, d\tau$$

$$\leq C(1+t)^{-1} N^2(0,t). \tag{3.3.96}$$

The estimate of $\| \int_0^{\frac{t}{2}} D^2 V_{i3} \left(\frac{\partial}{\partial \tau}(g_{32} v^2) \right) d\tau \|_{L^1}$ is the same as above. Thus we arrive at

$$\left\| \int_0^{\frac{t}{2}} D^2 G(t-\tau) * F(\tau) \, d\tau \right\|_{L^1} \leq C(1+t)^{-1} \left(N^2(0,t) + N^3(0,t) + N^4(0,t) \right). \tag{3.3.97}$$

By integration by parts and Theorem 2.2.1, we have

$$\left\| \int_{\frac{t}{2}}^t D^2 G(t-\tau) * F(\tau) \, d\tau \right\|_{L^1} = \left\| \int_{\frac{t}{2}}^t DG * DF \, d\tau \right\|_{L^1}$$

$$\leq C \int_{\frac{t}{2}}^t (1+t-\tau)^{-\frac{1}{2}} \|DF\|_{W^{1,1}} \, d\tau$$

$$\leq C \left(N^2(0,t) + N^3(0,t) + N^4(0,t) \right) \int_{\frac{t}{2}}^t (1+t-\tau)^{-\frac{1}{2}}(1+\tau)^{-\frac{3}{2}} \, d\tau$$

$$\leq C(1+t)^{-1} \left(N^2(0,t) + N^3(0,t) + N^4(0,t) \right). \tag{3.3.98}$$

It turns out from (3.3.92), (3.3.97)–(3.3.98) that (3.3.90) holds.

Combining the estimates in the previous three steps, we finally arrive at

$$N^2(0,t) \leq C \left(\|U_0\|_{W^{1,1}}^2 + \|U_0\|_{H^3}^2 + \sum_{i=3}^{10} N^i(0,t) \right). \tag{3.3.99}$$

For $3 < k < 10$, we use Young's inequality with $p = \frac{7}{10-k}, q = \frac{7}{k-3}$ to obtain

$$N^k(0,t) = N^{\frac{3(10-k)}{7}}(0,t) N^{\frac{10(k-3)}{7}}(0,t) \leq \frac{N^3(0,t)}{p} + \frac{N^{10}(0,t)}{q}. \tag{3.3.100}$$

Thus it turns out from (3.3.99) that

$$N^2(0,t) \leq C \left(\|U_0\|_{W^{1,1}}^2 + \|U_0\|_{H^3}^2 + N^3(0,t) + N^{10}(0,t) \right). \tag{3.3.101}$$

Let $\delta = \|U_0\|_{W^{1,1}}^2 + \|U_0\|_{H^3}^2$, $x = N^2(0,t)$, $f(x) = C(\delta + x^{\frac{3}{2}} + x^5) - x$. Then (3.3.101) implies that $f(x) \geq 0$, $f(0) = C\delta$ and

$$f'(x) = C(\frac{3}{2}x^{\frac{1}{2}} + 5x^4) - 1 \leq -\frac{1}{2}, \ as \ 0 \leq x \leq \min\left(1, \frac{1}{(13C)^2}\right). \tag{3.3.102}$$

It follows from $f(x) = f(0) + \int_0^x f'(x)dx$ that when $\delta \leq \delta_0 = \frac{1}{2C} \min(1, \frac{1}{(13C)^2})$, $f(x)$ must change its sign in $[0, 2C\delta]$. Thus we conclude that when $\|U_0\|_{W^{1,1}}^2 + \|U_0\|_{H^3}^2 \leq \varepsilon_1 = \min(\delta_0, \varepsilon_0)$,

$$N^2(0,t) \leq 2C\delta = 2C(\|U_0\|_{W^{1,1}}^2 + \|U_0\|_{H^3}^2). \tag{3.3.103}$$

Thus the proof of Theorem 3.3.1 is complete. □

The method used in the proof of Theorem 3.3.1 can also be applied to other classes of QHPCS. Consider the following initial value problem for the second class of one-dimensional QHPCS:

$$\begin{cases} \frac{\partial u_1}{\partial t} - \frac{\partial u_2}{\partial x} = 0, \\[2mm] \frac{\partial u_2}{\partial t} - \frac{\partial p(u_1, v, u_{2x})}{\partial x} = 0, \\[2mm] c_0(u_1, v)v_t - p_v(u_1, v, u_{2x})\frac{\partial u_2}{\partial x} - q(u_1, v, v_x)_x + d(u_1, v, v_x) = 0, \\[2mm] t = 0: \ u_1 = u_1^0(x), \ u_2 = u_2^0(x), \ v = v^0(x). \end{cases} \tag{3.3.104}$$

We make the following assumptions on the system (3.3.104):

$(H_1)'$ There exist positive numbers γ and a_0, a_1, a_2, k, M depending on γ such that when $|u_1|, |v|, |u_{2x}|, |v_x| \leq \gamma$, the following hold:

$$p_{u_1} = \frac{\partial p}{\partial u_1} \geq a_0 > 0, \ p_v \neq 0, \ c_0 \geq a_1 > 0, \ \frac{\partial p}{\partial u_{2x}} \geq a_2 > 0, \tag{3.3.105}$$

$$\frac{\partial q}{\partial v_x} \geq k > 0, \ |q_{u_1}|, |q_v| \leq M|v_x|. \tag{3.3.106}$$

$(H_2)'$ As in (3.3.5), d is a smooth higher-order term of its argument.

$(H_3)'$

$$p, q \in C^4, \quad c_0 \in C^3. \tag{3.3.107}$$

$(H_4)'$

$$U_0 = (u_1^0, u_2^0, v^0)^T \in H^3(\mathbb{R}) \bigcap W^{1,1}(\mathbb{R}). \tag{3.3.108}$$

Then we have (see Zheng & Shen [3])

Theorem 3.3.2 *Under the above assumptions $(H_1)'$–$(H_4)'$, there exists a small positive constant ε such that when $\|U_0\|_{H^3 \bigcap W^{1,1}} \leq \varepsilon$, the Cauchy problem (3.3.104) admits a unique global solution $U = (u_1, u_2, v)^T$ such that*

$u_1 \in C([0, +\infty); H^3 \bigcap W^{1,1}) \bigcap C^1([0, +\infty); H^2)$, $u_2, v \in C([0, +\infty); H^3 \bigcap W^{1,1}) \bigcap C^1([0, +\infty); H^1)$, $Du_2, Dv \in L^2([0, \infty); H^3)$.

Moreover, as $t \to \infty$,

$$
\begin{cases}
\|U(t)\|_{L^\infty} = O((1+t)^{-\frac{1}{2}}), \\
\|U(t)\| = O((1+t)^{-\frac{1}{4}}), \\
\|DU(t)\|_{L^1} = O((1+t)^{-\frac{1}{2}}).
\end{cases} \tag{3.3.109}
$$

The proof of Theorem 3.3.2 can be carried out in the same way as Theorem 3.3.1. However, we should use the results in Section 2.3.1 (Theorem 2.3.1) and the definition of $N(0, t)$ should be changed into

$$N(0, t) = N_1(0, t) + N_2(0, t) + \widetilde{N}_3(0, t) \tag{3.3.110}$$

with

$$\widetilde{N}_3(0, t) = \left(\int_0^t \left(\|DU(\tau)\|_{H^2}^2 + \tau \|D^2 U(\tau)\|^2 + \tau^2 \|D^3 U(\tau)\|^2 \right. \right.$$
$$\left. \left. + (1 + \tau)^2 \|D^4\{u_2, v\}\|^2 \right) d\tau \right)^{\frac{1}{2}}. \tag{3.3.111}$$

We refer to Zheng & Shen [3] for the details of the proof.

3.3.2 Applications

1. The system of one-dimensional nonlinear thermoelasticity.

In Lagrange coordinates the system of one-dimensional nonlinear thermoelasticity can be written as (see Dafermos [4])

$$
\begin{cases}
\dfrac{\partial u_1}{\partial t} - \dfrac{\partial u_2}{\partial x} = 0, \\[2mm]
\dfrac{\partial u_2}{\partial t} - \dfrac{\partial \sigma(u_1, \theta)}{\partial x} = 0, \\[2mm]
\dfrac{\partial \left(e(u_1, \theta) + \frac{u_2^2}{2}\right)}{\partial t} - \dfrac{\partial(\sigma u_2)}{\partial x} = \dfrac{\partial(K(u_1, \theta)\theta_x)}{\partial x}, \\[2mm]
t = 0: \quad u_1 = u_1^0(x), \quad u_2 = u_2^0(x), \quad \theta = \theta^0(x),
\end{cases}
\tag{3.3.112}
$$

where $u_1, u_2, e, \sigma, \theta$ denote deformation gradient, velocity, internal energy, stress and temperature in that order.

To comply with the second law of thermodynamics expressed by the Clausius–Duhem inequality, we must have

$$
\sigma(u_1, \theta) = \frac{\partial \psi(u_1, \theta)}{\partial u_1}, \quad \eta(u_1, \theta) = -\frac{\partial \psi(u_1, \theta)}{\partial \theta}
\tag{3.3.113}
$$

where $\psi = e - \theta \eta$ is the Helmholtz free energy and η is the specific entropy.

Let $\bar{u}_1, \bar{\theta}$ be any fixed positive constants. Suppose that there exist positive constants γ and a_0, a_1, k depending on γ such that when $|u_1 - \bar{u}_1| \leq \gamma, |\theta - \bar{\theta}| \leq \gamma$,

$$
\frac{\partial e}{\partial \theta} \geq a_1 > 0, \quad \frac{\partial \sigma}{\partial u_1} \geq a_0 > 0, \quad \frac{\partial \sigma}{\partial \theta} \neq 0, \quad K \geq k > 0.
\tag{3.3.114}
$$

Consider the problem (3.3.112). Then it is easy to follow from (3.3.113) that the reduced system for $\tilde{u}_1 = u_1 - \bar{u}_1$, $\tilde{u}_2 = u_2$, $\tilde{\theta} = \theta - \bar{\theta}$ is a special form of (3.3.1), (3.3.3) and we can apply Theorem 3.3.1 to conclude that if $(u_1^0 - \bar{u}_1, u_2^0, \theta^0 - \bar{\theta})$ is suitably small in $H^3 \cap W^{1,1}$, then problem (3.3.112) admits a unique global smooth solution converging to the equilibrium $(\bar{u}_1, 0, \bar{\theta})$ with the same decay rates as the heat equation.

2. The system of radiation hydrodynamics

In Lagrange coordinates the system of one-dimensional radiation hydrodynamics can be written as (see Li, Yu & Shen [1] and the references cited there)

$$
\begin{cases}
\dfrac{\partial u_1}{\partial t} - \dfrac{\partial u_2}{\partial x} = 0, \\[2ex]
\dfrac{\partial u_2}{\partial t} - \dfrac{\partial (p + p_\nu)}{\partial x} = 0, \\[2ex]
\dfrac{\partial (e + \frac{u_2^2}{2} + u_1 E_\nu)}{\partial t} + \dfrac{\partial (u_2(p + p_\nu) - \frac{d}{u_1}(E_\nu)_x)}{\partial x} = 0, \\[3ex]
t = 0,: \quad u_1 = u_1^0(x), \quad u_2 = u_2^0(x), \quad \theta = \theta^0(x),
\end{cases}
\tag{3.3.115}
$$

where u_1 is the specific volume, u_2 is the velocity and θ is the temperature; $p = \dfrac{R\theta}{u_1}, p_\nu = \dfrac{4\sigma}{3c}\theta^4, e = \dfrac{R\theta}{\gamma - 1}, E_\nu = \dfrac{4\sigma}{c}\theta^4, d = \dfrac{Ac}{3}\theta^\alpha$ with positive constants R, c, σ, A, α and $\gamma > 1$.

Let $\bar{u}_1, \bar{\theta}$ be positive constants and consider problem (3.3.115) near the equilibrium $(\bar{u}_1, 0, \bar{\theta})$. Then it is easy to see that the reduced system of (3.3.115) for $\tilde{u}_1 = u_1 - \bar{u}_1, \tilde{u}_2 = u_2, \tilde{\theta} = \theta - \bar{\theta}$ is a special form of (3.3.1), (3.3.3) and we can conclude from Theorem 3.3.1 the global existence and uniqueness of smooth solutions provided that $(u_1^0(x) - \bar{u}_1, u_2^0(x), \theta^0(x) - \bar{\theta})$ is suitably small in $H^3 \cap W^{1,1}$.

3. The Equations of motion of one-dimensional compressible viscous and heat-conductive fluids

Let ρ be the density, u the velocity and θ the temperature and let $\tau = \frac{1}{\rho}$. Then in Lagrange coordinates the equations of motion of one-dimensional compressible viscous and heat-conductive fluids can be written as (see Li, Yu & Shen [1])

$$
\begin{cases}
\dfrac{\partial \tau}{\partial t} - \dfrac{\partial u}{\partial x} = 0, \\[2ex]
\dfrac{\partial u}{\partial t} + \dfrac{\partial p(\tau, \theta)}{\partial x} = \dfrac{\partial (\frac{\nu}{\tau} u_x)}{\partial x}, \\[2ex]
\dfrac{\partial e}{\partial \theta}\dfrac{\partial \theta}{\partial t} + \theta \dfrac{\partial p}{\partial \theta}\dfrac{\partial u}{\partial x} = \dfrac{\partial (\frac{K}{\tau}\theta_x)}{\partial x}, \\[2ex]
t = 0 : \tau = \tau^0(x), \ u = u^0(x), \theta = \theta^0(x),
\end{cases}
\tag{3.3.116}
$$

where $e = e(\tau, \theta), p = p(\tau, \theta)$, are the internal energy and the pressure, respectively, and ν, K are given positive constants. Let $\bar{\tau}, \bar{\theta}$ be positive constants and suppose that in a neighbourhood of $(\bar{\tau}, \bar{\theta})$, $e_\theta \geq a_0 > 0$, $p_\tau \leq -a_1 < 0$, $p_\theta \neq 0$. Then it is easy to follow from (3.3.113) that (3.3.116) is a special form of (3.3.104). It follows from Theorem 3.3.2 that if $(\tau^0 - \bar{\tau}, u^0, \theta^0 - \bar{\theta})$ is small in $H^3 \cap W^{1,1}$, then the Cauchy problem (3.3.116) admits a unique global smooth solution converging to the equilibrium $(\bar{\tau}, 0, \bar{\theta})$ with the same decay rates as the heat equation.

4. The system of one-dimensional nonlinear thermoviscoelasticity

   In Lagrange coordinates this can be written as

$$
\begin{cases}
\dfrac{\partial u}{\partial t} - \dfrac{\partial v}{\partial x} = 0, \\[2mm]
\dfrac{\partial v}{\partial t} - \dfrac{\partial \sigma(u, \theta, v_x)}{\partial x} = 0, \\[2mm]
\dfrac{\partial(e + \frac{1}{2}v^2)}{\partial t} - \dfrac{\partial(\sigma v)}{\partial x} - \dfrac{\partial q(u, \theta, \theta_x)}{\partial x} = 0, \\[2mm]
t = 0: \quad u = u^0(x), \quad v = v^0(x), \quad \theta = \theta^0(x),
\end{cases}
\tag{3.3.117}
$$

where $u, v, e(u, \theta), \sigma(u, \theta, v_x), \theta, q(u, \theta, \theta_x)$ denote deformation gradient, velocity, internal energy, stress, temperature and heat flux in that order. Note that u, e, θ may only take positive value (see Dafermos [4]).

To comply with the second law of thermodynamics expressed by the Clausius–Duhem inequality (see Dafermos [4–5]), we must have

$$
\sigma(u, \theta, 0) = \frac{\partial \psi(u, \theta)}{\partial u}, \quad \eta = -\frac{\partial \psi}{\partial \theta}
\tag{3.3.118}
$$

where $\psi = e - \theta\eta$ is the Helmholtz free energy and η is the specific entropy.

Let $\bar{u}, \bar{\theta}$ be any fixed positive constants. Suppose that there exist positive constants γ and $a_0, \sigma_0, \sigma_1, q_0, M$ depending on γ such that when $|u - \bar{u}|, |\theta - \bar{\theta}|, |v_x|, |\theta_x| \leq \gamma$,

$$
\begin{cases}
e_\theta \geq a_0 > 0, \quad \sigma_u \geq \sigma_0 > 0, \quad \sigma_{v_x} \geq \sigma_1 > 0, \quad \sigma_\theta \neq 0, \\[2mm]
\dfrac{\partial q}{\partial \theta_x} \geq q_0 > 0,
\end{cases}
\tag{3.3.119}
$$

and

$$|q_u|, \ |q_\theta| \leq M|\theta_x|. \tag{3.3.120}$$

Then it is easy to follow from (3.3.118) that (3.3.117) is a special form of (3.3.104). We can conclude from Theorem 3.3.2 that if e, σ, $q \in C^4$ and $(u^0 - \bar{u}, v^0, \theta^0 - \bar{\theta})$ is suitably small in $H^3 \cap W^{1,1}$, then problem (3.3.117) admits a unique global smooth solution as precisely stated by Theorem 3.3.2.

3.4 Nonlinear Hyperbolic–Parabolic Coupled Systems: IBVP

In this section we consider a class of quasilinear hyperbolic–parabolic coupled system in one space dimension:

$$\begin{cases} u_{1t} + u_2 b_1(u_1, v)u_{1x} + b_2(u_1, v)u_{2x} = 0, \\[2mm] a(u_1, v)u_{2t} + b_2(u_1, v)u_{1x} + u_2 b_3(u_1, v)u_{2x} + \beta(u_1, v)v_x = 0, \\[2mm] c_0(u_1, v)v_t - (K(u_1, v)v_x)_x + u_2\alpha(u_1, v)v_x \\[1mm] \quad + \beta(u_1, v)u_{2x} + d(u_1, v, u_{1x}, v_x) = 0, \end{cases} \tag{3.4.1}$$

subject to the initial conditions and the boundary conditions

$$\begin{cases} t = 0: \quad u_1 = u_1^0(x), \ u_2 = u_2^0(x), \ v = v^0(x), \\[2mm] x = 0, 1: \quad u_2 = 0, \ v_x = 0. \end{cases} \tag{3.4.2}$$

We make the following assumptions on problem (3.4.1), (3.4.2):

(H_1) There exists positive constants γ and $a_0, k, \delta_0, k_\beta$ such that when $|u_1|, \ |v| \leq \gamma$,

$$b_2 \neq 0, \ a \geq a_0 \geq 0, \ K \geq k > 0, c_0 \geq \delta_0 > 0, \frac{\beta^2}{K} \geq k_\beta > 0. \tag{3.4.3}$$

(H_2)

$$d = d_1(u_1, v)u_{1x}v_x + d_2(u_1, v)(v_x)^2 \tag{3.4.4}$$

with d_1, d_2 being smooth functions.

(H_3) $K \in C^4$ and other coefficients in (3.4.1) belong to C^3.

(H_4) $u_1^0(x), u_2^0(x) \in H^3$, $v^0 \in H^4$.

Moreover, the following compatibility conditions are satisfied on the boundary $x = 0, 1$:

$$u_2^0 = \frac{du_1^0}{dx} = \frac{dv^0}{dx} = \frac{d^3 v^0}{dx^3} = \frac{d^2 u_2^0}{dx^2} = 0. \tag{3.4.5}$$

It can be easily seen that under assumption (H_1) the first two equations of (3.4.1) consist of a quasilinear symmetric hyperbolic system with respect to u_1, u_2 and the third equation of (3.4.1) is a quasilinear parabolic equation with respect to v. Several important equations can be reduced to (3.4.1). The one-dimensional nonlinear thermoelastic system in Euler coordinates can be written as (see Slemrod [1])

$$\begin{cases} u_{tt} - a(u_x, \theta)u_{xx} + b(u_x, \theta)\theta_x = 0, \\ c(u_x, \theta)\theta_t + b(u_x, \theta)u_{tx} - d(\theta)\theta_{xx} = 0. \end{cases} \tag{3.4.6}$$

If one introduces

$$u_1 = u_x, \ u_2 = u_t, \ v = \theta, \tag{3.4.7}$$

then it can be seen that (3.4.6) is reduced to (3.4.1). The physical meaning of boundary conditions (3.4.2) is: the rod is clamped and insulated at both ends.

The following example is more complicated. In Euler coordinates the system of radiation hydrodynamics can be written as (see Li, Yu & Shen [1] and Zheng [2])

$$\begin{cases} \rho_t + u\rho_x + \rho u_x = 0, \\[2mm] \rho(u_t + uu_x) + R\theta\rho_x + (R\rho + \frac{16\sigma}{3c}\theta^3)\theta_x = 0, \\[2mm] \left(\frac{R\rho}{\gamma - 1} + \frac{16\sigma}{c}\theta^2\right)\theta_t - \left(\frac{16\sigma A}{3}\theta^{3+\alpha}\theta_x\right)_x \\[2mm] + \left(\frac{R\rho}{\gamma - 1} + \frac{16\sigma}{c}\theta^3\right)u\theta_x + \left(R\rho\theta + \frac{16\sigma}{3c}\theta^4\right)u_x = 0 \end{cases} \tag{3.4.8}$$

where ρ is the density, u the velocity and θ the temperature, $R, \sigma, c, A, \gamma, \alpha$ are given positive constants with $\gamma > 1$.

Then it is easy to see that (3.4.8) can be rewritten as

$$
\begin{cases}
\rho_t + u\rho_x + \rho u_x = 0, \\[2ex]
\dfrac{\rho^2}{R\theta}(u_t + uu_x) + \rho\rho_x + \left(R + \dfrac{16\sigma}{3\rho c}\theta^3\right)\dfrac{\rho^2}{R\theta}\theta_x = 0, \\[2ex]
\dfrac{\rho}{R\theta^2}\left(\dfrac{R\rho}{\gamma-1} + \dfrac{16\sigma}{c}\theta^2\right)\theta_t - \dfrac{\rho}{R\theta^2}\left(\dfrac{16\sigma A}{3}\theta^{3+\alpha}\theta_x\right)_x \\[2ex]
\qquad + \dfrac{\rho}{R\theta^2}\left(\dfrac{R\rho}{\gamma-1} + \dfrac{16\sigma}{c}\theta^3\right)u\theta_x + \dfrac{\rho}{R\theta^2}\left(R\rho\theta + \dfrac{16\sigma}{3c}\theta^4\right)u_x = 0
\end{cases}
\tag{3.4.9}
$$

and this is a special form of (3.4.1) with $a = \dfrac{\rho^2}{R\theta}$, $b_1 = 1$, $b_2 = \rho$, $b_3 = \dfrac{\rho^2}{R\theta}$, $c_0 = \left(\dfrac{R\rho}{\gamma-1} + \dfrac{16\sigma}{c}\theta^2\right)\dfrac{\rho}{R\theta^2}$, $K = \dfrac{16\sigma A}{3R}\theta^{1+\alpha}\rho$, $\alpha = \left(\dfrac{R\rho}{\gamma-1} + \dfrac{16\sigma}{c}\theta^3\right)\dfrac{\rho}{R\theta^2}$, $\beta = \left(R + \dfrac{16\sigma}{3\rho c}\theta^3\right)\dfrac{\rho^2}{R\theta}$.

The physical meaning of boundary conditions (3.4.2) now is clear: the first one represents the condition that the velocity vanishes at a solid wall boundary and the second one means that the heat flux does not go through the boundary, i.e, the boundary is insulated.

Throughout this section, we use the following notation. Let

$$u = \{u_1, u_2\}, \{u, v\} = \{u_1, u_2, v\}, Du = \{u_t, u_x\}, Dv = \{v_t, v_x\}, D^2u = \{u_{tt}, u_{tx}, u_{xx}\},$$
$$D^2v = \{v_{tt}, v_{tx}, v_{xx}\}, |u|^2 = u_1^2 + u_2^2, |u|_1^2 = |u|^2 + |Du|^2, |u|_2^2 = |u|_1^2 + |D^2u|^2.$$

We now have (see Zheng [3])

Theorem 3.4.1 *Suppose that the assumptions (H_1)–(H_4) are satisfied. Then there is a small constant $\varepsilon > 0$ such that when $\|u_1^0\|_{H^3} + \|u_2^0\|_{H^3} + \|v^0\|_{H^4} \le \varepsilon$, problem (3.4.1)–(3.4.2) admits a unique global smooth solution $\{u_1, u_2, v\}$ such that*
$$(u_1, u_2) \in \bigcap_{k=0}^{2} C^k([0, +\infty); H^{2-k}), v \in C([0, +\infty); H^3) \cap C^1([0, +\infty); H^2) \cap C^2([0, +\infty);$$
$L^2), v_{xtt} \in L^2([0, \infty); L^2).$
Moreover, $\|u_2(t)\|, \|Du(t)\|, \|Dv(t)\|, \|D^2u(t)\|, \|D^2v(t)\|, \|v_{xxx}(t)\|, \|v_{xxt}(t)\|$ exponentially decay to zero as $t \to \infty$.

Proof. As before, we prove the theorem by combining a local existence and uniqueness theorem with uniform a priori estimates.

Let

$$M_0 = \|\{u, Du, D^2u\}|_{t=0}\|^2 + \|\{v, Dv, D^2v\}|_{t=0}\|^2 \tag{3.4.10}$$

where $\|\cdot\|$, as usual, denotes the L^2 norm in $[0,1]$ and $Du|_{t=0}, Dv|_{t=0}, D^2u|_{t=0}, D^2v|_{t=0}$ are obtained from (3.4.1) and the initial data. Since $u_1^0, u_2^0 \in H^3, v^0 \in H^4$, we have $M_0 < \infty$. Let M_1 be a positive constant such that when $\|\{u,v\}\|_{H^2} \leq M_1$, by the imbedding theorem, $\|\{u,v\}\|_{L^\infty} \leq \gamma$ where γ is the constant appearing in assumption (H_1). We now introduce the set

$$X_h(M_2, M_3) = \left\{ \{u,v\} \middle| \begin{array}{l} u \in \bigcap_{k=0}^{2} C^k([0,h]; H^{2-k}), v \in C([0,h]; H^3), \\[1mm] v_t \in C([0,h]; H^2), v_{tt} \in C([0,h]; L^2) \cap L^2([0,h]; H^1), \\[1mm] \sup_{0\leq t\leq h}\sum_{k=0}^{2}\|\{D^ku, D^kv\}\|^2 \leq M_2, \\[1mm] \sup_{0\leq t\leq h}(\|v_{xxx}(t)\|^2 + \|v_{xxt}(t)\|^2) + \int_0^h \|v_{xtt}\|^2 \, d\tau \leq M_3, \\[1mm] u_1|_{t=0} = u_1^0(x), u_2|_{t=0} = u_2^0(x), v|_{t=0} = v^0(x), \\[1mm] \text{on } x = 0,1 : u_2 = u_{1x} = v_x = 0. \end{array} \right\} \tag{3.4.11}$$

where $M_2 \leq M_1, M_3$ are positive constants specified in the course of the proof. We can now state the following local existence and uniqueness theorem.

Lemma 3.4.1 *Under assumptions (H_1)–(H_4), there exist positive constants ε_0 and t_0 depending only on ε_0 such that when $M_0 \leq \varepsilon_0$, problem (3.4.1), (3.4.2) admits a unique local solution $\{u,v\} \in X_{t_0}(C_1M_0, C_2M_0)$ with C_1, C_2 being positive constants depending only on γ.*

Local existence and uniqueness results on initial boundary value problems for quasilinear hyperbolic–parabolic coupled systems have been obtained in the literature (see, e.g., Zheng [2], D. Li [2-3] and Li, Yu & Shen [1-5]). For the convenience of the reader we briefly describe the major steps of the proof here.

Step 1. For any $\{\tilde{u}, \tilde{v}\} \in X_h(M_2, M_3)$, consider the auxiliary linear problems

$$\begin{cases} \alpha_0(\tilde{u}_1, \tilde{v})u_t + \alpha_1(\tilde{u}, \tilde{v})u_x + F = 0, \\ u_2|_{x=0,1} = 0, \quad u|_{t=0} = (u_1^0, u_2^0)^T \end{cases} \tag{3.4.12}$$

and

$$\begin{cases} c_0(\tilde{u}_1, \tilde{v})v_t - (K(\tilde{u}_1, \tilde{v})v_x)_x + \tilde{u}_2\alpha(\tilde{u}_1, \tilde{v})\tilde{v}_x \\ +\beta(\tilde{u}_1, \tilde{v})\tilde{u}_{2x} + D(\tilde{u}_1, \tilde{v}, \tilde{u}_{1x}, \tilde{v}_x) = 0, \\ v_x|_{x=0,1} = 0, \quad v|_{t=0} = v^0(x), \end{cases} \tag{3.4.13}$$

where we have used the matrix notation:

$$\alpha_0(u_1, v) = \begin{pmatrix} 1 & 0 \\ 0 & a(u_1, v) \end{pmatrix}, \quad \alpha_1(u, v) = \begin{pmatrix} u_2 b_1(u_1, v) & b_2(u_1, v) \\ b_2(u_1, v) & u_2 b_3(u_1, v) \end{pmatrix} \tag{3.4.14}$$

and

$$F = \begin{pmatrix} 0 \\ \beta(\tilde{u}_1, \tilde{v})\tilde{v}_x \end{pmatrix}. \tag{3.4.15}$$

Problem (3.4.12) is a linear symmetric hyperbolic system of first order with respect to u. By (3.4.3) and (3.4.11), α_1 has rank 1 on the boundary. Therefore, the boundary condition $u_2|_{x=0,1} = 0$ is admissible in the Friedrichs sense (see Section 1.3). Since the coefficients belong to $\bigcap_{k=0}^{2} C^k([0, h]; H^{2-k})$ and $D^2 \left(\beta \frac{\partial \tilde{v}}{\partial x}\right) \in L^2([0, h]; L^2)$, and by (3.4.5) the compatibility conditions up to $s - 1$ order ($s = 2$) are satisfied, by Theorem 1.3.10, problem (3.4.12) admits a unique solution $u \in \bigcap_{k=0}^{2} C^k([0, h]; H^{2-k})$ satisfying the following energy inequality

$$\|\{u, Du, D^2u\}(t)\|^2 \leq C \left(\|\{u, Du, D^2u\}|_{t=0}\|^2 + \sum_{k=0}^{1} \|D_t^k f_1|_{t=0}\|_{H^{1-k}}^2 \right.$$
$$\left. + Ct \int_0^t \sum_{l=0}^{2} \|D^l f_1\|^2 \, d\tau \right) e^{CM_2 t}, \quad 0 \leq t \leq h \tag{3.4.16}$$

where

$$f_1 = \beta(\tilde{u}_1, \tilde{v})\frac{\partial \tilde{v}}{\partial x} \tag{3.4.17}$$

and C is a positive constant depending on γ.

Concerning the initial boundary value problem for linear parabolic equation (3.4.13), we deduce from the energy method that the unique solution v satisfies the following estimate:

$$\|\{v, Dv, D^2v\}(t)\|^2 + \int_0^t \left(\|v_x\|^2 + \|v_{xt}\|^2 + \|v_{xtt}\|^2 + \|v_{xx}\|^2 + \|v_{xxt}\|^2 \right) d\tau$$
$$\leq C e^{CM_2 t} \left(\|\{v, Dv, D^2v\}|_{t=0}\|^2 + M_2 t \right), \quad 0 \leq t \leq h. \tag{3.4.18}$$

$$\|v_{xxx}(t)\|^2 + \|v_{xxt}(t)\|^2 \leq C(M_2) \left(1 + \|\{Dv, D^2v\}(t)\|^2 \right), \quad 0 \leq t \leq h \tag{3.4.19}$$

where $C(M_2)$ is a positive constant depending on M_2.

<u>Step 2.</u> It can be seen from the above estimates that appropriately choosing M_2 first, then appropriately choosing M_3, for h small enough, the nonlinear operator defined by (3.4.12), (3.4.13) maps $X_h(M_2, M_3)$ into itself.

<u>Step 3.</u> Applying the energy estimates again yields that the iterative sequence $\{u_n, v_n\}$ starting from any point in $X_h(M_2, M_3)$ converges in $C([0, h]; H^1) \cap C^1([0, h]; L^2) \times C([0, h]; H^2) \cap C^1([0, h]; L^2) \cap L^2([0, h]; H^3)$ and the limit point $\{u, v\}$ satisfies (3.4.1). Moreover, by $\{u_n, v_n\} \in X_h(M_2, M_3)$, $\{u, v\}$ belongs to the class: for $0 \leq k \leq 2$, $\{D^k u, D^k v\} \in L^\infty([0, h]; L^2)$, $v_{xxx}, v_{xxt} \in L^\infty([0, h]; L^2), v_{xtt} \in L^2([0, h]; L^2)$ and is a unique pair of solutions to problem (3.4.1)–(3.4.2) in this class.

<u>Step 4.</u> Substituting this $\{u, v\}$ into the coefficients in (3.4.1), we deduce that the reduced linearized problems (3.4.11), (3.4.13) admit a unique pair of solutions in $X_h(M_2, M_3)$. By uniqueness, this pair of solutions must be $\{u, v\}$, i.e, $\{u, v\} \in X_h(M_2, M_3)$. □

We now proceed to prove global existence. It remains to establish uniform a priori estimates. Before going into tedious and technical estimates, let us explain the idea of the proof first. In order to obtain uniform a priori estimates, in the following we introduce auxiliary energy functions $E(t)$ and $\tilde{E}(t)$ which are equivalent to $\sum_{j=0}^{2} \|\{D^j u, D^j v\}\|^2$ and satisfy the following inequalities

$$E(t) + \tilde{C}_3 \int_0^t G(\tau)d\tau \leq \tilde{C}_4 \left(E(0) + \nu \int_0^t G(\tau)d\tau \right) \tag{3.4.20}$$

and

$$\frac{d\tilde{E}(t)}{dt} + \tilde{C}_5 \tilde{E}(t) \leq \tilde{C}_6 \nu \tilde{E}(t) \tag{3.4.21}$$

where $\tilde{C}_i, (i = 3, \cdots, 6)$ are positive constants independent of the solution and t; ν is the boundedness of a certain norm of the solution in $Q_T = (0, 1) \times (0, T)$ and G is a certain energy function. Then it follows from (3.4.20) that as long as

$$\nu \leq \frac{\tilde{C}_3}{\tilde{C}_4}, \tag{3.4.22}$$

$E(t)$ is bounded by $\tilde{C}_4 E(0)$. When initial data are small, by the local existence result, the solution in the interval $[0, t_0]$ satisfies (3.4.22). It turns out that $E(t)$ is bounded by $\tilde{C}_4 E(0)$. Thus the solution can be extended into $[0, 2t_0]$ and (3.4.22) is satisfied provided that the initial data are small. Therefore, $E(t)$ is bounded by $\tilde{C}_4 E(0)$ in $[0, 2t_0]$. Since t_0 depends only on $E(t)$, in this way the solution can be extended step by step to become a global one.

Applying the same argument to (3.4.21) yields the exponential decay of $\tilde{E}(t)$.

Let $T > 0$ be an arbitrary constant. In what follows we derive a priori estimates of the solution $\{u, v\}$. Hereafter we denote by C a positive constant, which may vary in different places and is independent of u, v, T. We denote by $(\ ,\ )$ the inner product in $\mathbb{R}^2$. Let

$$
\begin{cases}
E_1(t) = \frac{1}{2} \int_0^1 \left((u, \alpha_0 u) + c_0 v^2 \right) dx, \\
E_2(t) = \frac{1}{2} \int_0^1 \left((u_t, \alpha_0 u_t) + c_0 v_t^2 \right) dx, \\
E_3(t) = \frac{1}{2} \int_0^1 \left((u_x, \alpha_0 u_x) + c_0 v_x^2 \right) dx, \\
E_4(t) = \frac{1}{2} \int_0^1 \left((u_{tt}, \alpha_0 u_{tt}) + c_0 v_{tt}^2 \right) dx, \\
E_5(t) = \frac{1}{2} \int_0^1 \left((u_{xt}, \alpha_0 u_{xt}) + c_0 v_{xt}^2 \right) dx.
\end{cases}
\tag{3.4.23}
$$

We also introduce the auxiliary functions $F_i(x)$:

$$
\begin{cases}
F_1(t) = \displaystyle\int_0^1 u_2 u_{2t} dx, \\[2mm]
F_2(t) = \displaystyle\int_0^1 u_{2x} \frac{\beta c_0}{K} v \, dx, \\[2mm]
F_3(t) = \displaystyle\int_0^1 u_{2xx} \frac{\beta c_0}{K} v_x dx, \\[2mm]
F_4(t) = \displaystyle\int_0^1 u_{2xt} \frac{\beta c_0}{K} v_t dx, \\[2mm]
F_5(t) = \displaystyle\int_0^1 u_{2t} u_{2tt} dx,
\end{cases}
\tag{3.4.24}
$$

and the auxiliary energy functions $E_0(t), E_{01}(t), E_{02}(t)$:

$$\begin{cases} E_0(t) = \dfrac{1}{2} \displaystyle\int_0^1 (u_x, \alpha_0 u_x) dx, \\[3mm] E_{01}(t) = \dfrac{1}{2} \displaystyle\int_0^1 (u_{xx}, \alpha_0 u_{xx}) dx, \\[3mm] E_{02}(t) = \dfrac{1}{2} \displaystyle\int_0^1 (u_{xt}, \alpha_0 u_{xt}) dx. \end{cases} \qquad (3.4.25)$$

Clearly, $E_i(t)\,(i = 1, \cdots, 5), E_0(t), E_{01}(t), E_{02}(t)$ should be part of the auxiliary energy function $E(t)$. However, because of the feature of hyperbolic–parabolic coupled systems, we have to introduce auxiliary functions $F_i(t), (i = 1, \cdots, 5)$ to obtain the term $\int_0^t G(\tau) d\tau$ in (3.4.20) with G being another auxiliary energy function (also see Section 2.3.2).

From now on, we always assume that $u \in \bigcap\limits_{k=0}^{2} C^k([0,T]; H^{2-k})$, $v \in C([0,T]; H^3)$, $v_t \in C([0,T]; H^2)$, $v_{tt} \in C([0,T]; L^2) \cap L^2([0,T]; H^1)$ and $\{u, v\}$ is a pair of solutions to problem (3.4.1)–(3.4.2) in Q_T. Using the standard energy method we can obtain the following results.

Lemma 3.4.2 *Suppose*

$$|u|_1 + |v|_1 + |v_{xx}| + |v_{xt}| \le \nu, \ \ in \ \bar{Q}_T = [0,1] \times [0,T]. \qquad (3.4.26)$$

Then for $\nu < 1$ appropriately small, the following estimates hold for $0 \le t \le T$:

$$E_1(t) - E_1(0) + \frac{1}{2}\int_0^t\int_0^1 Kv_x^2 dx\, d\tau \le C\nu\int_0^t\int_0^1 (u_2^2 + u_x^2)dx\, d\tau, \quad (3.4.27)$$

$$E_2(t) - E_2(0) + \frac{1}{2}\int_0^t\int_0^1 Kv_{xt}^2 dx d\tau \le C\nu\int_0^t\int_0^1 (|Du|^2 + |Dv|^2)dx d\tau, \quad (3.4.28)$$

$$E_3(t) - E_3(0) + \frac{1}{2}\int_0^t\int_0^1 Kv_{xx}^2 dx\, d\tau \le C\nu\int_0^t\int_0^1 |Du|^2 dx\, d\tau, \qquad (3.4.29)$$

$$E_4(t) - E_4(0) + \frac{1}{2}\int_0^t\int_0^1 Kv_{xtt}^2 dx\, d\tau$$
$$\le C\nu\int_0^t\int_0^1 (|u_t|^2 + |v_t|^2 + |u_{xt}|^2 + v_{xt}^2 + |u_{tt}|^2 + v_{tt}^2)dx\, d\tau, \qquad (3.4.30)$$

$$E_5(t) - E_5(0) + \frac{1}{2} \int_0^t \int_0^1 K v_{xxt}^2 dx \, d\tau$$
$$\leq C\nu \int_0^t \int_0^1 (|u_x|^2 + v_x^2 + |D^2 u|^2 + |D^2 v|^2) dx \, d\tau. \tag{3.4.31}$$

Proof. Acting with $\frac{\partial^k}{\partial t^k}$ $(k = 0, 1, 2)$ on the equations in (3.4.1) and then multiplying the first two equations by $\frac{\partial^k u}{\partial t^k}$ and the third equation by $\frac{\partial^k v}{\partial t^k}$, adding up and integrating with respect to x and t, using the boundary conditions in (3.4.2) and (3.4.26), then applying the Young inequality and the Hölder inequality, we can get (3.4.27), (3.4.28) and (3.4.30). To prove (3.4.29) and (3.4.31), in the same manner as before, we act with $\frac{\partial}{\partial x}$ and $\frac{\partial^2}{\partial x \partial t}$ on the equations in (3.4.1) respectively. Then we multiply them by $\frac{\partial u}{\partial x}, \frac{\partial v}{\partial x}$ and $\frac{\partial^2 u}{\partial x \partial t}, \frac{\partial^2 v}{\partial x \partial t}$ respectively. It follows from the second equation in (3.4.1) and the boundary conditions (3.4.2) that $\frac{\partial u_1}{\partial x}$ vanishes at the boundary. Therefore, when we use integration by parts, the boundary integral terms vanish. Thus, (3.4.29) and (3.4.31) can also be derived. □

Lemma 3.4.3 *Under the same assumptions as Lemma 3.4.2, when $\nu < 1$ is suitably small, for $0 \leq t \leq T$ the following estimates hold:*

$$\int_0^t \int_0^1 v_t^2 dx \, d\tau \leq \int_0^t \int_0^1 v_{xt}^2 dx \, d\tau + C\nu \int_0^t \int_0^1 (u_2^2 + v_x^2) dx \, d\tau, \tag{3.4.32}$$

$$\int_0^t \int_0^1 v_{tt}^2 dx \, d\tau$$
$$\leq \int_0^t \int_0^1 v_{xtt}^2 dx \, d\tau + C\nu \int_0^t \int_0^1 (|u_x|^2 + v_x^2 + |u_{xt}|^2 + v_{xt}^2) dx \, d\tau. \tag{3.4.33}$$

Proof. By the Poincaré inequality, we have

$$\int_0^1 v_t^2 dx \leq \left(\int_0^1 v_t dx \right)^2 + \int_0^1 v_{xt}^2 dx. \tag{3.4.34}$$

We can easily deduce from the third equation of (3.4.1), the boundary conditions (3.4.2) and the Hölder inequality, that

$$\left(\int_0^1 v_t dx \right)^2 \leq C\nu \int_0^1 (v_x^2 + u_2^2) dx. \tag{3.4.35}$$

Combining (3.4.34) with (3.4.35) and integrating with respect to t yields (3.4.32). The estimate (3.4.33) can be derived in the same way. □

Lemma 3.4.4 *When $\nu < 1$ is suitably small, for $0 \leq t \leq T$, the following estimates hold:*

$$F_1(t) - F_1(0) \geq \int_0^t \int_0^1 u_{2t}^2 dx\, d\tau - C \left(\int_0^t \int_0^1 (u_{2x}^2 + u_{1t}^2 + v_t^2) dx\, d\tau \right.$$
$$\left. + \nu \int_0^t \int_0^1 (u_2^2 + u_{2t}^2 + u_{1x}^2 + v_x^2) dx\, d\tau \right), \tag{3.4.36}$$

$$F_2(t) - F_2(0) + E_0(t) - E_0(0) + \int_0^t \int_0^1 \frac{\beta^2}{K} u_{2x}^2 dx\, d\tau$$
$$\leq C \left(\frac{1}{\delta} \int_0^t \int_0^1 v_x^2 dx\, d\tau + \delta \int_0^t \int_0^1 u_{1x}^2 dx\, d\tau \right.$$
$$\left. + \nu \int_0^t \int_0^1 (|u_x|^2 + |u_t|^2 + v_t^2) dx\, d\tau \right), \tag{3.4.37}$$

$$F_3(t) - F_3(0) + E_{01}(t) - E_{01}(0) + \int_0^t \int_0^1 \frac{\beta^2}{K} u_{2xx}^2 dx\, d\tau$$
$$\leq C \left(\frac{1}{\delta} \int_0^t \int_0^1 v_{xx}^2 dx\, d\tau + \delta \int_0^t \int_0^1 u_{1xx}^2 dx\, d\tau \right.$$
$$\left. + \nu \int_0^t \int_0^1 (|Du|^2 + |Dv|^2 + |u_{xx}|^2 + |u_{xt}|^2) dx\, d\tau \right), \tag{3.4.38}$$

$$F_4(t) - F_4(0) + E_{02}(t) - E_{02}(0) + \int_0^t \int_0^1 \frac{\beta^2}{K} u_{2xt}^2 dx\, d\tau$$
$$\leq C \left(\frac{1}{\delta} \int_0^t \int_0^1 v_{xt}^2 dx\, d\tau + \delta \int_0^t \int_0^1 u_{1xt}^2 dx\, d\tau \right.$$
$$\left. + \nu \int_0^t \int_0^1 (|Du|^2 + |Dv|^2 + u_{1xt}^2 + v_{xt}^2) dx\, d\tau \right), \tag{3.4.39}$$

$$F_5(t) - F_5(0) \geq \int_0^t \int_0^1 u_{2tt}^2 dx\, d\tau - C \left(\int_0^t \int_0^1 (u_{2xt}^2 + u_{1tt}^2 + v_{tt}^2) dx\, d\tau \right.$$
$$\left. + \nu \int_0^t \int_0^1 (u_{2t}^2 + |u_{tt}|^2 + u_{1xt}^2 + v_{xt}^2) dx\, d\tau \right) \tag{3.4.40}$$

where δ is an arbitrary positive constant.

Proof. Notice that when u_2 is a smoother function,

$$F_1(t) - F_1(0) = \int_0^t \frac{dF_1}{d\tau} d\tau = \int_0^t \int_0^1 (u_{2t}^2 + u_2 u_{2tt}) dx\, d\tau \tag{3.4.41}$$

and

$$F_5(t) - F_5(0) = \int_0^t \frac{dF_5}{d\tau} d\tau = \int_0^t \int_0^1 (u_{2tt}^2 + u_{2t} u_{2ttt}) dx\, d\tau. \tag{3.4.42}$$

Differentiating the second equation in (3.4.1) with respect to t, then multiplying the resulting equation by $\frac{u_2}{a}$, integrating with respect to x and t, using integration by parts and (3.4.26), we get

$$\left| \int_0^t \int_0^1 u_2 u_{2tt} dx \, d\tau \right|$$
$$\leq C \left(\int_0^t \int_0^1 (u_{2x}^2 + u_{1t}^2 + v_t^2) dx \, d\tau \right.$$
$$\left. + \nu \int_0^t \int_0^1 (u_2^2 + u_{2t}^2 + u_{1x}^2 + v_x^2) dx \, d\tau \right). \tag{3.4.43}$$

Similarly, we obtain

$$\left| \int_0^t \int_0^1 u_{2t} u_{2ttt} dx \, d\tau \right|$$
$$\leq C \left(\int_0^t \int_0^1 (u_{2xt}^2 + u_{1tt}^2 + v_{tt}^2) \, dx \, d\tau \right.$$
$$\left. + \nu \int_0^t \int_0^1 (u_{2t}^2 + u_{2tt}^2 + u_{1xt}^2 + v_{xt}^2) dx \, d\tau \right). \tag{3.4.44}$$

Combining (3.4.43), (3.4.44) with (3.4.41), (3.4.42), respectively, we immediately get (3.4.36) and (3.4.40). For $\{u, v\}$ in the class as it is, the estimates are justified by the usual density argument.

Now we differentiate the first two equations of (3.4.1) respect to x, then multiply by u_x to obtain

$$_{,}(u_x, \alpha_0 u_{xt}) + (u_x, \alpha_{0x} u_t) + (u_x, (\alpha_1 u_x)_x)$$
$$+ u_{2x} \beta v_{xx} + u_{2x} \beta_x v_x = 0. \tag{3.4.45}$$

We also differentiate the second equation with respect to x, then multiply it by $\frac{\beta c_0}{K a} v$ to obtain

$$v \frac{\beta c_0}{K} u_{2xt} + v \frac{\beta c_0}{K a} (b_2 u_{1x} + u_2 b_3 u_{2x})_x + v \frac{\beta c_0}{K a} a_x u_{2t} + v \frac{\beta c_0}{K a} (\beta v_x)_x = 0. \tag{3.4.46}$$

Multiplying the third equation of (3.4.1) by $\frac{\beta}{K} u_{2x}$, adding up with (3.4.45) and (3.4.46), integrating with respect to x and t and then making estimates term by term, we can get (3.4.37). The estimates (3.4.38), (3.4.39) can be derived in the same way. $\qquad \square$

Lemma 3.4.5 *When $\nu < 1$ is suitably small, for $0 \leq t \leq T$, the following estimates hold:*

$$\int_0^t \int_0^1 u_{1x}^2 dx \, d\tau \leq C \left(\int_0^t \int_0^1 (u_{2t}^2 + v_x^2) dx \, d\tau + \nu \int_0^t \int_0^1 u_{2x}^2 dx \, d\tau \right), \quad (3.4.47)$$

$$\int_0^t \int_0^1 u_{1t}^2 dx \, d\tau \leq C \left(\int_0^t \int_0^1 u_{2x}^2 dx \, d\tau + \nu \int_0^t \int_0^1 (u_{2t}^2 + v_x^2) dx \, d\tau \right), \quad (3.4.48)$$

$$\int_0^t \int_0^1 u_{1xt}^2 dx \, d\tau \leq C \int_0^t \int_0^1 (u_{2tt}^2 + v_{xt}^2) dx \, d\tau$$
$$+ C\nu \int_0^t \int_0^1 (|Du|^2 + |Dv|^2 + u_{xxt}^2) dx \, d\tau, \quad (3.4.49)$$

$$\int_0^t \int_0^1 u_{1xx}^2 dx \, d\tau \leq C \int_0^t \int_0^1 (u_{2xt}^2 + v_{xx}^2) dx \, d\tau$$
$$+ C\nu \int_0^t \int_0^1 (|Du|^2 + |Dv|^2 + u_{2xx}^2) dx \, d\tau, \quad (3.4.50)$$

$$\int_0^t \int_0^1 u_{1tt}^2 dx \, d\tau \leq C \int_0^t \int_0^1 u_{2xt}^2 dx \, d\tau$$
$$+ C\nu \int_0^t \int_0^1 (u_{2tt}^2 + v_{xt}^2 + |Du|^2 + |Dv|^2) dx \, d\tau. \quad (3.4.51)$$

Proof. The estimate (3.4.47) follows from the second equation of (3.4.1) and the fact that $|b_2| \geq Const. > 0$. We can easily deduce (3.4.48) from the first equation of (3.4.1). The estimates (3.4.49), (3.4.50) can be obtained by differentiating the second equation of (3.4.1) with respect to t and x, respectively. Differentiating the first equation of (3.4.1) yields (3.4.51). $\qquad \square$

Lemma 3.4.6 *When $\nu < 1$ is suitably small, for $0 \leq t \leq T$, the following estimates hold:*

$$\int_0^1 v_{xx}^2 dx \leq C \int_0^1 (|Dv|^2 + u_{2x}^2) dx, \quad (3.4.52)$$

$$\int_0^1 v_{xxx}^2 dx$$
$$\leq C \int_0^1 (v_{xt}^2 + |u_{xx}|^2 + v_{xx}^2 + |Du|^2 + |Dv|^2) dx, \quad (3.4.53)$$

$$\int_0^1 v_{xxt}^2 dx$$
$$\leq C \int_0^1 (v_{tt}^2 + v_{xt}^2 + |u_{xt}|^2 + |Du|^2 + |Dv|^2) dx, \tag{3.4.54}$$

$$\int_0^t \int_0^1 v_{xxx}^2 dx \, d\tau \leq C \int_0^t \int_0^1 (v_{xt}^2 + u_{2xx}^2) dx \, d\tau$$
$$+ C\nu \int_0^t \int_0^1 (|Du|^2 + |Dv|^2 + v_{xx}^2 + |u_{xx}|^2) dx \, d\tau. \tag{3.4.55}$$

Proof. These estimates easily follow from the third equation of (3.4.1). $\qquad \square$

Let

$$E(t) = N \sum_{i=1}^5 E_i(t) - \eta_1 F_1(t) + \eta_2(F_2(t) + E_0(t)) + \eta_3(F_3(t) + E_{01}(t))$$
$$+ \eta_4(F_4(t) + E_{02}(t)) - \eta_5 F_5(t) \tag{3.4.56}$$

where N is a large positive constant and η_i $(i = 1, \cdots, 5)$ are small positive constants specified later. It is easy to see that for any given positive constants $N, \eta_i (i = 1, \cdots, 5)$

$$E(t) \leq C_7 \sum_{j=0}^2 \|\{D^j u(t), D^j v(t)\}\|^2. \tag{3.4.57}$$

In what follows we show that for η_i, N chosen appropriately, $E(t)$ satisfies

$$E(t) \geq C_8 \sum_{j=0}^2 \|\{D^j u(t), D^j v(t)\}\|^2, \tag{3.4.58}$$

and (3.4.20) with a certain energy function $G(t)$. It turns out from Lemma 3.4.2– Lemma 3.4.5 that there exists $0 < \nu_1 < 1$, independent of T and the solution, such that when $0 < \nu \leq \nu_1$,

$$E(t) - E(0) \leq -\frac{N}{2} \int_0^t \int_0^1 K(v_x^2 + v_{xt}^2 + v_{xx}^2 + v_{xtt}^2 + v_{xxt}^2) dx \, d\tau$$

$$-\eta_1 \left(\int_0^t \int_0^1 u_{2t}^2 dx \, d\tau - C \int_0^t \int_0^1 (u_{2x}^2 + u_{1t}^2 + v_t^2) dx \, d\tau \right)$$

$$-\eta_2 \left(\int_0^t \int_0^1 \frac{\beta^2}{K} u_{2x}^2 dx \, d\tau - C \int_0^t \int_0^1 \left(\frac{1}{\delta} v_x^2 + \delta u_{1x}^2 \right) dx \, d\tau \right)$$

$$-\eta_3 \left(\int_0^t \int_0^1 \frac{\beta^2}{K} u_{2xx}^2 dx\, d\tau - C \int_0^t \int_0^1 \left(\frac{1}{\delta} v_{xx}^2 + \delta u_{1xx}^2 \right) dx\, d\tau \right)$$

$$-\eta_4 \left(\int_0^t \int_0^1 \frac{\beta^2}{K} u_{2xt}^2 dx\, d\tau - C \int_0^t \int_0^1 \left(\frac{1}{\delta} v_{xt}^2 + \delta u_{1xt}^2 \right) dx\, d\tau \right)$$

$$-\eta_5 \left(\int_0^t \int_0^1 u_{2tt}^2 dx\, d\tau - C \int_0^t \int_0^1 \left(u_{2xt}^2 + u_{1tt}^2 + v_{tt}^2 \right) dx\, d\tau \right)$$

$$+ \tilde{C}\nu \int_0^t \int_0^1 (u_2^2 + |Du|^2 + |Dv|^2 + |D^2 u|^2 + |D^2 v|^2) dx\, d\tau$$

$$\leq \left(-\frac{N}{2}k + \frac{\eta_2 C}{\delta} + C^2 \delta \eta_2 \right) \int_0^t \int_0^1 v_x^2 dx\, d\tau$$

$$+ \left(-\frac{N}{2}k + \eta_1 C + \frac{\eta_4 C}{\delta} + \eta_4 C^2 \delta \right) \int_0^t \int_0^1 v_{xt}^2 dx\, d\tau$$

$$+ \left(-\frac{N}{2}k + \frac{\eta_3 C}{\delta} + C^2 \eta_3 \delta \right) \int_0^t \int_0^1 v_{xx}^2 dx\, d\tau$$

$$+ \left(-\frac{N}{2}k + \eta_5 C \right) \int_0^t \int_0^1 v_{xtt}^2 dx\, d\tau - \frac{N}{2}k \int_0^t \int_0^1 v_{xxt}^2 dx\, d\tau$$

$$+ \left(-\eta_1 + C^2 \eta_2 \delta \right) \int_0^t \int_0^1 u_{2t}^2 dx\, d\tau + \left(-\eta_2 k_\beta + C(C+1)\eta_1 \right) \int_0^t \int_0^1 u_{2x}^2 dx\, d\tau$$

$$-\eta_3 k_\beta \int_0^t \int_0^1 u_{2xx}^2 dx\, d\tau + \left(-\eta_4 k_\beta + C^2 \eta_3 \delta + \eta_5 C(C+1) \right) \int_0^t \int_0^1 u_{2xt}^2 dx\, d\tau$$

$$+ \left(-\eta_5 + C^2 \eta_4 \delta \right) \int_0^t \int_0^1 u_{2tt}^2 dx\, d\tau$$

$$+ \tilde{C}_1 \nu \int_0^t \int_0^1 \left(u_2^2 + |Du|^2 + |Dv|^2 + |D^2 u|^2 + |D^2 v|^2 \right) dx\, d\tau \qquad (3.4.59)$$

where $\tilde{C}, \tilde{C}_1$ are positive constants depending on $C \geq 1, N$ and η_i; k, k_β are positive constants appearing in assumption (H_1). Taking $\eta_2 = \eta_3 = \eta_4$ any positive constant

and

$$\eta_1 = \eta_5 = \frac{k_\beta \eta_2}{2C(C+1)}, \quad \delta = \frac{k_\beta}{4C^3(C+1)}, \quad N \geq \frac{4(\eta_1 C + \frac{\eta_4 C}{\delta} + \eta_4 C^2 \delta)}{k} \quad (3.4.60)$$

yields that the coefficients of all the terms on the right-hand side of (3.4.59) except the last term are negative. Thus

$$E(t) + \delta_2 \int_0^t \int_0^1 \left(|Du_2|^2 + |D^2 u_2|^2 + v_x^2 + v_{xt}^2 + v_{xx}^2 + v_{xtt}^2 + v_{xxt}^2 \right) dx\, d\tau$$

$$\leq E(0) + \tilde{C}_1 \nu \int_0^t \int_0^1 \left(u_2^2 + |Du|^2 + |Dv|^2 + |D^2 u|^2 + |D^2 v|^2 \right) dx\, d\tau \quad (3.4.61)$$

where δ_2 is a constant independent of u, v and T. Moreover, by the assumption (H_1), we have

$$\begin{cases} E_1 \geq \frac{1}{2}(\|u_1\|^2 + a_0\|u_2\|^2 + \delta_0\|v\|^2), \\[2mm] E_2 \geq \frac{1}{2}(\|u_{1t}\|^2 + a_0\|u_{2t}\|^2 + \delta_0\|v_t\|^2), \\[2mm] E_3 \geq \frac{1}{2}(\|u_{1x}\|^2 + a_0\|u_{2x}\|^2 + \delta_0\|v_x\|^2), \\[2mm] E_4 \geq \frac{1}{2}(\|u_{1tt}\|^2 + a_0\|u_{2tt}\|^2 + \delta_0\|v_{tt}\|^2), \\[2mm] E_5 \geq \frac{1}{2}(\|u_{1xt}\|^2 + a_0\|u_{2xt}\|^2 + \delta_0\|v_{xt}\|^2), \end{cases} \quad (3.4.62)$$

and

$$\begin{cases} E_0 \geq \frac{1}{2}(\|u_{1x}\|^2 + a_0\|u_{2x}\|^2), \\[2mm] E_{01} \geq \frac{1}{2}(\|u_{1xx}\|^2 + a_0\|u_{2xx}\|^2), \\[2mm] E_{02} \geq \frac{1}{2}(\|u_{1xt}\|^2 + a_0\|u_{2xt}\|^2). \end{cases} \quad (3.4.63)$$

Choosing

$$\eta_2 = \eta_3 = \eta_4 = \frac{a_0 \delta_0}{2\delta_1^2}, \quad \eta_1 = \eta_5 = \frac{k_\beta \eta_2}{2C(C+1)} \quad (3.4.64)$$

and N large enough where δ_1 is a constant such that when $|u|, |v| \leq \gamma$, $\left| \frac{\beta c_0}{K} \right| \leq \delta_1$, we get

$$E(t) \geq \delta_3 \int_0^1 \left(|u|_2^2 + |v|_1^2 + v_{xt}^2 + v_{tt}^2 \right) dx \quad (3.4.65)$$

where δ_3 is a positive constant independent of u, v and T. Combining (3.4.65) and (3.4.52), we get (3.4.58). Thus $E(t)$ is equivalent to $\sum_{j=0}^{2} \|\{D^j u(t), D^j v(t)\}\|^2$.

To prove (3.4.20), we take

$$G(t) = \int_0^1 (u_2^2 + |Du|^2 + |Dv|^2 + |D^2 u|^2 + |D^2 v|^2) dx. \tag{3.4.66}$$

Combining (3.4.61), (3.4.65) yields that there exists $\delta_4 > 0$ independent of u, v and T such that

$$\delta_4 \left(\int_0^1 \left(|u|_2^2 + |v|_1^2 + v_{xt}^2 + v_{tt}^2 \right) dx \right.$$
$$+ \int_0^t \int_0^1 \left(|Du_2|^2 + |D^2 u_2|^2 + v_x^2 + v_{xt}^2 + v_{xx}^2 + v_{xtt}^2 + v_{xxt}^2 \right) dx \, d\tau \right)$$
$$\leq E(0) + \tilde{C}_1 \nu \int_0^t \int_0^1 \left(u_2^2 + |Du|^2 + |Dv|^2 + |D^2 u|^2 + |D^2 v|^2 \right) dx \, d\tau. \tag{3.4.67}$$

Using Lemma 3.4.3, Lemma 3.4.5 and Lemma 3.4.6 again, we obtain

$$\delta_5 \left(\int_0^1 (|u|_2^2 + |v|_2^2 + v_{xxx}^2 + v_{xxt}^2) dx \right.$$
$$+ \int_0^t \int_0^1 (u_2^2 + |Du|^2 + |Dv|^2 + |D^2 u|^2 + |D^2 v|^2 + v_{xxx}^2 + v_{xxt}^2 + v_{xtt}^2) dx \, d\tau \right)$$
$$\leq E(0) + \tilde{C}_1 \nu \int_0^t \int_0^1 \left(u_2^2 + |Du|^2 + |Dv|^2 + |D^2 u|^2 + |D^2 v|^2 \right) dx \, d\tau. \tag{3.4.68}$$

Thus when $\nu \leq \nu_2 = \min \left(\nu_1, \dfrac{\delta_5}{2\tilde{C}_1} \right)$, the solution $\{u, v\}$ of problem (3.4.1)–(3.4.2) in $[0,1] \times [0, T]$ satisfies the following uniform a priori estimate:

$$\int_0^1 \left(|u|_2^2 + |v|_2^2 + v_{xxx}^2 + v_{xxt}^2 \right) dx$$
$$+ \int_0^t \int_0^1 \left(u_2^2 + |Du|^2 + |Dv|^2 + |D^2 u|^2 + |D^2 v|^2 + v_{xxx}^2 + v_{xxt}^2 + v_{xtt}^2 \right) dx \, d\tau$$
$$\leq C_3 M_0 \tag{3.4.69}$$

where C_3 is a positive constant independent of u, v, T and M_0.

Now we can combine the local existence and uniqueness result with the uniform a priori estimates to conclude the global existence and uniqueness result.

Indeed, when $M_0 \leq \varepsilon_0$, there exists t_0 such that problem (3.4.1)–(3.4.2) admits a unique smooth solution in $X_{t_0}(C_1 M_0, C_2 M_0)$. By the imbedding theorem,

$$|u|_1 + |v|_1 + |v_{xx}| + |v_{xt}| \leq C_4 M_0, \quad in \ [0,1] \times [0, t_0]. \tag{3.4.70}$$

Let

$$\varepsilon = \min\left(\frac{\nu_2}{C_4}, \frac{\nu_2}{C_3 C_4}, \varepsilon_0, \frac{\varepsilon_0}{C_3}\right). \tag{3.4.71}$$

Then when $M_0 \leq \varepsilon \leq \varepsilon_0$, problem (3.4.1)–(3.4.2) admits a unique solution in $X_{t_0}(C_1 M_0, C_2 M_0)$. Moreover, it follows from $C_4 M_0 \leq \nu_2$ that we can apply Lemma 3.4.2–Lemma 3.4.6 to conclude that the solution $\{u, v\}$ satisfies (3.4.69) in $[0, 1] \times [0, t_0]$. It follows from (3.4.69) that

$$\int_0^1 (|u(t_0)|_2^2 + |v(t_0)|_2^2) dx \leq C_3 M_0 \leq C_3 \varepsilon \leq \varepsilon_0. \tag{3.4.72}$$

By the local exsitence and uniqueness results, the solution can be extended to $[t_0, 2t_0]$. Moreover,

$$|u|_1 + |v|_1 + |v_{xx}| + |v_{xt}| \leq C_4(C_3 M_0), \quad in \ [0, 1] \times [0, 2t_0]. \tag{3.4.73}$$

Since $C_3 C_4 M_0 \leq \nu_2$, the previous argument shows that (3.4.69) holds in $[0, 1] \times [0, 2t_0]$. Thus we can repeat the above argument so that the local solution $\{u, v\}$ can be extended step by step to get a unique global solution.

It remains to prove the exponential decay of $\|u_2\|$, $\|Du\|$, $\|Dv\|$, $\|D^2 u\|$, $\|D^2 v\|$, $\|D^3 v\|$ and $\|v_{xxt}\|$. For this purpose, we introduce another auxiliary energy function

$$\tilde{E}(t) = \tilde{N} \sum_{i=2}^{5} E_i(t) - \tilde{\eta}_1 F_1(t) + \tilde{\eta}_3(F_3(t) + E_{01}(t))$$
$$+ \tilde{\eta}_4(F_4(t) + E_{02}(t)) - \tilde{\eta}_5 F_5(t) + \tilde{\eta}_6 E_{11}(t) \tag{3.4.74}$$

where

$$E_{11}(t) = \frac{1}{2} \int_0^1 a u_2^2 dx. \tag{3.4.75}$$

In the same manner as before, we can find suitable constants $\tilde{N}, \tilde{\eta}_i \ (i = 1, 3, 4, 5, 6)$ and $\delta_6, \delta_7 > 0$ such that for all $0 \leq t \leq T$

$$\frac{d\tilde{E}}{dt} + \delta_6 \int_0^1 \left(u_{2t}^2 + u_{2xx}^2 + u_{2xt}^2 + u_{2tt}^2 + v_{xt}^2 + v_{xx}^2 + v_{xtt}^2 + v_{xxt}^2\right) dx$$
$$\leq C\nu \int_0^1 \left(|Du|^2 + |Dv|^2 + |D^2 u|^2 + |D^2 v|^2\right) dx \tag{3.4.76}$$

and

$$\tilde{E}(t) \geq \delta_7 \int_0^1 \left(u_2^2 + |Du|^2 + |Dv|^2 + |D^2 u|^2 + v_{xt}^2 + v_{tt}^2\right) dx. \tag{3.4.77}$$

On the other hand, by the Poincaré inequality and the second and third equations of (3.4.1), we have

$$
\int_0^1 \left(u_2^2 + u_{2x}^2 \right) dx \le C \int_0^1 u_{2x}^2 dx \le \int_0^1 \left(v_t^2 + v_{xx}^2 + v_x^2 \right) dx
$$
$$
\le C \left(\int_0^1 \left(v_{xt}^2 + v_{xx}^2 \right) dx + \nu \int_0^1 \left(|Du|^2 + |Dv|^2 \right) dx \right). \tag{3.4.78}
$$

Combining (3.4.76) with (3.4.78) yields that there exists $\delta_8 > 0$ such that for $0 \le t \le T$,

$$
\frac{d\tilde{E}}{dt} + \delta_8 \int_0^1 \left(u_2^2 + |Du|^2 + |Dv|^2 + |D^2u|^2 + |D^2v|^2 + v_{xtt}^2 + v_{xxt}^2 \right) dx
$$
$$
\le C\nu \int_0^1 \left(|Du|^2 + |Dv|^2 + |D^2u|^2 + |D^2v|^2 \right) dx. \tag{3.4.79}
$$

It easily follows from the definition of $\tilde{E}(t)$ and the system (3.4.1) that

$$
\tilde{E}(t) \le C_5 \int_0^1 \left(u_2^2 + |Du|^2 + |Dv|^2 + |D^2u|^2 + |D^2v|^2 + v_{xtt}^2 + v_{xxt}^2 \right) dx. \tag{3.4.80}
$$

On the other hand, by (3.4.77) and the system (3.4.1) we have

$$
\tilde{E}(t) \ge \delta_9 \int_0^1 \left(u_2^2 + |Du|^2 + |Dv|^2 + |D^2u|^2 + |D^2v|^2 + v_{xxx}^2 + v_{xxt}^2 \right) dx. \tag{3.4.81}
$$

We can deduce from (3.4.79) and (3.4.80) that

$$
\frac{d\tilde{E}}{dt} + \frac{\delta_8}{2C_5} \tilde{E}(t) \le 0 \tag{3.4.82}
$$

provided that ν is suitably small, which corresponds to the suitable smallness of the initial data.

Thus it follows from (3.4.82) and the Gronwall inequality that $\tilde{E}(t)$ and, therefore, the terms on the right-hand side of (3.4.81) decay exponentially as $t \to \infty$. The proof of Theorem 3.4.1 is complete. $\square$

As an application of Theorem 3.4.1, we consider the system of radiation hydrodynamics (3.4.8) subject to the following boundary conditions and initial conditions:

$$
\begin{cases}
u|_{x=0,1} = 0, \quad \theta_x|_{x=0,1} = 0, \\
t = 0 : \rho = \rho_0(x), \ u = u_0(x), \ \theta = \theta_0(x).
\end{cases} \tag{3.4.83}
$$

Let

$$\rho = \bar{\rho}_0(1 + \rho'), \ u = u, \ \theta = T_0(1 + \theta') \tag{3.4.84}$$

where $\bar{\rho}_0$, T_0 are given positive constants. For the new dependent functions ρ', u, θ', we deduce from Theorem 3.4.1 that when $\left\| \dfrac{\rho_0(x) - \bar{\rho}_0}{\bar{\rho}_0} \right\|_{H^3}$, $\|u_0\|_{H^3}$, $\left\| \dfrac{\theta_0(x) - \bar{T}_0}{T_0} \right\|_{H^4}$ are suitably small, problem (3.4.8), (3.4.83) admits a unique global smooth solution which has the corresponding exponential decay.

Theorem 3.4.1 can also be applied to the one-dimensional nonlinear thermoelastic system and we leave the details to the reader.

3.5   Nonexistence of Global Solutions

In this section we are concerned with nonexistence of global solution to the initial boundary value problems for nonlinear parabolic equations with *small* initial data. The results presented in this section show that for *small* initial data, the question whether solutions to the initial boundary value problems for nonlinear parabolic equations globally exist or blow-up in finite time is very closely related to whether the corresponding elliptic operators subject to the boundary conditions are positive definite. It turns out that this question is also closely related to whether solutions to the linearized problems have exponential decay rates. The same question for the nonlinear hyperbolic equations is also discussed in this section.

In the past forty years a large literature has developed concerning the nonexistence of global solution or blow-up in finite time. For instance, to name just a few, see Levine [1], [2], Kaplan [1], Glassay [1], Ball [1], [2] and the references cited there. We refer to Ball [1], [2], for the distinction between two different concepts: nonexistence of global solutions and blow-up in finite time. Most conclusions concerning nonexistence of global solution or blow-up in finite time were drawn under the assumption on initial data that the appropriate 'energy function' is negative at $t = 0$. We will show that this assumption amounts to the initial data being 'large'. Typically the following problem is considered:

$$\begin{cases} u_t - \Delta u = u^p, & (x,t) \in \Omega \times I\!R^+, \\ u|_\Gamma = 0, \\ u|_{t=0} = u_0(x), & u_0 \geq 0, x \in \Omega \end{cases} \tag{3.5.1}$$

where Γ is a smooth boundary of a bounded domain $\Omega \in I\!R^n$ and $p > 1$ is a given constant. Let

$$E(u(t)) = \int_\Omega (\frac{1}{2}|\nabla u|^2 - \frac{1}{p+1}|u|^{p+1})dx. \tag{3.5.2}$$

Then it is well known (see the references mentioned before) that $E(u_0) \leq 0$ implies the nonexistence of global smooth solutions no matter how smooth the initial data u_0 are (see Ball [1], [2] for the blow-up results under additional restrictions on p).

Let $u_0 = \varepsilon\varphi(x)$ with $\varphi(x) \neq 0, \varphi \in H_o^1$ being a given function. Then it is easy to see that for sufficiently small ε, we have $E(u_0) > 0$. Therefore, roughly speaking, $E(u_0) \leq 0$ amounts to assuming that the initial data are 'large'.

On the other hand, because the solution to the linearized problem of (3.5.1) has exponential decay rate (see Section 2.1.2), by the so-called potential well method or by Theorem 3.2.1 with $p > 1$, integer, $u_0(x) \geq 0$ small and suitably smooth, we can conclude the global existence of solutions to the problem (3.5.1).

In contrast, for the following problem

$$\begin{cases} u_t - \Delta u = u^2, \\ \dfrac{\partial u}{\partial n}\Big|_\Gamma = 0, \\ u|_{t=0} = u_0, u_0 \geq 0, u_0 \neq 0, \end{cases} \tag{3.5.3}$$

the operator $-\Delta$ subject to the Neumann boundary condition is not positive definite and the solution to the corresponding linearized problem does not decay to zero as time goes to infinity. It turns out that the problem (3.5.3) does not have a global classical solution no matter how small and smooth the initial data u_0 are. Indeed, this assertion can be proved as follows.

Integrating the equation (3.5.3) with respect to x, using the boundary condition, we obtain

$$\frac{d}{dt}\int_\Omega u dx = \int_\Omega u^2 dx \geq \frac{1}{mes(\Omega)}\left(\int_\Omega u dx\right)^2. \tag{3.5.4}$$

Thus we can easily deduce from (3.5.4) that as $t \to \dfrac{mes(\Omega)}{\int_\Omega u_0 dx}$,

$$F(t) = \int_\Omega u dx \geq \frac{1}{\dfrac{1}{\int_\Omega u_0 dx} - \dfrac{t}{mes(\Omega)}} \to \infty. \tag{3.5.5}$$

Concerning the *initial value problem* for nonlinear parabolic equations with *small* initial data, it is well known (see Fujita [1], Weissler [1]) that global existence or nonexistence of a classical solution heavily depends on the decay rate of the solution to the linearized problem. It can be seen from the previous discussions about problems (3.5.1) and (3.5.3) that as far as the *initial boundary value problem* for nonlinear parabolic equations with *small* initial data is concerned, global existence or nonexistence depends on whether the solution to the linearized problem has exponential decay rate. We now extend the above results on problem (3.5.3) to more general initial boundary value problems for nonlinear parabolic equations (and also hyperbolic equations) (see Zheng & Chen [2]).

More precisely, we consider the following initial boundary value problem for nonlinear parabolic equations

$$\begin{cases} u_t - \displaystyle\sum_{i,j=1}^n \frac{\partial}{\partial x_i}\left(a_{ij}(x)\frac{\partial u}{\partial x_j}\right) - cu = f(x,t,u), \\ Bu|_\Gamma = 0, \\ u|_{t=0} = u_0(x). \end{cases} \tag{3.5.6}$$

Here

$$Au = -\sum_{i,j=1}^n \frac{\partial}{\partial x_i}\left(a_{ij}(x)\frac{\partial u}{\partial x_j}\right) \tag{3.5.7}$$

is a self-adjoint elliptic operator with smooth coefficients $a_{ij}(x)$ $(i,j = 1, \cdots, n)$, and the boundary operator Bu denotes u or $\displaystyle\sum_{i,j=1}^n a_{ij}(x)\frac{\partial u}{\partial x_j}\cos(n, x_i) + \sigma(x)u$ with smooth $\sigma \geq 0$ and n being the exterior normal to the smooth boundary Γ of a bounded domain $\Omega \subseteq \mathbb{R}^n$.

Let λ_1^B be the first eigenvalue of the following eigenvalue problem:

$$\begin{cases} A\varphi = \lambda\varphi, \\ B\varphi|_{\Gamma=0} = 0. \end{cases} \tag{3.5.8}$$

It is well-known (see e.g., Courant & Hilbert [1]) that the corresponding eigenfunctions span one-dimensional subspace and we have corresponding eigenfunction φ_1^B such that

$$\begin{cases} \varphi_1^B(x) > 0, \quad x \in \Omega, \\[2mm] \displaystyle\int_\Omega \varphi_1^B \, dx = 1. \end{cases} \tag{3.5.9}$$

Now we have the following nonexistence result.

Theorem 3.5.1 *Suppose that $c = c(t) \geq \lambda_1^B$, $f(x,t,u) \geq b|u|^p$ with given constants $b > 0, p > 1$. Suppose u is a classical solution to problem (3.5.6) in $\bar\Omega \times [0, T_{max})$. Then when $\int_\Omega u_0 \varphi_1^B \, dx > 0$, no matter how small the initial data u_0 are, we must have $T_{max} < \infty$, i.e., nonexistence of a global classical solution.*

Proof. Multiplying the equation and the initial condition in (3.5.6) by φ_1^B and integrating with respect to x, we obtain

$$\begin{cases} v_t(t) \geq (c(t) - \lambda_1^B)v(t) + b\displaystyle\int_\Omega |u|^p \varphi_1^B \, dx, \\[2mm] v(0) = \displaystyle\int_\Omega u_0(x)\varphi_1^B \, dx > 0 \end{cases} \tag{3.5.10}$$

where

$$v(t) = \int_\Omega u\varphi_1^B \, dx. \tag{3.5.11}$$

Applying the Jensen inequality (see Section 1.6), we get

$$\int_\Omega |u|^p \varphi_1^B \, dx \geq \left(\int_\Omega |u| \varphi_1^B \, dx\right)^p = \left(\int_\Omega |u\varphi_1^B| \, dx\right)^p$$

$$\geq \left|\int_\Omega u\varphi_1^B \, dx\right|^p = v^p. \tag{3.5.12}$$

Combining (3.5.10) with (3.5.12) yields

$$\begin{cases} v_t \geq (c(t) - \lambda_1^B)v(t) + b|v|^p, \\[2mm] v(0) = \displaystyle\int_\Omega u_0\varphi_1^B \, dx > 0. \end{cases} \tag{3.5.13}$$

Therefore, by (3.5.13) we have

$$v(t) \geq v(0) > 0, \quad for \; t \in [0, T_{max}) \tag{3.5.14}$$

and

$$v_t \geq b v^p > 0, \quad for \ t \in [0, T_{max}). \tag{3.5.15}$$

Since $p > 1$, we deduce from (3.5.15) that $T_{max} < \infty$, i.e., nonexistence of a global classical solution. □.

Remark 3.5.1 *It can be easily seen that the assumption $c \geq \lambda_1^B$ implies that the corresponding elliptic operator $- \sum\limits_{i,j=1}^{n} \dfrac{\partial}{\partial x_i}(a_{ij}(x) \dfrac{\partial u}{\partial x_j}) - cu$ subject to the boundary condition $Bu|_\Gamma = 0$ is not positive definite. It turns out that the solution to the linearized problem does not decay to zero.*

Remark 3.5.2 *The above result is sharp in the following sense that if $c = Const. < \lambda_1^B$, then it follows from Theorem 3.2.1 that problem (3.5.6) with $f = f(u)$ being smooth, $f(u) = O(u^2)$ near $u = 0$, must have a global smooth solution provided that the initial data are small.*

Similarly, we can discuss the initial boundary value problem for nonlinear hyperbolic equations:

$$\begin{cases} u_{tt} - \sum\limits_{i,j=1}^{n} \dfrac{\partial}{\partial x_i}(a_{ij}(x) \dfrac{\partial u}{\partial x_j}) - c(t)u + \alpha u_t = f(x, t, u, u_t, \nabla u), \\ Bu|_\Gamma = 0, \\ u|_{t=0} = u_0(x), \quad u_t|_{t=0} = u_1(x). \end{cases} \tag{3.5.16}$$

The assumptions on the operators A, B and Ω are the same as before. We also assume that c, f and initial data u_0, u_1 are smooth given functions and α is a positive constant. The dissipative term αu_t usually is a good one to stabilize the system (see e.g., Matsumura [1]). However, the following result shows that when the corresponding elliptic operator is not positive definite, the dissipative term αu_t is not strong enough to prevent the solution from developing a singularity in the finite time even if the initial data are small and smooth.

Theorem 3.5.2 *Suppose that $c = c(t) \geq \lambda_1^B, \alpha > 0, f \geq b_1|u|^{p_1} + b_2|u_t|^{p_2}$ with constants $b_i > 0, p_i > 1 \ (i = 1, 2)$. Suppose that $u(x, t)$ is a classical solution in $\Omega \times [0, T_{max})$. Then when $\int_\Omega u_0 \varphi_1^B dx > 0$, $\int_\Omega u_1 \varphi_1^B dx > 0$, no matter how small and smooth the initial data are, we must have $T_{max} < +\infty$.*

Proof. In the same manner as before, we get

$$
\begin{cases}
v_{tt}(t) - (c(t) - \lambda_1^B)v(t) + \alpha v_t(t) \geq b_1|v|^{p_1} + b_2|v_t(t)|^{p_2}, \\
\\
v(0) = \displaystyle\int_\Omega u_0 \varphi_1^B \, dx > 0, \quad v_t(0) = \displaystyle\int_\Omega u_1 \varphi_1^B \, dx > 0.
\end{cases}
\tag{3.5.17}
$$

We first claim that

$$
v(t) \geq v(0) > 0, \quad v_t(t) > 0, \quad t \in [0, T_{max}).
\tag{3.5.18}
$$

Indeed, it follows from $v(0) > 0, v_t(0) > 0$ that there is a neighbourhood of $t = 0$ such that $v(t) > 0, v_t(t) > 0$. If (3.5.18) is not true, then there is a $t^* \in [0, T_{max})$ such that $v_t(t) > 0$, for $t \in [0, t^*)$ and $v_t(t^*) = 0$. Thus, it follows from (3.5.17) that

$$
v_{tt} + \alpha v_t(t) > 0, \quad t \in [0, t^*).
\tag{3.5.19}
$$

Therefore,

$$
\frac{d}{dt}(e^{\alpha t} v_t) > 0, \quad t \in [0, t^*),
\tag{3.5.20}
$$

$$
v_t(t) > v_t(0)e^{-\alpha t^*} > 0, \quad t \in [0, t^*),
\tag{3.5.21}
$$

$$
v_t(t^*) \geq v_t(0)e^{-\alpha t^*} > 0,
\tag{3.5.22}
$$

a contradiction.

To prove the theorem, we need the following comparison lemma (see Beckenbach & Bellman [1], p. 139, Theorem 5).

Lemma 3.5.1 *Suppose that $p(t), q(t) \in C, u, v \in C^2$ and satisfy*
(i) $u_{tt} + p(t)u_t - q(t)u > 0, \quad t \geq 0$;
(ii) $v_{tt} + p(t)v_t - q(t)v = 0, \quad t \geq 0$;
(iii) $q(t) \geq 0, \quad t \geq 0$;
(iv) $u(0) = v(0), u_t(0) = v_t(0)$.
Then $u(t) > v(t)$, for $t > 0$.

It follows from (3.5.17), (3.5.18) and $c(t) \geq \lambda_1^B$ that

$$
v_{tt} + \alpha v_t - b_1 v^{p_1-1}(0)v \geq b_2 v_t^{p_2} > 0.
\tag{3.5.23}
$$

Let $w(t)$ be the solution to the following problem

$$\begin{cases} w_{tt} + \alpha w_t - b_1 v^{p_1-1}(0)w = 0, \\ w(0) = v(0), \quad w_t(0) = v_t(0). \end{cases} \tag{3.5.24}$$

Thus

$$w(t) = C_1 e^{\mu_1 t} + C_2 e^{\mu_2 t} \tag{3.5.25}$$

where

$$\begin{cases} \mu_1 = \dfrac{1}{2}(-\alpha + \sqrt{\alpha^2 + 4b_1 v^{p_1-1}(0)}) > 0, \\[3mm] \mu_2 = \dfrac{1}{2}(-\alpha - \sqrt{\alpha^2 + 4b_1 v^{p_1-1}(0)}) < 0 \end{cases} \tag{3.5.26}$$

and

$$\begin{cases} C_1 = \dfrac{v_t(0) - \mu_2 v(0)}{\sqrt{\alpha^2 + 4b_1 v^{p_1-1}(0)}} > 0, \\[4mm] C_2 = \dfrac{\mu_1 v(0) - v_t(0)}{\sqrt{\alpha^2 + 4b_1 v^{p_1-1}(0)}}. \end{cases} \tag{3.5.27}$$

Therefore, we conclude from (3.5.25) that

$$w(t) \to +\infty, \quad as \ t \to \infty. \tag{3.5.28}$$

By Lemma 3.5.1, we have

$$v(t) > w(t), \quad \forall \, t > 0. \tag{3.5.29}$$

We now use a contradiction argument to prove that $T_{max} < \infty$.

If $T_{max} = \infty$, then by (3.5.29) we get

$$v(t) \to \infty, \quad as \ t \to \infty. \tag{3.5.30}$$

By the Young inequality, we get for any $\delta > 0$,

$$v_t \le \frac{\delta^{p_2}}{p_2} v_t^{p_2} + \frac{1}{p_2' \delta^{p_2'}} \tag{3.5.31}$$

where $p_2' > 1$, $\frac{1}{p_2} + \frac{1}{p_2'} = 1$. Taking δ small enough, we have

$$\alpha v_t \le \frac{b_2}{2} v_t^{p_2} + C_\delta \tag{3.5.32}$$

where $C_\delta > 0$ is a constant depending on δ. Combining (3.5.23) with (3.5.32) yields

$$v_{tt} \geq \frac{b_2}{2} v_t^{p_2} + b_1 v^{p_1-1}(0)v - C_\delta. \tag{3.5.33}$$

On the other hand, it follows from (3.5.30), (3.5.33) that there exists $t_0 \in \mathbb{R}^+$ such that

$$v_{tt} \geq \frac{b_2}{2} v_t^{p_2}, \quad \forall t \geq t_0. \tag{3.5.34}$$

Since $b_2 > 0, p_2 > 1$, the function v satisfying (3.5.34) must blow up in a finite time. This contradicts $T_{max} = \infty$. Thus the proof is complete . $\Box$.

Remark 3.5.3 *The above result is sharp in the following sense that if the corresponding elliptic operator is positive definite (i.e., $c = Const < \lambda_1^B$), as shown by Matsumura [1], problem (3.5.16) admits a unique global solution provided that the initial data are smooth, small and the right-hand side term f is smooth and is at least second order about u and its derivatives near the origin. For more details, see Matsumura [1].*

Chapter 4

Global Existence for Large Initial Data

In this chapter we discuss global existence and uniqueness of solutions to some non-linear evolution equations arising from the study of phase transitions with arbitrary (large) initial data. It turns out that the technique used in Chapter 3 is no longer useful and one has to derive uniform a priori estimates of solutions based on the special structure conditions of the equations.

In the first section of this chapter we deal with a nondiagonal system of nonlinear parabolic equations of second order, namely the phase-field equations. Section 4.2 is devoted to a nonlinear system of partial differential equations arising from the study of phase transitions in shape memory alloys. This is the coupled system of a nonlinear heat equation and a nonlinear beam equation. In Section 4.3 we are concerned with a coupled system of nonlinear parabolic equations of second order and fourth order: the coupled Cahn–Hilliard equations. The spirit of the method deriving uniform a priori estimates, especially the L^∞ norm estimate, is illustrated through these examples. These systems of nonlinear partial differential equations have some interesting features: 1. The phase-field equations and the coupled Cahn–Hilliard equations are non-diagonal nonlinear parabolic systems. 2. In the phase-field equations of the Penrose–Fife model and in the coupled Cahn–Hilliard equations, the unusual nonlinear term, namely $\frac{1}{\theta}$ and its derivatives with θ being an unknown

145

function, appear in the systems. It turns out that one has to bound θ from above and also from below. 3. At the very begining, for an unknown function, the temperatute θ, only a very weak a priori estimate, the L^1 norm (see Sections 4.1.2, 4.2, 4.3) or L^2 norm (see Section 4.1.1) estimate, is available. Then the key issue is to derive further L^∞ norm and higher-order norm estimates.

The methods used in this chapter include an extension of the technique by Alikakos [1]. A special technique developed in Section 4.1.2 and Section 4.3 is proved to be useful to deal with the unusual nonlinearity mentioned above. We also establish an important lemma in analysis (Lemma 4.3.1) which is not only useful for getting uniform a priori estimates, but also useful for studying the asymptotic behaviour of the solution as time goes to infinity.

4.1   Phase-field Equations

4.1.1   The Caginalp Model

The following phase-field equations

$$
\begin{cases}
\alpha\varphi_t = \xi^2\Delta\varphi + \frac{1}{2}(\varphi - \varphi^3) + 2u, \\
\\
u_t + \frac{l}{2}\varphi_t = K\Delta u,
\end{cases}
\tag{4.1.1}
$$

was proposed by Caginalp [1] to describe the phase transition with finite thickness of the interface in a material, which may be in either of two phases, e.g., solid or liquid, occupying a domain $\Omega \subset I\!R^n, (n \leq 3)$ with a smooth boundary Γ. The function $u(x,t)$ represents the distribution of the (reduced) temperature in $\Omega \times (0,\infty)$ such that $u = 0$ is the equilibrium melting temperature, i.e., the temperature at which the solid and liquid can coexist independently in equilibrium separated by an interface. The function $\varphi(x,t)$, which is also called the order parameter, represents the phase distribution, which is near $+1$ and -1, respectively, in the liquid and solid states. The positive parameters α, ξ, l, K represent the relaxation time, a length scale, which, at the microscope level, is a measure of the strength of the bonding, the latent heat

and the thermal diffusivity, respectively. We also refer to the paper of Penrose &
Fife [1] for the historic background of the model and the derivation of a more general
thermodynamically consistent model. The global existence of solutions, among other
things, to equations (4.1.1) subject to the Dirichlet boundary conditions

$$\varphi|_\Gamma = \varphi_\Gamma(x), \quad u|_\Gamma = u_\Gamma(x) \tag{4.1.2}$$

and initial conditions

$$\varphi|_{t=0} = \varphi_0(x), \quad u|_{t=0} = u_0(x) \tag{4.1.3}$$

was proved by Caginalp [1] under the restriction on the coefficients:

$$\frac{\xi^2}{\alpha} < K. \tag{4.1.4}$$

Elliott & Zheng [2] used a different method to prove global existence for problem
(4.1.1)–(4.1.3) so that they were able to drop the restriction (4.1.4). They also considered other boundary conditions, e.g.,

$$\frac{\partial \varphi}{\partial n}\bigg|_\Gamma = 0, \quad \frac{\partial u}{\partial n}\bigg|_\Gamma = 0, \tag{4.1.5}$$

or

$$\varphi|_\Gamma = \varphi_\Gamma(x), \quad \frac{\partial u}{\partial n}\bigg|_\Gamma = 0, \tag{4.1.6}$$

or

$$\frac{\partial \varphi}{\partial n}\bigg|_\Gamma = 0, \quad u|_\Gamma = u_\Gamma(x). \tag{4.1.7}$$

For results with initial data in different settings of spaces, see Elliott & Zheng [2],
Bates & Zheng [1], and also Brochet, Chen & Hilhorst [1].

In what follows we present the results only for the Dirichlet initial boundary value
problem (4.1.1)–(4.1.3) with u_Γ and $\varphi_\Gamma(x)$ being given functions only in x, which is
exactly the problem considered by Caginalp [1].

Let $\bar{u}(x), \bar{\varphi}(x)$ be harmonic functions in Ω such that $\bar{u}|_\Gamma = u_\Gamma$ and $\bar{\varphi}|_\Gamma = \varphi_\Gamma$,
respectively. Then $\psi = \varphi - \bar{\varphi}, v = u - \bar{u}$ vanish on Γ and satisfy similar equations to
(4.1.1). Therefore, in what follows, without loss of generality, we always assume that
$\varphi_\Gamma = u_\Gamma = 0$. Now we have

Theorem 4.1.1 Let $\Omega \subset \mathbb{R}^n (n \leq 3)$ be a bounded domain with smooth boundary Γ. Suppose $u_0 \in L^2(\Omega), \varphi_0 \in H_o^1(\Omega)$. Then problem (4.1.1)–(4.1.3) admits a unique global solution (φ, u) such that $\varphi \in C([0, \infty); H_o^1), u \in C([0, \infty); L^2)$; for any $T > 0$, $\varphi \in L^2([0, T]; H^2), \varphi_t \in L^2([0, T]; L^2), u \in L^2([0, T]; H^1)$ and the associated nonlinear semigroup $S(t)$ is a dynamical system.

Moreover, $u, \varphi \in C^\infty((0, \infty); C^\infty(\bar{\Omega}))$ and for any $\varepsilon > 0$, the orbit $t \in [\varepsilon, +\infty) \mapsto (\varphi(\cdot, t), u(\cdot, t))$ is compact in $H_o^1 \times L^2$.

Remark 4.1.1 The restriction $n \leq 3$ is not necessary and for general n, the solution (φ, u) will belong to $(H^1 \cap L^4) \times L^2$ for all $t > 0$ provided the initial data are in the same space.

Proof of Theorem 4.1.1 The proof consists of four steps.

1. We use the contraction mapping theorem to prove the local existence and uniqueness of the solution to problem (4.1.1)–(4.1.3) ($\varphi_\Gamma = u_\Gamma = 0$). Let

$$X_\delta(M_1, M_2) = \left\{ (\varphi, u) \,\middle|\, \begin{array}{l} \varphi \in C([0, \delta]; H_o^1),\ \varphi_t \in L^2([0, \delta]; L^2),\ u \in C([0, \delta]; L^2), \\ \varphi|_{t=0} = \varphi_0,\ u|_{t=0} = u_0,\ \max_{0 \leq t \leq \delta} \|\varphi(t)\|_{H^1}^2 \leq 2\|\varphi_0\|_{H^1}^2, \\ \max_{0 \leq t \leq \delta} \|u(t)\|^2 \leq M_1,\ \int_0^\delta \|\varphi_t\|^2 d\tau \leq M_2 \end{array} \right\}$$

(4.1.8)

with M_1, M_2 being positive constants to be specified later.

For $(\psi, v) \in X_\delta(M_1, M_2)$ consider the following auxiliary linear problems

$$\begin{cases} \alpha \varphi_t = \xi^2 \Delta \varphi + f_1(\psi, v), \\ \varphi|_\Gamma = 0, \\ \varphi|_{t=0} = \varphi_0(x) \end{cases}$$

(4.1.9)

and

$$\begin{cases} u_t = K \Delta u + f_2(\psi), \\ u|_\Gamma = 0, \\ u|_{t=0} = u_0(x) \end{cases}$$

(4.1.10)

where

$$f_1(\psi, v) = \frac{1}{2}(\psi - \psi^3) + 2v,$$

(4.1.11)

$$f_2(\psi) = -\frac{l}{2}\psi_t. \tag{4.1.12}$$

It is easy to see that f_1, f_2 belong to $L^2([0,\delta]; L^2)$. We deduce from Theorem 1.3.1 that there is a unique pair of solutions (φ, u) such that $\varphi \in C([0,\delta]; H_o^1) \cap L^2([0,\delta]; H^2)$, $\varphi_t \in L^2([0,\delta]; L^2)$, $u \in C([0,\delta]; L^2) \cap L^2([0,\delta]; H_o^1)$, $u_t \in L^2([0,\delta]; H^{-1})$. Moreover, for $0 \le t \le \delta$,

$$\|\varphi\|_{H^1}^2 + \frac{\alpha}{\xi^2}\int_0^t \|\varphi_t\|^2 d\tau \le \|\varphi_0\|_{H^1}^2 + C_1 \int_0^t \|f_1\|^2 d\tau, \tag{4.1.13}$$

$$\|u\|^2 + K\int_0^t \|\nabla u\|^2 d\tau \le \|u_0\|^2 + C_2 \int_0^t \|f_2\|^2 d\tau \tag{4.1.14}$$

where C_1, C_2 are positive constants depending only on α, ξ, l, K. It turns out from the Sobolev imbedding theorem and (4.1.11)–(4.1.14) that for $M_2 \ge \frac{2\xi^2}{\alpha}\|\varphi_0\|^2$, $M_1 \ge \|u_0\|^2 + \frac{C_2 l^2}{4}M_2$, there exists $\delta_1 > 0$ depending only on C_1, C_2, M_1, M_2 such that when $\delta \le \delta_1$ and $(\psi, v) \in X_\delta(M_1, M_2)$, $(\varphi, u) \in X_\delta(M_1, M_2)$, i.e., the mapping $(\psi, v) \mapsto (\varphi, u)$ maps $X_\delta(M_1, M_2)$ into itself.

For $(\psi_i, v_i) \in X_\delta(M_1, M_2)$ $(i = 1, 2)$, we have by letting $\psi = \psi_1 - \psi_2, v = v_1 - v_2, u = u_1 - u_2, \varphi = \varphi_1 - \varphi_2$

$$\begin{cases} \alpha\varphi_t = \xi^2\Delta\varphi + f_1(\psi_1, v_1) - f_1(\psi_2, v_2), \\ \varphi|_\Gamma = 0, \\ \varphi|_{t=0} = 0 \end{cases} \tag{4.1.15}$$

and

$$\begin{cases} u_t = K\Delta u - \frac{l}{2}\psi_t, \\ u|_\Gamma = 0, \\ u|_{t=0} = 0. \end{cases} \tag{4.1.16}$$

Using estimates (4.1.13), (4.1.14) again, we obtain

$$\|\varphi\|_{H^1}^2 + \frac{\alpha}{\xi^2}\int_0^t \|\varphi_t\|^2 d\tau \le C_1 \int_0^t \|f_1(\psi_1, v_1) - f_1(\psi_2, v_2)\|^2 d\tau, \tag{4.1.17}$$

$$\|u\|^2 + K\int_0^t \|\nabla u\|^2 d\tau \le \frac{C_2 l^2}{4}\int_0^t \|\psi_t\|^2 d\tau. \tag{4.1.18}$$

Thus if we define the norm of (φ, u) in $X_\delta(M_1, M_2)$:

$$\|(\varphi, u)\|_{X_\delta} = \left(\max_{0 \leq t \leq \delta} \|\varphi(t)\|_{H^1}^2 + \frac{\alpha}{\xi^2} \int_0^\delta \|\varphi_t\|^2 d\tau + \eta \max_{0 \leq t \leq \delta} \|u\|^2 \right)^{\frac{1}{2}} \quad (4.1.19)$$

with $\eta = \frac{2\alpha}{C_2 l^2 \xi^2}$, then we deduce from (4.1.17), (4.1.18) that there exists δ_2 depending only on M_1, M_2 and the coefficients such that when $\delta \leq \delta_0 = \min(\delta_1, \delta_2)$,

$$\|(\varphi, u)\|_{X_\delta}^2 \leq \frac{1}{2} \|(\psi, v)\|_{X_\delta}^2. \quad (4.1.20)$$

Thus the contraction mapping theorem yields a unique local solution (φ, u) to problem (4.1.1), (4.1.2) and (4.1.3) such that $\varphi \in C([0, \delta_0]; H_o^1) \cap L^2([0, \delta_0]; H^2)$, $\varphi_t \in L^2([0, \delta_0]; L^2)$, $u \in C([0, \delta_0]; L^2) \cap L^2([0, \delta_0]; H_o^1)$, $u_t \in L^2([0, \delta_0]; H^{-1})$.

2. To prove global existence, we need only prove uniform boundedness of $\|\varphi\|_{H^1}$, and $\|u\|$. Let

$$E(t) = \int_\Omega \left(\frac{\xi^2}{2} |\nabla \varphi|^2 + \frac{1}{8} \varphi^4 - \frac{1}{4} \varphi^2 + \frac{2}{l} u^2 \right) dx. \quad (4.1.21)$$

By equations (4.1.1) we can easily obtain

$$\frac{dE}{dt} + \alpha \|\varphi_t\|^2 + \frac{4K}{l} \|\nabla u\|^2 = 0, \ \forall t > 0. \quad (4.1.22)$$

Then integrating (4.1.22) with respect to t yields the uniform boundedness of $\|\varphi\|_{H^1}^2$ and $\|u\|^2$.

Thus global existence follows and the nonlinear semigroup $S(t)$ associated with problem (4.1.1)–(4.1.3) is a C_o-semigroup in $H_o^1 \times L^2$: for each $(\varphi_0, u_0) \in H_o^1 \times L^2$, $t \mapsto S(t)(\varphi_0, u_0)$ is continuous.

3. To prove that the nonlinear semigroup $S(t)$ is a dynamical system, by Definition 1.5.1 we have to prove that for each $t \geq 0, S(t)$ is continuous from $H_o^1 \times L^2$ to $H_o^1 \times L^2$. Let $(\varphi^{(1)}, u^{(1)}), (\varphi^{(2)}, u^{(2)})$ be a pair of solutions corresponding to $(\varphi_0^{(1)}, u_0^{(1)})$ and $(\varphi_0^{(2)}, u_0^{(2)})$. We also denote

$$\varphi = \varphi^{(1)} - \varphi^{(2)}, \quad u = u^{(1)} - u^{(2)}, \quad (4.1.23)$$

$$\varphi_0 = \varphi_0^{(1)} - \varphi_0^{(2)}, \quad u_0 = u_0^{(1)} - u_0^{(2)}. \quad (4.1.24)$$

Then φ and u satisfy

$$\alpha \varphi_t = \xi^2 \Delta \varphi + \frac{1}{2}(\varphi - \varphi f(\varphi^{(1)}, \varphi^{(2)})) + 2u, \quad (4.1.25)$$

$$u_t + \frac{l}{2}\varphi_t = K\Delta u, \tag{4.1.26}$$

$$\varphi|_\Gamma = u|_\Gamma = 0, \tag{4.1.27}$$

$$\varphi|_{t=0} = \varphi_0(x), \quad u|_{t=0} = u_0(x) \tag{4.1.28}$$

where

$$f(\varphi^{(1)}, \varphi^{(2)}) = (\varphi^{(1)})^2 + \varphi^{(1)}\varphi^{(2)} + (\varphi^{(2)})^2. \tag{4.1.29}$$

Multiplying (4.1.25) by φ_t, (4.1.26) by $\frac{4}{l}u$ and adding together, then integrating with respect to x and t yields

$$\frac{\xi^2}{2}\|\nabla\varphi(t)\|^2 + \frac{2}{l}\|u(t)\|^2 + \alpha\int_0^t\|\varphi_t\|^2 d\tau + \frac{4K}{l}\int_0^t\|\nabla u\|^2 d\tau$$

$$= \frac{1}{2}\int_0^t\int_\Omega \varphi_t\varphi(f-1)dx d\tau + \frac{\xi^2}{2}\|\nabla\varphi_0\|^2 + \frac{2}{l}\|u_0\|^2$$

$$\leq \frac{1}{2}\int_0^t\|\varphi_t\|\,\|\varphi\|_{L^6}\|(f-1)\|_{L^3}d\tau + \frac{\xi^2}{2}\|\nabla\varphi_0\|^2 + \frac{2}{l}\|u_0\|^2. \tag{4.1.30}$$

By the Hölder inequality,

$$\|f-1\|_{L^3} \leq C\left(1 + \|\varphi^{(1)}\|_{L^6}^2 + \|\varphi^{(2)}\|_{L^6}^2\right)$$

$$\leq C\left(1 + \|\varphi_0^{(1)}\|_{H^1}^2 + \|\varphi_0^{(2)}\|_{H^1}^2 + \|u_0^{(1)}\|^2 + \|u_0^{(2)}\|^2\right) \tag{4.1.31}$$

with C being a positive constant.

It follows from (4.1.30)–(4.1.31) that

$$\frac{\xi^2}{2}\|\nabla\varphi(t)\|^2 + \frac{2}{l}\|u(t)\|^2 + \alpha\int_0^t\|\varphi_t\|^2 d\tau + \frac{4K}{l}\int_0^t\|\nabla u\|^2 d\tau$$

$$\leq \frac{\alpha}{2}\int_0^t\|\varphi_t\|^2 d\tau + \tilde{C}\int_0^t\|\varphi\|_{L^6}^2 d\tau + \frac{\xi^2}{2}\|\nabla\varphi_0\|^2 + \frac{2}{l}\|u_0\|^2 \tag{4.1.32}$$

where $\tilde{C} > 0$ is a constant depending on $\|\varphi_0^{(i)}\|_{H^1}, \|u_0^{(i)}\|$, $(i = 1, 2)$. Thus the assertion follows from (4.1.32), the Sobolev imbedding theorem and the Gronwall inequality.

4. It remains to prove that $\varphi, u \in C^\infty((0, \infty); C^\infty(\bar{\Omega}))$ and for any $\varepsilon > 0$, the orbit starting from ε is compact in $H^1 \times L^2$. First, as has been proved in Elliott & Zheng [2], if $(\varphi_0, u_0) \in H^2 \cap H_o^1 \times H^2 \cap H_o^1$, then there is a unique global solution (φ, u) such that $\varphi, u \in C([0, \infty); H^2 \cap H_o^1), \varphi_t, u_t \in C([0, \infty); L^2) \cap L^2([0, \infty); H_o^1)$.

Moreover, by (4.1.1), we have

$$\frac{K}{2}\frac{d}{dt}\|\nabla u\|^2 + \|u_t\|^2 = -\frac{l}{2}\int_\Omega \varphi_t u_t dx \le \frac{1}{2}\|u_t\|^2 + \frac{l^2}{8}\|\varphi_t\|^2, \qquad (4.1.33)$$

$$\frac{\alpha}{2}\frac{d}{dt}\|\varphi_t\|^2 + \xi^2\|\nabla\varphi_t\|^2 + \frac{3}{2}\int_\Omega \varphi^2\varphi_t^2 dx = \frac{1}{2}\|\varphi_t\|^2 + 2\int_\Omega u_t\varphi_t dx, \quad (4.1.34)$$

$$\frac{1}{2}\frac{d}{dt}\|u_t\|^2 + K\|\nabla u_t\|^2 + \frac{l}{\alpha}\|u_t\|^2 - \frac{\xi^2 l}{2\alpha}\int_\Omega \nabla u_t\nabla\varphi_t dx$$

$$+\frac{l}{4\alpha}\int_\Omega u_t(\varphi_t - 3\varphi^2\varphi_t)dx = 0. \qquad (4.1.35)$$

It easily follows from (4.1.22), (4.1.33)–(4.1.35) and equations (4.1.1) that

$$\|\nabla u(t)\|^2 + \int_0^t \|u_t\|^2 d\tau \le C_3, \quad \forall t \ge 0, \qquad (4.1.36)$$

$$\|\varphi_t(t)\|^2 + \|\varphi(t)\|_{H^2(\Omega)}^2 + \int_0^t \|\nabla\varphi_t\|^2 d\tau \le C_4, \quad \forall t \ge 0, \qquad (4.1.37)$$

$$\|u_t(t)\|^2 + \|u(t)\|_{H^2}^2 + \int_0^t \|\nabla u_t\|^2 d\tau \le C_4, \quad \forall t \ge 0 \qquad (4.1.38)$$

where C_3, C_4 are positive constants depending only on the H^1 norm of (φ_0, u_0) and H^2 norm of (φ_0, u_0), respectively.

Multiplying (4.1.33)–(4.1.34) by t, (4.1.35) by t^2, respectively, then integrating with respect to t over $[0, T]$, using (4.1.22) we obtain that

$$t\|\nabla u(t)\|^2 + \int_0^T \tau\|u_t\|^2 d\tau \le C_T, \quad t \in [0, T], \qquad (4.1.39)$$

$$t\|\varphi_t(t)\|^2 + \int_0^T \tau\|\nabla\varphi_t\|^2 d\tau \le C_T, \quad t \in [0, T], \qquad (4.1.40)$$

$$t^2\|u_t(t)\|^2 + \int_0^T \tau^2(\|\nabla u_t\|^2 + \|u_t\|^2)d\tau$$

$$\leq C\left(\int_0^T (\tau\|u_t\|^2 + \tau^2(\|\nabla\varphi_t\|^2 + \|u_t\|\,\|\varphi\|_{L^6}^2\|\varphi_t\|_{L^6}))d\tau\right)$$

$$\leq C\left(\int_0^T \tau\|u_t\|^2 d\tau + T\int_0^T \tau\|\nabla\varphi_t\|^2 d\tau + \int_0^T (\tau\|u_t\|^2 + \|\nabla\varphi_t\|^2)d\tau\right)$$

$$\leq C_T \tag{4.1.41}$$

where $C_T > 0$ is a constant depending only on $\|\varphi_0\|_{H^1}, \|u_0\|$ and T. Thus it follows from (4.1.39)–(4.1.41) and equations (4.1.1) that for any small $\varepsilon > 0$,

$$\|\varphi(\varepsilon)\|_{H^2} + \|u(\varepsilon)\|_{H^2} \leq C_{T,\varepsilon} \tag{4.1.42}$$

where $C_{T,\varepsilon} > 0$ depends only on $\|\varphi_0\|_{H^1}, \|u_0\|, \varepsilon$ and T.

By the standard density argument, for any $(\varphi_0, u_0) \in H_o^1 \times L^2$, there is a sequence $(\varphi_0^{(n)}, u_0^{(n)}) \in C_o^\infty \times C_o^\infty$ such that $\varphi_0^{(n)} \to \varphi_0$ in H_o^1, $u_0^{(n)} \to u_0$ in L^2. Then it follows that for $t \in (0, T], S(t)(\varphi_0^{(n)}, u_0^{(n)})$ converges to $S(t)(\varphi_0, u_0)$ in $H^2 \times H^2$. Moreover, for any $\varepsilon > 0, (\varphi(\varepsilon), u(\varepsilon)) \in H^2 \times H^2$. Combining this results with (4.1.37)–(4.1.38) yields that for any ε, for each $(\varphi_0, u_0) \in H_o^1 \times L^2$, the orbit $\bigcup_{t \geq \varepsilon} S(t)(\varphi_0, u_0)$ is bounded in $H^2 \times H^2$.

We now use the standard bootstrap argument to prove that $(\varphi, u) \in C^\infty((0, +\infty); C^\infty(\bar{\Omega})) \times C^\infty((0, +\infty); C^\infty(\bar{\Omega}))$. Differentiating equations (4.1.1) with respect to t successively, then making similar energy estimates to (4.1.33)–(4.1.35), we obtain the uniform boundedness of the orbit $\bigcup_{t \geq \varepsilon} S(t)(\varphi_0, u_0)$ in $H^{2m} \times H^{2m}, (m = 1, 2, \cdots)$. Thus the assertion follows from the Sobolev imbedding theorem. The proof of theorem is complete. $\qquad\square$

Accordingly, for $\varphi \in H^2 \cap H_o^1, u_0 \in H_o^1$, we have

Remark 4.1.2 *Let $\Omega \subset \mathbb{R}^n (n \leq 3)$ be a bounded domain with smooth boundary Γ. Suppose $\varphi_0 \in H^2 \times H_o^1, u_0 \in H_o^1$. Then problem (4.1.1)–(4.1.3) admits a unique global solution (φ, u) such that $\varphi \in C([0, +\infty); H^2 \cap H_o^1), u \in C([0, +\infty); H_o^1), \varphi_t \in$*

$C([0,+\infty); L^2)$, *for any* $T > 0$, $\varphi_t \in L^2([0,T]; H^1)$, $u \in L^2([0,T]; H^2 \cap H_o^1)$, $u_t \in L^2([0,T]; L^2)$ *and the associated semigroup* $S(t)$ *is a dynamical system. Moreover,* φ, $u \in C((0,+\infty); C^\infty)$ *and for any* $\varepsilon > 0$, *the orbit* $t \in [\varepsilon, +\infty) \mapsto (\varphi(\cdot,t), u(\cdot,t))$ *is compact in* $H^2 \cap H_o^1 \times H_o^1$.

The proof of this result can be carried out in the same manner as for Theorem 4.1.1 and we can omit the details here.   □

4.1.2   The Penrose–Fife Model

Penrose and Fife in their paper [1] proposed a thermodynamically consistent model based on the idea that the value of the entropy functional cannot decrease along solutions paths, which is in agreement with what one expects from the second law of thermodynamics.

In particular, for systems with non-conserved order parameter, it turns out that the order paramerter φ and the absolute temperature θ satisfy the following coupled system of partial differential equations (see Penrose & Fife [1], Section 6):

$$\begin{cases} \varphi_t = K_1\left(k_1\Delta\varphi + s'(\varphi) + \frac{\lambda(\varphi)}{\theta}\right), \\[3mm] \theta_t - \lambda(\varphi)\varphi_t = -M_2\Delta(\frac{1}{\theta}) \end{cases} \tag{4.1.43}$$

with positive constants K_1, k_1, M_2, a double well function $-s(\varphi)$ and a linear function $\lambda(\varphi)$. As has been observed in Penrose & Fife [1], if the temperature θ stays close to some average value θ_0, if the equations in (4.1.43) are linearized about $u = \theta - \theta_0$, and if the dependence of $\lambda(\varphi)$ on φ is ignored, then system (4.1.43) is reduced to the phase-field equations (4.1.1).

We now consider the initial boundary value problem for system (4.1.43) in three space dimensions. More precisely, we consider the system

$$\begin{cases} \varphi_t = K_1(k_1\Delta\varphi + s'(\varphi) + \frac{a\varphi}{\theta}), \\[3mm] \theta_t - a\varphi\varphi_t = -M_2\Delta(\frac{1}{\theta}) + g(x,t) \end{cases} \tag{4.1.44}$$

with positive constants K_1, a, k_1, M_2 and Ω being a boundary domain in $\mathbb{R}^3$ with smooth boundary Γ.

System (4.1.44) is supplemented by the initial and boundary conditions

$$\varphi|_{t=0} = \varphi_0(x), \quad \theta|_{t=0} = \theta_0(x), \quad x \in \bar{\Omega}, \tag{4.1.45}$$

$$\frac{\partial \varphi}{\partial n}\bigg|_\Gamma = 0, \quad \left(\frac{\partial \theta}{\partial n} + \alpha(\theta - \theta_\Gamma)\right)\bigg|_\Gamma = 0. \tag{4.1.46}$$

Here n denotes the outer unit normal at Γ. The functions g and θ_Γ are given and can be considered as distributed heat source and outside temperature, respectively. Hence we require that $\theta_\Gamma(x,t) \geq \beta > 0$ and the constant α be positive. The Neumann boundary condition for φ is a variational boundary condition. As mentioned before, the function $-s$ in (4.1.44) is a double-well function. Therefore, we assume:

(A) $s \in C^3(\mathbb{R})$ and there exist positive constants $\tilde{C}_0 > \frac{a}{2}$, $\tilde{C}_1$, $\tilde{C}_2$ such that for all $\varphi \in \mathbb{R}$

$$\tilde{C}_0 \varphi^2 - \tilde{C}_1 \leq -s(\varphi), \quad s''(\varphi) \leq \tilde{C}_2. \tag{4.1.47}$$

Then we have the following main result (see Sprekels and Zheng [2])

Theorem 4.1.2 *Let assumption (A) be satisfied. Suppose that $\varphi_0 \in H^4(\Omega)$ and $\theta_0 \in H^3(\Omega)$ satisfy the compatibility condition and suppose that $\theta_0 > 0$ in $\bar{\Omega}$. Furthermore, suppose that for any $T > 0, g \in C^1([0,T]; L^2) \cap C([0,T]; H^2(\Omega))$, $\theta_\Gamma \in C^1([0,T]; H^{\frac{3}{2}}(\Gamma)) \cap H^2([0,T]; L^2(\Gamma))$ and $\theta_\Gamma \geq \beta > 0$ on $\Gamma \times [0,\infty)$. Then system (4.1.44), (4.1.45) and (4.1.46) has a unique global solution (φ, θ) such that $\varphi \in C([0,T]; H^4) \cap C^1([0,T]; H^2) \cap C^2([0,T]; L^2)$, $\theta \in C([0,T]; H^3) \cap C^1([0,T]; H^1) \cap H^2([0,T]; L^2) \cap L^2([0,T]; H^4)$. Moreover, $\theta(x,t) > 0$ for all $(x,t) \in \bar{\Omega} \times [0,\infty)$.*

Remark 4.1.3 *We refer to Zheng [13] for the one-dimensional case. In that paper, the global existence, uniqueness and asymptotic behaviour of smooth solutions to the initial boundary value problem with both φ and θ satisfying the Neumann boundary conditions have been proved.*

Before giving the proof, we first notice that system (4.1.44) is not a diagonal parabolic system. Indeed, it is a triangular parabolic system as in the case of the Caginalp model. Moreover, the right-hand side of the second equation in (4.1.44) is not a uniformly elliptic operator of θ. To establish global existence, one has to get the uniform boundedness of θ and $\frac{1}{\theta}$ in $\bar{\Omega} \times [0, T]$ for any $T > 0$. This consists of the essential part of the proof.

We first establish a local existence and uniqueness result (also refer to Amann [4–7] for general results about local existence and uniqueness).

Lemma 4.1.1 *Under the same assumption as in Theorem 4.1.2 there exists $t^* > 0$ depending only on $\|\varphi_0\|_{H^4}, \|\theta_0\|_{H^3}$ and $\min\limits_{\bar{\Omega}}\theta_0(x)$ such that problem (4.1.44), (4.1.45)– (4.1.46) admits in $\bar{\Omega} \times [0, t^*]$ a unique solution (φ, θ) such that*

$\varphi \in C([0, t^*]; H^4(\Omega)) \cap C^1([0, t^*]; H^2(\Omega)) \cap C^2([0, t^*]; L^2(\Omega))$ *and* $\theta \in C([0, t^*]; H^3(\Omega))$
$\cap C^1([0, t^*]; H^1(\Omega)) \cap L^2([0, t^*]; H^4(\Omega)) \cap H^2([0, t^*]; L^2(\Omega)).$
Moreover, $\theta(x, t) \geq \frac{1}{2}\min\limits_{x \in \bar{\Omega}}(\theta_0(x), \beta) > 0$ *holds in* $\bar{\Omega} \times [0, t^*].$

Sketch of the proof of Lemma 4.1.1 The strategy of the proof of Lemma 4.1.1 is still to use the contraction mapping theorem.

Let us consider the linear auxiliary problems:

$$
\begin{cases}
\varphi_t - k_1 \Delta \varphi = f_1, \\[2mm]
\left.\dfrac{\partial \varphi}{\partial n}\right|_\Gamma = 0, \\[2mm]
\varphi|_{t=0} = \varphi_0(x)
\end{cases}
\tag{4.1.48}
$$

and

$$
\begin{cases}
\theta_t - \tilde{a}\Delta\theta = f_2, \\[2mm]
\left.\left(\dfrac{\partial \theta}{\partial n} + \alpha(\theta - \theta_\Gamma)\right)\right|_\Gamma = 0, \\[2mm]
\theta|_{t=0} = \theta_0(x)
\end{cases}
\tag{4.1.49}
$$

with

$$f_1 = s_1'(\tilde{\varphi}) + \frac{a\tilde{\varphi}}{\tilde{\theta}}, \quad \tilde{a} = \frac{M_2}{\tilde{\theta}^2}, \quad f_2 = a\tilde{\varphi}\tilde{\varphi}_t - \frac{2M^2}{\tilde{\theta}^3}|\nabla\tilde{\theta}|^2 + g. \tag{4.1.50}$$

For any $h > 0$, we define the set

$$X_h(M_0, \cdots, M_4) = \left\{ (\varphi, \theta) \,\middle|\, \begin{array}{l} \varphi \in C([0,h]; H^4) \cap C^1([0,h]; H^2) \cap C^2([0,h]; L^2), \\ \theta \in C([0,h]; H^3) \cap C^1([0,h]; H^1) \cap L^2([0,h]; H^4) \\ \cap H^2([0,h]; L^2), \\ \varphi(x,0) = \varphi_0(x), \ \theta(x,0) = \theta_0(x), \ \forall x \in \bar{\Omega} \\ \max_{0\leq t\leq h} \|\varphi(t)\|_{H^2} \leq M_0, \ \max_{0\leq t\leq h} \|\theta(t)\|_{H^2} \leq M_2, \\ \max_{0\leq t\leq h}(\|\varphi(t)\|_{H^4} + \|\varphi_t(t)\|_{H^2} + \|\varphi_{tt}(t)\|) \leq M_1, \\ \max_{0\leq t\leq h}(\|\theta(t)\|_{H^3} + \|\theta_t(t)\|_{H^1}) \leq M_3, \\ \int_0^h (\|\theta_t\|_{H^2}^2 + \|\theta_{tt}\|^2 + \|\theta\|_{H^4}^2)\,d\tau \leq M_4, \\ \min_{\bar{\Omega}\times[0,h]} \theta(x,t) \geq \lambda = \frac{1}{2}\min_{\bar{\Omega}\times[0,T]}(\theta_0(x), \beta) > 0 \end{array} \right\}$$

$$\tag{4.1.51}$$

where the positive constants M_i $(i = 0, 1, \cdots, 4)$ are specified during the course of the proof. Then for $(\tilde{\varphi}, \tilde{\theta}) \in X_h(M_0, \cdots, M_4)$, it follows from the standard theory of linear parabolic equations that with suitable choices of M_i $(i = 0, \cdots, 4)$ and small $h = t^*$ that the nonlinear operator defined by the auxiliary problems (4.1.48), (4.1.49) maps X_{t^*} into itself. Moreover, the operator is a strict contraction. Thus local existence and uniqueness follow. Since the proof is essentially the same as in Theorem 4.1.1, we can omit the details here. We refer the interested reader to Sprekels & Zheng [2] for the details of the proof.   □

Proof of Theorem 4.1.2 In order to prove global existence, we have to establish uniform a priori estimates of $\|\theta(t)\|_{H^3}, \|\varphi(t)\|_{H^4}$ and $\min_{x\in\bar{\Omega}}\theta(x,t)$, respectively. As mentioned before, the key point is to find positive upper and lower bounds for the temperature θ. In what follows, without loss of generality, we assume that $K_1 = k_1 = M_2 = \alpha = 1$. Let

$$u = \frac{1}{\theta}. \tag{4.1.52}$$

Then the pair (φ, u) solves the system

$$\varphi_t = \Delta\varphi + s'(\varphi) + au\varphi, \tag{4.1.53}$$

$$\frac{1}{u^2}u_t + a\varphi\varphi_t = \Delta u - g, \tag{4.1.54}$$

$$\left.\frac{\partial\varphi}{\partial n}\right|_\Gamma = 0, \quad \left.\frac{\partial u}{\partial n}\right|_\Gamma = (u - \theta_\Gamma u^2)|_\Gamma, \tag{4.1.55}$$

$$\varphi|_{t=0} = \varphi_0(x), \quad u|_{t=0} = \frac{1}{\theta_0}. \tag{4.1.56}$$

We now have

Lemma 4.1.2 *There exist $C_T > 0$ and $C'_T > 0$ depending only on T and the initial data such that for $0 \le t \le T$ the following estimates hold:*

$$\|\varphi(t)\|_{H^1} \le C_T, \tag{4.1.57}$$

$$\int_0^t (\|\varphi_t\|^2 + \|\nabla u\|^2)d\tau \le C_T, \tag{4.1.58}$$

$$\int_0^t \int_\Gamma u^3 dx d\tau + \int_0^t \|u\|^2_{H^1} d\tau \le C_T, \tag{4.1.59}$$

$$0 < C'_T \le \int_\Omega \frac{1}{u} dx \le C_T. \tag{4.1.60}$$

Proof. We multiply (4.1.53) by φ_t and (4.1.54) by u, respectively, integrate over Ω and add up the resulting equations to obtain the identity

$$\frac{d}{dt}\int_\Omega (\frac{1}{2}|\nabla\varphi|^2 - s(\varphi) + \ln u)dx + \|\nabla u\|^2 + \|\varphi_t\|^2$$

$$+ \int_\Gamma (\theta_\Gamma u^3 - u^2)dx = -\int_\Omega gu dx. \tag{4.1.61}$$

Integrating with respect to t and using the Young inequality, we obtain

$$\int_\Omega \left(\frac{1}{2}|\nabla\varphi|^2 - s(\varphi)\right)dx + \int_0^t (\|\nabla u\|^2 + \|\varphi_t\|^2)d\tau + \frac{\beta}{2}\int_0^t \int_\Gamma u^3 dx d\tau$$

$$\leq C + C \int_0^t \|g\|^2 d\tau + \int_\Omega \ln \frac{1}{u} dx. \tag{4.1.62}$$

Hereafter we denote by C, C_T positive constants which may vary in different places and depend on T and the initial data, but not on t.

On the other hand, integrating (4.1.54) with respect to x and t yields

$$\int_\Omega \frac{1}{u} dx = \frac{a}{2} \|\varphi\|^2 + \int_0^t \int_\Gamma (\theta_\Gamma u^2 - u) dx d\tau + \int_0^t \int_\Omega g dx d\tau + C. \tag{4.1.63}$$

Combining (4.1.62) and (4.1.63) and using assumption (A), the Young inequality and the elementary inequality $\ln \frac{1}{u} \leq \frac{1}{u}$, we conclude that (4.1.57)–(4.1.59) are valid. Then it follows from (4.1.63) that

$$\int_\Omega \frac{1}{u} dx \leq C_T. \tag{4.1.64}$$

Integrating (4.1.61) with respect to t yields the boundedness of $\int_\Omega \ln u dx$ from above. Applying the Jensen inequality to the convex function $-\ln u$ yields

$$-\frac{1}{|\Omega|} \int_\Omega \ln \frac{1}{u} dx \geq -\ln(\frac{1}{|\Omega|} \int_\Omega \frac{1}{u} dx). \tag{4.1.65}$$

Thus the left-hand side inequality in (4.1.60) follows from (4.1.65) and the boundedness of $\int_\Omega \ln u dx$ from above. The proof of the lemma is complete.   □

Lemma 4.1.3 *There exists $C_T > 0$ such that for $0 \leq t \leq T$ the following estimates hold:*

$$\|\varphi_t(t)\| + \|u(t)\|_{H^1} \leq C_T, \tag{4.1.66}$$

$$\max_{0\leq\tau\leq t} \int_\Gamma u^3 ds \leq C_T, \tag{4.1.67}$$

$$\int_0^t \left\| \frac{u_t}{u} \right\|^2 d\tau + \int_0^t \|\nabla \varphi_t\|^2 d\tau \leq C_T, \tag{4.1.68}$$

$$\|\varphi(t)\|_{H^2} \leq C_T, \quad \|\varphi(t)\|_{L^\infty} \leq C_T. \tag{4.1.69}$$

Proof. Differentiating equation (4.1.53) with respect to t and multiplying the result by φ_t, then multiplying equation (4.1.54) by u_t and adding up the resulting equations, and finally integrating over Ω and using integration by parts, we obtain

$$\frac{d}{dt}\left(\frac{1}{2}\|\varphi_t\|^2 + \frac{1}{2}\|\nabla u\|^2\right) + \|\nabla\varphi_t\|^2 + \int_\Omega s''(\varphi)\varphi_t^2 dx + \left\|\frac{u_t}{u}\right\|^2$$

$$= \|\varphi_t\|^2 + a\int_\Omega u\varphi_t^2 dx + \int_\Gamma u_t(u - \theta_\Gamma u^2)ds - \int_\Omega u_t g dx. \qquad (4.1.70)$$

Hence,

$$\frac{d}{dt}\left(\frac{1}{2}\|\varphi_t\|^2 + \frac{1}{2}\|\nabla u\|^2 + \frac{1}{3}\int_\Gamma \theta_\Gamma u^3 ds - \frac{1}{2}\int_\Gamma u^2 ds\right)$$

$$+\|\nabla\varphi_t\|^2 + \left\|\frac{u_t}{u}\right\|^2 + \int_\Omega s''(\varphi)\varphi_t^2 dx$$

$$= a\int_\Omega u\varphi_t^2 dx + \frac{1}{3}\int_\Gamma \frac{\partial\theta_\Gamma}{\partial t}u^3 ds - \int_\Omega u_t g dx. \qquad (4.1.71)$$

We now estimate the terms on the right-hand side of (4.1.71). By the Hölder inequality and (4.1.59), we have

$$\left|\int_0^t\int_\Omega u_t g dx d\tau\right| \leq \delta\int_0^t\left\|\frac{u_t}{u}\right\|^2 d\tau + C\int_0^t\int_\Omega u^2 g^2 dx d\tau$$

$$\leq \delta\int_0^t\left\|\frac{u_t}{u}\right\|^2 d\tau + C\max_{\bar\Omega\times[0,T]}|g(x,t)|^2\int_0^t\|u\|^2 d\tau$$

$$\leq \delta\int_0^t\left\|\frac{u_t}{u}\right\|^2 d\tau + C. \qquad (4.1.72)$$

with constant $\delta > 0$. It follows from (4.1.58), (4.1.59) and the Nirenberg inequality that

$$\left|\int_0^t\int_\Omega au\varphi_t^2 dx d\tau\right| \leq \int_0^t a\|u\|_{L^6}\|\varphi_t\|_{L^{\frac{12}{5}}}^2 d\tau$$

$$\leq a\left(\int_0^t\|u\|_{L^6}^2 d\tau\right)^{\frac{1}{2}}\left(\int_0^t\|\varphi_t\|_{L^{\frac{12}{5}}}^4 d\tau\right)^{\frac{1}{2}}$$

$$\leq C \left(\int_0^t \|u\|_{H^1}^2 d\tau \right)^{\frac{1}{2}} \left(\int_0^t (\|\nabla \varphi_t\| \|\varphi_t\|^3 + \|\varphi_t\|^4) d\tau \right)^{\frac{1}{2}}$$

$$\leq C \sup_{0 \leq \tau \leq t} \|\varphi_t(\tau)\| \left(\int_0^t \|\nabla \varphi_t\| \|\varphi_t\| d\tau + \int_0^t \|\varphi_t\|^2 d\tau \right)^{\frac{1}{2}}$$

$$\leq C \sup_{0 \leq \tau \leq t} \|\varphi_t(\tau)\| \left(1 + \left(\int_0^t \|\nabla \varphi_t\|^2 d\tau \right)^{\frac{1}{4}} \right)$$

$$\leq \delta \sup_{0 \leq \tau \leq t} \|\varphi_t(\tau)\|^2 + \delta \int_0^t \|\nabla \varphi_t\|^2 d\tau + C. \tag{4.1.73}$$

Combining (4.1.71)–(4.1.73) yields that

$$\frac{1}{2}\|\varphi_t(t)\|^2 + \frac{1}{2}\|\nabla u(t)\|^2 + \frac{1}{3}\int_\Gamma \theta_\Gamma u^3 ds + \int_0^t \int_\Omega s''(\varphi)\varphi_t^2 dx d\tau$$

$$+ (1-\delta) \int_0^t \|\nabla \varphi_t\|^2 d\tau + (1-\delta) \int_0^t \left\| \frac{u_t}{u} \right\|^2 d\tau$$

$$\leq C + \delta \sup_{0 \leq \tau \leq t} \|\varphi_t\|^2 + \frac{\beta}{6} \int_\Gamma u^3 ds + C \int_0^t \int_\Gamma u^3 ds d\tau. \tag{4.1.74}$$

Taking the supremum with respect to t in (4.1.74) and choosing δ sufficiently small, we deduce (4.1.66)–(4.1.68) from the Gronwall inequality and assumption (A). Finally, we deduce (4.1.69) from (4.1.57), (4.1.66) and the standard elliptic estimates of (4.1.53). The proof of the lemma is complete.   □

Now we derive further estimates for φ.

Lemma 4.1.4 *For any $q \in (1, 3)$, there exists a constant $C_{q,T} > 0$ depending on q and T such that*

$$\max_{0 \leq t \leq T} \|\varphi_t(t)\|_{L^q} \leq C_{q,T}. \tag{4.1.75}$$

Proof. It follows from (4.1.53) that $\psi = \varphi_t$ satisfies

$$\psi_t - \Delta \psi + \psi = f, \tag{4.1.76}$$

$$\left. \frac{\partial \psi}{\partial n} \right|_\Gamma = 0, \tag{4.1.77}$$

$$\psi|_{t=0} = \psi_0 \tag{4.1.78}$$

where

$$f = \varphi_t + s''(\varphi)\varphi_t + au_t\varphi + au\varphi_t \tag{4.1.79}$$

and

$$\psi_0 = \Delta\varphi_0 + s'(\varphi_0) + au_0\varphi_0 \in H^2(\Omega). \tag{4.1.80}$$

We can easily deduce from Lemma 4.1.2 and Lemma 4.1.3 that

$$\int_0^T \|f\|_{L^{\frac{3}{2}}}^2 dt \leq C. \tag{4.1.81}$$

Let

$$A = -\Delta + I, \; D(A) = \left\{ u \,\middle|\, u \in W^{2,\frac{3}{2}}(\Omega), \left.\frac{\partial u}{\partial n}\right|_{\Gamma} = 0 \right\}, \; X = L^{\frac{3}{2}}(\Omega). \tag{4.1.82}$$

As we know from Chapter 1, A is a sectorial operator with $Re\,\sigma(A) > 0$ and

$$\psi = e^{-At}\psi_0 + \int_0^t e^{-A(t-\tau)} f(\tau) d\tau. \tag{4.1.83}$$

Moreover, for any $\alpha \in (0, \frac{1}{2})$, there exists a constant $C_\alpha > 0$ such that

$$\|A^\alpha\psi\|_X \leq \|A^\alpha e^{-At}\psi_0\|_X + \int_0^t \|A^\alpha e^{-A(t-\tau)} f(\tau)\|_X d\tau$$

$$\leq C\|\psi_0\|_{H^2} + \int_0^t C_\alpha(t-\tau)^{-\alpha}\|f\|_X d\tau$$

$$\leq C + C_\alpha \left(\int_0^t (t-\tau)^{-2\alpha} d\tau\right)^{\frac{1}{2}} \left(\int_0^t \|f\|_X^2 d\tau\right)^{\frac{1}{2}}$$

$$\leq C, \;\; \forall t \in [0, T]. \tag{4.1.84}$$

Now, let $q \in (1,3)$ be fixed,. Then we can choose $\alpha \in (0, \frac{1}{2})$ such that

$$2 - \frac{3}{q} < 2\alpha. \tag{4.1.85}$$

Denoting $X^\alpha = D(A^\alpha)$, we can infer from the Sobolev imbedding theorem (see Henry [1], Theorem 1.6.1) with $n = 3$ that X^α is continuously imbedded in $L^q(\Omega)$. Thus the proof of the lemma is complete. $\square$

After these preparations we can use a technique by Alikakos [1] or J. Moser to derive the L^∞ norm bounds of θ and u (see also Nakao [1]). We now need an important lemma (see Laurencot [1]):

Lemma 4.1.5 *Let $a > 1, b \geq 0,\ c \in \mathbb{R}, c_0 \geq 1,\ c_1 \geq 1$ and $\delta_0 > 0$ be given numbers and $\{\delta_k\}_{k\geq 0},\ \{\gamma_k\}_{k\geq 0}$ be two sequences of positive real numbers such that*

$$\delta_0 + \frac{c}{a-1} > 0,$$

$$\delta_{k+1} = a\delta_k + c, \quad k \in N. \tag{4.1.86}$$

$$\gamma_0 \leq c_1^{\delta_0}, \tag{4.1.87}$$

$$\gamma_{k+1} \leq c_0 \delta_{k+1}^b \max(c_1^{\delta_k}, \gamma_k^a). \tag{4.1.88}$$

Then

$$\gamma_k^{\frac{1}{\delta_k}} \leq M, \quad \forall k \in N \tag{4.1.89}$$

with M being a positive constant depending on $c_0, c_1, a, b, \delta_0, c$.

Proof. A straightforward calculation gives

$$\delta_{k+1} - \delta_k = a(\delta_k - \delta_{k-1}), \tag{4.1.90}$$

$$\delta_{k+1} - \delta_0 = \sum_{i=0}^{k} a^i(\delta_1 - \delta_0) = \frac{a^{k+1} - 1}{a-1}(\delta_1 - \delta_0), \tag{4.1.91}$$

$$\delta_{k+1} = \frac{(\delta_1 - \delta_0)a^{k+1} + (\delta_0 a - \delta_1)}{a-1} \leq \frac{\delta_1 - \delta_0 + |c|}{a-1} a^{k+1} \tag{4.1.92}$$

and

$$\frac{\delta_{k+1}}{a^{k+1}} \to \frac{\delta_1 - \delta_0}{a-1}, \ as \ k \to \infty. \tag{4.1.93}$$

Therefore, without loss of generality, we may assume that $\delta_k \geq 1$ for all $k \geq 0$. It is easy to prove by induction that

$$\gamma_{k+1} \leq c_0^{\frac{a^{k+1}-1}{a-1}} \delta_{k+1}^b \delta_k^{ab} \cdots \delta_1^{a^k b} c_1^{(\delta_{k+1} + \frac{|c|(a^{k+1}-1)}{a-1})}$$

$$\leq c_0^{\frac{a^{k+1}-1}{a-1}} \left(\frac{\delta_1 - \delta_0 + |c|}{a-1} \right)^{b\frac{(a^{k+1}-1)}{a-1}} a^{b(k+1+ak+a^2(k-1)+\cdots+a^k)}$$

$$\times c_1^{(\delta_{k+1} + \frac{|c|(a^{k+1}-1)}{a-1})}. \tag{4.1.94}$$

To calculate $\sum_{j=0}^{k}(j+1)x^j$, we define

$$f(x) = \sum_{j=0}^{k} x^{j+1} = \frac{x^{k+2} - x}{x-1}, \ for\ x \neq 1. \tag{4.1.95}$$

Then

$$f'(x) = \sum_{j=0}^{k}(j+1)x^j = \frac{(k+1)x^{k+2} - (k+2)x^{k+1} + 1}{(x-1)^2}. \tag{4.1.96}$$

Therefore

$$s_k = k+1+ak+\cdots+a^k = a^k \sum_{j=0}^{k}(j+1)(\frac{1}{a})^j$$

$$= a^k \frac{(k+1)(\frac{1}{a})^{k+2} - (k+2)(\frac{1}{a})^{k+1} + 1}{(\frac{1}{a} - 1)^2}$$

$$= \frac{(k+1) - a(k+2) + a^{k+2}}{(a-1)^2}. \tag{4.1.97}$$

Combining (4.1.94), (4.1.97) yields

$$\gamma_{k+1}^{\frac{1}{\delta_{k+1}}} \leq c_0^{\frac{a^{k+1}-1}{\delta_{k+1}(a-1)}} \left(\frac{\delta_1 - \delta_0 + |c|}{a-1} \right)^{\frac{b(a^{k+1}-1)}{(a-1)\delta_{k+1}}} a^{\frac{s_k}{\delta_{k+1}}} c_1 \left(1 + \frac{|c|a^{k+1}}{(a-1)\delta_{k+1}} \right). \tag{4.1.98}$$

It turns out from (4.1.93) and (4.1.97) that the term on the right-hand side of (4.1.98) converges to

$$c_0^{\frac{1}{\delta_1-\delta_0}} \left(\frac{\delta_1 - \delta_0 + |c|}{a-1} \right)^{\frac{b}{\delta_1-\delta_0}} a^{\frac{a}{(\delta_1-\delta_0)(a-1)}} c_1^{1+\frac{|c|}{\delta_1-\delta_0}}. \tag{4.1.99}$$

Thus the proof of Lemma 4.1.5 is complete. $\qquad\square$

Lemma 4.1.6 *There exists $C_T > 0$ such that for $0 \leq t \leq T$ the following estimates are satisfied:*

$$\|\theta(t)\|_{L^\infty} \leq C_T, \quad \|u(t)\|_{L^\infty} \leq C_T. \tag{4.1.100}$$

Proof. Using the idea of Alikakos [1], we prove (4.1.100) by first deriving a uniform bound for the L^{p_k} norm and then passing to the limit. Without loss of generality, we shall henceforth assume that $vol(\Omega) = 1$.

It follows from (4.1.44) that for $p > 2$, the equation satisfied by $\theta > 0$ can be rewritten in the form

$$\theta^p \theta_t - \theta^{p-2} \Delta \theta + 2\theta^{p-3}|\nabla \theta|^2 = \tilde{g}\theta^p, \tag{4.1.101}$$

$$\left.\frac{\partial \theta}{\partial n}\right|_\Gamma = (\theta_\Gamma - \theta)|_\Gamma \tag{4.1.102}$$

where, as shown in Lemma 4.1.2–Lemma 4.1.4, $\tilde{g} = g + a\varphi\varphi_t$ is bounded in both $L^2([0,T];L^6(\Omega))$ and $L^\infty([0,T];L^q(\Omega))$ for $1 < q < 3$. Integrating (4.1.101) with respect to x and using the Young inequality, we obtain that

$$\frac{1}{p+1}\frac{d}{dt}\int_\Omega \theta^{p+1}dx + p\int_\Omega \theta^{p-3}|\nabla\theta|^2 dx + \int_\Gamma \theta^{p-1}ds$$

$$= \int_\Gamma \theta_\Gamma \theta^{p-2}ds + \int_\Omega \tilde{g}\theta^p dx$$

$$\leq \frac{p-2}{p-1}\int_\Gamma \theta^{p-1}ds + \frac{1}{p-1}\int_\Gamma \theta_\Gamma^{p-1}ds + \int_\Omega \tilde{g}\theta^p dx. \tag{4.1.103}$$

We first prove the uniform boundedness of $\max_{0 \leq t \leq T}\|\theta(t)\|_{L^4}$ by taking $p = 3$. Indeed, for $p = 3$, we have

$$\left|\int_0^t \int_\Omega \tilde{g}\theta^3 dx d\tau\right| \leq \int_0^t \|\tilde{g}\|_{L^6}\|\theta\|_{L^{\frac{18}{5}}}^3 d\tau$$

$$\leq C \max_{0 \leq \tau \leq t}\|\theta(\tau)\|_{L^{\frac{18}{5}}}^3 \leq C \max_{0 \leq \tau \leq t}\|\theta(\tau)\|_{L^4}^3. \tag{4.1.104}$$

Integrating (4.1.103) with respect to t and using the Young inequality, we obtain

$$\max_{0 \leq t \leq T}\|\theta(t)\|_{L^4} \leq \tilde{C}. \tag{4.1.105}$$

Let $\psi = \theta^{\frac{p-1}{2}}$. Then (4.1.103) can be rewritten as

$$\frac{d}{dt}\int_\Omega \theta^{p+1}dx + \left(\int_\Omega |\nabla\psi|^2 dx + \int_\Gamma \psi^2 ds\right)$$

$$\leq CR^{p-1} + (p+1)\int_\Omega |\tilde{g}||\theta|^p dx \qquad (4.1.106)$$

with $R = \max\left(1, \tilde{C}, \max_{\Gamma\times[0,T]}|\theta_\Gamma|, \frac{1}{\beta}, \|\theta_0\|_{L^\infty}, \|\frac{1}{\theta_0}\|_{L^\infty}\right)$. By the imbedding theorem we deduce from (4.1.106) that

$$\frac{d}{dt}\int_\Omega \theta^{p+1}dx + \delta\left(\int_\Omega \theta^{3(p-1)}dx\right)^{\frac{1}{3}}$$

$$\leq CR^{p-1} + (p+1)\int_\Omega |\tilde{g}||\theta|^p dx \qquad (4.1.107)$$

with $\delta > 0$ being a constant depending only on Ω.

Now we apply the Hölder and Young inequality to get

$$(p+1)\int_\Omega |\tilde{g}||\theta|^p dx \leq (p+1)\|\tilde{g}\|_{L^6}\left(\int_\Omega \theta^{3(p-1)}dx\right)^{\frac{1}{6}}\left(\int_\Omega |\theta|^{\frac{3(p+1)}{4}}dx\right)^{\frac{2}{3}}$$

$$\leq \frac{\delta}{2}\left(\int_\Omega |\theta|^{3(p-1)}dx\right)^{\frac{1}{3}} + C(p+1)^2\|\tilde{g}\|_{L^6}^2\left(\int_\Omega |\theta|^{\frac{3(p+1)}{4}}dx\right)^{\frac{4}{3}}. \qquad (4.1.108)$$

Combining (4.1.107), (4.1.108) and integrating with respect to t yields for $t \in [0,T]$

$$\int_\Omega \theta^{p+1}dx \leq C_T\left(R^{p+1} + (p+1)^2\sup_{0\leq t\leq T}\left(\int_\Omega |\theta|^{\frac{3(p+1)}{4}}dx\right)^{\frac{4}{3}}\right)$$

$$\leq C_T(p+1)^2\max\left(R^{p+1}, \sup_{0\leq t\leq T}\left(\int_\Omega |\theta|^{\frac{3(p+1)}{4}}dx\right)^{\frac{4}{3}}\right) \qquad (4.1.109)$$

where we may assume $C_T > 1$. Let

$$p_0 = 4, \quad p_{k+1} = \frac{4}{3}p_k, \quad k \in N. \qquad (4.1.110)$$

Then $p_k \to \infty$, as $k \to \infty$.

Taking $p = p_{k+1} - 1$, we deduce from (4.1.109) that

$$\int_\Omega |\theta|^{p_{k+1}}dx \leq C_T p_{k+1}^2\max\left(R^{p_{k+1}}, \sup_{0\leq t\leq T}\left(\int_\Omega |\theta|^{p_k}dx\right)^{\frac{4}{3}}\right). \qquad (4.1.111)$$

Appling Lemma 4.1.5 to (4.1.111) with $a = \frac{4}{3}$, $b = 2$, $c = 0$, $c_0 = C_T$, $c_1 = R^{\frac{4}{3}}$, $\delta_0 = 4$, $\delta_{k+1} = p_{k+1}$, $\gamma_{k+1} = \sup\limits_{0 \le t \le T} \int_\Omega |\theta|^{p_{k+1}} dx$ yields

$$\sup_{0 \le t \le T} \|\theta\|_{L^{p_k}} \le C_T. \tag{4.1.112}$$

Passing to the limit yields the first inequality in (4.1.100).

We now proceed to prove the second inequality in (4.1.100). For $p > 1$ we infer from (4.1.54), (4.1.55) that

$$\frac{1}{p}\frac{d}{dt}\int_\Omega u^p dx + (p+1)\int_\Omega u^p |\nabla u|^2 dx + \int_\Gamma \theta_\Gamma u^{p+3} ds$$

$$= \int_\Gamma u^{p+2} ds - \int_\Omega \tilde{g} u^{p+1} dx \tag{4.1.113}$$

with $\tilde{g} = g + a\varphi\varphi_t$. Let $v = u^{\frac{p+1}{2}}$. Then by the Young inequality, we get

$$\frac{d}{dt}\int_\Omega u^p dx + \frac{4p}{p+1}\int_\Omega u|\nabla v|^2 dx + \frac{p}{p+3}\int_\Gamma u^2 \theta_\Gamma v^2 ds$$

$$\le CR^p + p\int_\Omega |\tilde{g}|u^{p+1} dx \tag{4.1.114}$$

It follows from the first inequality in (4.1.100) that for $(x,t) \in \bar{\Omega} \times [0,T]$,

$$u \ge \bar{C} > 0. \tag{4.1.115}$$

From the Sobolev imbedding theorem and (4.1.114), (4.1.115) we deduce that there exists $\bar{\delta} > 0$ depending only on T, Ω, β such that

$$\frac{d}{dt}\int_\Omega u^p dx + \bar{\delta}\left(\int_\Omega u^{3(p+1)} dx\right)^{\frac{1}{3}}$$

$$\le CR^p + p\|\tilde{g}\|_{L^6}\left(\int_\Omega u^{3(p+1)} dx\right)^{\frac{1}{6}}\left(\int_\Omega u^{\frac{3(p+1)}{4}} dx\right)^{\frac{2}{3}}$$

$$\le CR^p + \frac{\bar{\delta}}{2}\left(\int_\Omega u^{3(p+1)} dx\right)^{\frac{1}{3}} + Cp^2\|\tilde{g}\|_{L^6}^2\left(\int_\Omega u^{\frac{3(p+1)}{4}} dx\right)^{\frac{4}{3}}. \tag{4.1.116}$$

Therefore, integrating (4.1.116) with respect to t yields for $t \in [0,T]$

$$\int_\Omega u^p dx \le C_T\left(R^p + p^2 \sup_{0 \le t \le T}\left(\int_\Omega u^{\frac{3(p+1)}{4}} dx\right)^{\frac{4}{3}}\right)$$

$$\le C_T' p^2 \max\left(R^p, \sup_{0 \le t \le T}\left(\int_\Omega u^{\frac{3(p+1)}{4}} dx\right)^{\frac{4}{3}}\right). \tag{4.1.117}$$

Let

$$q_0 = 6, \quad q_{k+1} = \frac{4}{3}q_k - 1, \quad \forall k \in N \tag{4.1.118}$$

and

$$p = q_{k+1}. \tag{4.1.119}$$

Then it follows from (4.1.117) that

$$\int_\Omega u^{q_{k+1}} dx \leq C_T' q_{k+1}^2 \max\left(R^{q_{k+1}}, \sup_{0 \leq t \leq T} \left(\int_\Omega u^{q_k} dx \right)^{\frac{4}{3}} \right) \tag{4.1.120}$$

where we may assume $C_T' > 1$ again. Moreover, we deduce from (4.1.66) that

$$\|u(t)\|_{L^6} \leq C_T'. \tag{4.1.121}$$

Then applying Lemma 4.1.5 again with $a = \frac{4}{3}$, $b = 2$, $c = -1$, $c_0 = C_T'$, $c_1 = max(R^{\frac{4}{3}}, C_T')$, $\delta_0 = 6$, $\delta_k = q_k$, $\gamma_k = \sup_{0 \leq t \leq T} \int_\Omega |u|^{q_k} dx$ yields the second inequality in (4.1.100). The proof now is complete. □

With the uniform L^∞ estimates for u and θ established in Lemma 4.1.6 , we can conclude the proof of Theorem 4.1.2 by further deriving higher-order estimates or by a direct application of Theorem 5.2 in Amann [8]. Indeed, it can be easily verified that the nonlinear parabolic system (4.1.53), (4.1.54) is triangular and normal parabolic in the sense of Amann [8]. Moreover, the corresponding growth conditions are satisfied. Hence we can infer from Theorem 5.2 in Amann [8] that the unique local classical solution can be extended to a global one. Thus the proof of Theorem 4.1.2 is complete.

□

Remark 4.1.4 *¿From the proof of Lemma 4.1.6 we can see that the bounds of the L^∞ norm of θ and u depend on $\|\varphi_0\|_{H^2}, \|\theta_0\|_{L^\infty}$ and $\|\frac{1}{\theta_0}\|_{L^\infty}$. Therefore the assumptions on the initial data in Theorem 4.1.2 can be weakened to the condition that $\varphi_0 \in H^2, \theta_0 \in H^1 \cap L^\infty, \theta_0 > 0$, a.e, $\frac{1}{\theta_0} \in L^\infty$ (see Laurencot [1]).*

We refer the readers to Horn, Sprekels and Zheng [1] and Laurencot [1] for problems with $s(\varphi)$ defined only on a finite interval in IR or with constraint.

4.2 Systems of PDEs from Shape Memory Alloys

In this section we consider the system of partial differential equations arising from the study of phase transitions in shape memory alloys. A particular class of materials, where both stress and temperature induced first-order phase transitions occur and lead to a rather spectacular hysterestic behaviour, are the so-called shape memory alloys. In these materials the metallic lattice is deformed by shear, and the assumption of a constant density is justified. The shape memory effect itself is due to martensitic phase transitions between different configurations of the crystal lattice, namely austenite and martensitic twins. For an account of the physical properties of shape memory alloys we refer to the review paper: Delaey, Krishnam, Tas & Warlimont [1]. To study the thermomechanics of shape memory alloys in one space dimension, Falk [1,2] has proposed a Ginzburg–Landau theory, using the strain $\varepsilon = u_x$ as order parameter and assuming that the Helmholtz free energy density F is a potential of Ginzburg–Landau form, i.e.,

$$F = F(\varepsilon, \varepsilon_x, \theta) \tag{4.2.1}$$

where θ is the absolute temperature.

The simplest form for the free energy density F that accounts quite well for the experimentally observed behaviour and takes couple stresses into account is (see Falk [1,2])

$$F(\varepsilon, \varepsilon_x, \theta) = F_0(\theta) + \alpha_1(\theta - \theta_1)\varepsilon^2 - \alpha_2\varepsilon^4 + \alpha_3\varepsilon^6 + \frac{\gamma}{2}\varepsilon_x^2 \tag{4.2.2}$$

with positive constants $\theta_1, \gamma, \alpha_i$ $(i = 1, \cdots, 3)$. A typical form for $F_0(\theta)$ is given by

$$F_0(\theta) = -C_\nu\theta \ln(\frac{\theta}{\theta_2}) + C_\nu\theta + \tilde{C} \tag{4.2.3}$$

where $C_\nu, \theta_2 > 0$ and $\tilde{C}$ are given constants. Observe that in the interesting range of temperature where θ is close to θ_1, F is not a convex function of the shear strain ε and it may have up to three minima, corresponding to the austenitic and the two martensitic phases.

Let $\Omega = (0, 1)$ and, for simplicity, constant density $\rho \equiv 1$. Then the balance laws of momentum and energy read

$$u_{tt} - \sigma_x + \mu_{xx} = f(x,t), \tag{4.2.4}$$

$$U_t + q_x - \sigma\varepsilon_t - \mu\varepsilon_{xt} = g(x,t) \tag{4.2.5}$$

with $u, \varepsilon, \sigma, U, q, f, g$ being the displacement, strain, stress, internal energy, heat flux, density of distributed loads, and density of heat sources, respectively. Moreover, we have the constitutive relations

$$\varepsilon = u_x, \ \sigma = \frac{\partial F}{\partial\varepsilon}, \ \mu = \frac{\partial F}{\partial\varepsilon_x}, \ U = F - \theta\frac{\partial F}{\partial\theta}. \tag{4.2.6}$$

For the heat flux q, we assume that the Fourier law is satisfied:

$$q = -\kappa\theta_x \tag{4.2.7}$$

with $\kappa > 0$ a given constant.

In what follows we assume that F is in a somewhat more general form than (4.2.2):

$$F(\varepsilon, \varepsilon_x, \theta) = F_0(\theta) + \alpha_1\theta F_1(\varepsilon) + F_2(\varepsilon) + \frac{\gamma}{2}\varepsilon_x^2 \tag{4.2.8}$$

satisfying the following hypotheses:

(H_1) $F_0 \in C^4(0,\infty)$, $F_1, F_2 \in C^4(\mathbb{R})$.

(H_2) The mapping $\theta \mapsto -\theta F_0''(\theta)$ extends onto $[0,\infty)$ as an element of $C^2[0,\infty)$.

(H_3) $-\theta F_0''(\theta) \geq \bar{C}_1 > 0$, $\forall\theta \in [0,\infty)$.

(H_4) $F_2(\varepsilon) \geq \bar{C}_2|\varepsilon| - \bar{C}_3$, $\forall\varepsilon \in \mathbb{R}$ with given positive constants $\bar{C}_2, \bar{C}_3$.

Inserting (4.2.6), (4.2.7) into (4.2.4), (4.2.5) yields

$$u_{tt} - \left(\frac{\partial F}{\partial\varepsilon}\right)_x + \gamma D^4 u = f, \tag{4.2.9}$$

$$-\theta F_0''(\theta)\theta_t - \alpha_1\theta F_1'\varepsilon_t - \kappa\theta_{xx} = g \tag{4.2.10}$$

i.e.,

$$u_{tt} - (\alpha_1 F_1''\theta + F_2'')D^2 u + \gamma D^4 u - \alpha_1 F_1'D\theta = f, \tag{4.2.11}$$

$$-\theta F_0''(\theta)\theta_t - \alpha_1\theta F_1'u_{xt} - \kappa\theta_{xx} = g \tag{4.2.12}$$

where, as usual, $D^i = \dfrac{\partial^i}{\partial x^i}$.

In addition, we prescribe the initial and boundary conditions

$$u|_{x=0,1} = D^2 u|_{x=0,1} = 0, \tag{4.2.13}$$

$$D\theta|_{x=0} = 0, \ (\kappa\theta_x + \beta(\theta - \theta_\Gamma(t))|_{x=1} = 0, \tag{4.2.14}$$

$$u|_{t=0} = u_0(x), \ u_t|_{t=0} = u_1(x), \theta|_{t=0} = \theta_0(x). \tag{4.2.15}$$

Here $\theta_\Gamma(t) > 0$ is the temperature of the surrounding medium and β is a given positive constant.

Equations $(4.2.11)$–$(4.2.12)$ can be considered as a coupled system of a nonlinear beam equation with respect to u and a nonlinear parabolic equation with respect to θ.

We impose the following assumptions on the data of our system:

(H_5) For any $T > 0, f \in H^1([0,T]; H^1(\Omega)), f_{tt} \in L^2([0,T]; L^2(\Omega))$;

$\quad g \in L^2([0,T]; H^2(\Omega)) \cap H^1([0,T]; H^1(\Omega)), \ g(x,t) \geq 0$, on $\bar\Omega \times [0,\infty)$;

$\quad \theta_\Gamma \in H^2[0,T], \ \theta_\Gamma > 0$ on $[0,\infty)$.

(H_6) $u_0 \in H_E^5 = \{u | u \in H^5(\Omega), u|_{x=0,1} = D^2 u|_{x=0,1} = 0\}$;

$\quad u_1 \in H_E^3 = \{u | u \in H^3(\Omega), u|_{x=0,1} = D^2 u|_{x=0,1} = 0\}$;

$\quad \theta_0 \in H_B^3 = \{\theta | \theta \in H^3(\Omega), D\theta|_{x=0} = 0, \ (\kappa D\theta + \beta(\theta - \theta_\Gamma))|_{x=1} = 0\}$,

$\quad \theta_0(x) > 0$, on $\bar\Omega$.

Then the main result in this section (see Sprekels and Zheng [2]) is the following

Theorem 4.2.1 *Suppose (H_1)–(H_6) are satisfied. Then problem $(4.2.11)$–$(4.2.15)$ admits a unique global classical solution (u, θ) such that for any $T > 0, u \in C([0,T]; H^5)$ $\cap C^1([0,T]; H^3) \cap C^2([0,T]; H^1), \theta \in C([0,T]; H^3) \cap C^1([0,T]; H^1) \cap L^2([0,T]; H^4)$ and $D^4 u, D^2 u_t, u_{tt}, \theta_t, \theta_{xx}$ are all in $C^{\alpha,\frac{\alpha}{2}}(Q_T)$ for some $\alpha, 0 < \alpha < 1$. Moreover, $\theta > 0$ in Q_∞.*

Remark 4.2.1 *The method used here also applies to other boundary conditions such as*

$$Du|_{x=0,1} = D^3u|_{x=0,1} = D\theta|_{x=0,1} = 0. \tag{4.2.16}$$

(see Zheng [12]).

Proof of Theorem 4.2.1. For simplicity of exposition, without loss of generality, in the remaining part of this section, we always assume that $f \equiv g \equiv \theta_\Gamma \equiv 0$. The whole proof consists of several lemmas. We first establish a local existence and uniqueness result.

Lemma 4.2.1 *Suppose (H_1)–(H_6) are satisfied. Then there exists $t^* > 0$ depending only on $\|u_0\|_{H^5}, \|u_1\|_{H^3}, \|\theta_0\|_{H^3}$ such that problem (4.2.11)–(4.2.15) admits a unique solution (u, θ) in $\bar{\Omega} \times [0, t^*]$ such that*

$u \in C([0,t^*]; H^5) \cap C^1([0,t^*]; H^3) \cap C^2([0,t^*]; H^1), \ \theta \in C([0,t^*]; H^3) \cap C^1([0,t^*]; H^1)$
$\cap L^2([0,t^*]; H^4), \ \theta(x,t) > 0 \ in \ \bar{\Omega} \times [0,t^*].$

Moreover, for $t \in [0,t^*]$,

$$\|u(t)\|_{H^5} + \|u_t(t)\|_{H^3} + \|\theta(t)\|_{H^3} \le C\left(\|u_0\|_{H^5} + \|u_1\|_{H^3} + \|\theta_0\|_{H^3}\right) \tag{4.2.17}$$

where $C > 0$ depends only on $\|u_0\|_{H^5}, \|u_1\|_{H^3}, \|\theta_0\|_{H^3}$.

Proof of Lemma 4.2.1

Let

$$X_h(M_0, M_1) = \left\{ (u,\theta) \; \middle| \; \begin{array}{l} u \in C([0,h]; H^5) \cap C^1([0,h]; H^3) \cap C^2([0,h]; H^1), \\ \theta \in C([0,h]; H^3) \cap C^1([0,h]; H^1) \cap L^2([0,h]; H^4), \\ \theta > 0, \ in \ \bar{\Omega} \times [0,h], \\ u|_{t=0} = u_0(x), u_t|_{t=0} = u_1(x), \theta|_{t=0} = \theta_0(x), \\ \max_{0 \le t \le h}\left(\|u(t)\|_{H^5}^2 + \|u_t(t)\|_{H^3}^2 + \|u_{tt}(t)\|_{H^1}^2\right) \le M_0, \\ \max_{0 \le t \le h}\left(\|\theta(t)\|_{H^3}^2 + \|\theta_t\|_{H^1}^2\right) \\ + \int_0^h \left(\|\theta\|_{H^4}^2 + \|\theta_t\|_{H^2}^2 + \|\theta_{tt}\|^2\right) d\tau \le M_1 \end{array} \right\}$$

where M_0, M_1 are positive constants. For $(\tilde{u}, \tilde{\theta}) \in X_h(M_0, M_1)$, we consider the following linear auxiliary problems:

$$u_{tt} + \gamma D^4 u = \alpha_1 \tilde{\theta} F_1''(\tilde{u}_x)\tilde{u}_{xx} + F_2''(\tilde{u}_x)\tilde{u}_{xx} + \alpha_1 F_1'(\tilde{u}_x)\tilde{\theta}_x, \tag{4.2.18}$$

$$u|_{x=0,1} = D^2 u|_{x=0,1} = 0, \tag{4.2.19}$$

$$u|_{t=0} = u_0, u_t|_{t=0} = u_1 \tag{4.2.20}$$

and

$$c_0(\tilde{\theta})\theta_t - \alpha_1 \theta F_1'(\tilde{u}_x)\tilde{u}_{xt} - \kappa\theta_{xx} = 0, \tag{4.2.21}$$

$$\theta_x|_{x=0} = 0, \ (\kappa\theta_x + \beta\theta)|_{x=1} = 0, \tag{4.2.22}$$

$$\theta|_{t=0} = \theta_0(x) \tag{4.2.23}$$

where $c_0(\theta) = -\theta F_0''(\theta) \geq \bar{C}_1 > 0$. Since the operator $A = \gamma D^4 : D(A) = \{u \in H^4| \ u|_{x=0,1} = D^2 u|_{x=0,1} = 0\} \mapsto L^2$ is positive definite and $f_1 =$ the term on the right-hand side of (4.2.18) belongs to $H^1([0, h]; H^1)$, we can deduce from Theorem 1.3.3 that problem (4.2.18)–(4.2.20) has a unique solution $u \in C([0, h]; H^5) \cap C^1([0, h]; H^3) \cap C^2([0, h]; H^1)$. Moreover, the usual energy method gives

$$\|u_t\|^2 + \gamma\|D^2 u\|^2 = \|u_1\|^2 + \gamma\|D^2 u_0\|^2 + 2\int_0^t \int_0^1 f_1 u_t \, dx \, d\tau \tag{4.2.24}$$

and

$$\|u_{tt}\|^2 + \gamma\|D^2 u_t\|^2 = \gamma\|D^2 u_1\|^2$$

$$+ \| -\gamma D^4 u_0 + \alpha_1 \theta_0 F_1''(u_{0x})D^2 u_0 + F_2''(u_{0x})D^2 u_0 + \alpha_1 F_1'(u_{0x})\theta_{0x} \|^2$$

$$+ \int_0^t \int_0^1 f_{1t} u_{tt} \, dx \, d\tau. \tag{4.2.25}$$

Thus,

$$\|u_t(t)\|^2 + \gamma\|D^2 u(t)\|^2 \leq e^t \left(\|u_1\|^2 + \gamma\|D^2 u_0\|^2 + \int_0^t \|f_1\|^2 d\tau \right), \tag{4.2.26}$$

$$\|u_{tt}\|^2 + \gamma\|D^2u_t\|^2 \le e^t \left(\gamma\|D^2u_1\|^2\right.$$

$$\left. + \| - \gamma D^4u_0 + \alpha_1\theta_0 F_1''(u_{0x})D^2u_0 + F_2''(u_{0x})D^2u_0 + \alpha_1 F_1'(u_{0x})\theta_{0x}\|^2 + \int_0^t \|f_{1t}\|^2 d\tau\right)$$

$$\le e^t \left(M_2 + \int_0^t \|f_{1t}\|^2 d\tau\right), \tag{4.2.27}$$

where $M_2 > 0$ depends only on $\|u_0\|_{H^5}, \|u_1\|_{H^3}, \|\theta_0\|_{H^3}$. Differentiating (4.2.18) with respect to x and t, then multiplying it with u_{ttx}, in the same manner as before, we obtain

$$\|u_{tt}\|^2_{H^1} + \gamma\|D^2u_t\|^2_{H^1} \le e^t \left(M_3 + \int_0^t \|f_{1t}\|^2_{H^1} d\tau\right). \tag{4.2.28}$$

Hereafter $M_i, (i = 3, \cdots, 7)$ are positive constants depending on $\|u_0\|_{H^5}, \|u_1\|_{H^3}$ and $\|\theta_0\|_{H^3}$.

Combining (4.2.18) with (4.2.24)–(4.2.28) yields

$$\|u(t)\|^2_{H^5} + \|u_t(t)\|^2_{H^3} + \|u_{tt}(t)\|^2_{H^1}$$

$$\le e^t \left(M_4 + \int_0^t (\|f_1\|^2 + \|f_{1t}\|^2_{H^1}) d\tau\right)$$

$$\le e^t (M_4 + C(M_0, M_1)t) \tag{4.2.29}$$

where $C(M_0, M_1)$ is a positive constant depending on M_0, M_1.

By the energy method we can also deduce that the unique solution θ to the linear problem (4.2.21)–(4.2.23) satisfies the following estimates

$$\frac{\kappa}{2}\|D\theta\|^2 + \frac{\beta}{2}\theta^2|_{x=1} + \int_0^t\int_0^1 c_0\theta_t^2 dxd\tau$$

$$\le \frac{\kappa}{2}\|D\theta_0\|^2 + \frac{\beta}{2}\theta_0^2|_{x=1} + \int_0^t\int_0^1 |\theta_t\theta f_2| dxd\tau \tag{4.2.30}$$

and

$$\frac{\kappa}{2}\|\theta_{xt}\|^2 + \frac{\beta}{2}\theta_t^2|_{x=1} + \int_0^t\int_0^1 c_0\theta_{tt}^2 dxd\tau$$

$$\le M_5 + \int_0^t\int_0^1 |c_0'(\tilde\theta)\tilde\theta_t\theta_t\theta_{tt}| dxd\tau + \int_0^t\int_0^1 |f_2\theta_t + f_{2t}\theta||\theta_{tt}| dxd\tau. \tag{4.2.31}$$

where $f_2 = \alpha_1 F_1'(\tilde{u}_x)\tilde{u}_{xt}$.

Combining (4.2.30) with (4.2.31) yields

$$\|\theta\|_{H^1}^2 + \|\theta_t\|_{H^1}^2 + \int_0^t (\|\theta_t\|^2 + \|\theta_{tt}\|^2)d\tau \leq e^{C(M_0,M_1)t} M_6. \tag{4.2.32}$$

Using equation (4.2.21), we obtain

$$\|\theta\|_{H^3}^2 + \|\theta_t\|_{H^1}^2 + \int_0^t \left(\|\theta_{tt}\|^2 + \|\theta_t\|_{H^2}^2 + \|\theta\|_{H^4}^2 \right) d\tau$$

$$\leq e^{C(M_0,M_1)t} M_7 + C(M_0, M_1)t \tag{4.2.33}$$

where $C(M_0, M_1) > 0$ is a constant depending on M_0, M_1.

By the maximum principle, θ is always positive in $\bar{\Omega} \times [0, h]$. We take $M_0 = 2M_4, M_1 = 2M_7$. Then there exists $t^* > 0$ such that for $h \leq t^*$, the nonlinear operator defined by the auxiliary linear problems maps $X_h(M_0, M_1)$ into itself. A similar argument to that used before yields that this nonlinear operator is a strict contraction provided that t^* is small enough. We leave the details to the reader. Thus the proof of Lemma 4.2.1 is complete.   □

In order to complete the proof of Theorem 4.2.1, it remains to obtain uniform a priori estimates for any $T > 0$.

Lemma 4.2.2 *For any $T > 0$, there exists $C_T > 0$ such that for $0 \leq t \leq T$ the following estimates hold*

$$\|\theta(t)\|_{L^1(\Omega)} + \|u_t(t)\|^2 + \|u_{xx}(t)\|^2 \leq C_T, \tag{4.2.34}$$

$$\|u_x(t)\|_{L^\infty} = \|\varepsilon(t)\|_{L^\infty} \leq C_T. \tag{4.2.35}$$

Proof. Multiplying (4.2.9) by u_t, adding the result to (4.2.10) and integrating over Ω, we obtain

$$\frac{d}{dt}\left(\frac{1}{2}\|u_t(t)\|^2 + \int_0^1 U(t)dx\right) + \beta\theta|_{x=1} = 0. \tag{4.2.36}$$

By hypotheses (H_3), (H_4), we have

$$U \geq C_1 \left(\theta + |\varepsilon| + \varepsilon_x^2 \right) - C_2. \tag{4.2.37}$$

Thus integrating (4.2.36) with respect to t and using (4.2.37), we get (4.2.34). By the Sobolev imbedding theorem, we obtain (4.2.35).   □

Lemma 4.2.2. shows that at the begining we only have an L^1 norm estimate of θ. In order to obtain L^∞ norm and higher-order norm estimates of θ, the following lemma plays a very crucial role in bounding the superlinear terms in the equations.

Lemma 4.2.3 *For any $T > 0$, there exists $C_T > 0$ such that for $0 \leq t \leq T$, the following estimates hold*

$$\|\theta(t)\|^2 + \int_0^t \theta^2|_{x=1} d\tau + \int_0^t \|\theta_x\|^2 d\tau \leq C_T, \tag{4.2.38}$$

$$\int_0^t \|\theta(\tau)\|_{L^\infty}^4 d\tau \leq C_T. \tag{4.2.39}$$

Proof. Multiplying (4.2.10) by θ and integrating with respect to x and t yields

$$\int_0^t \int_0^1 -\theta^2 F_0''(\theta)\theta_t dx d\tau - \alpha_1 \int_0^t \int_0^1 \theta^2 F_1'(u_x)u_{xt} dx d\tau$$

$$+\kappa \int_0^t \int_0^1 \theta_x^2 dx d\tau + \beta \int_0^t \theta^2|_{x=1} d\tau = 0. \tag{4.2.40}$$

By hypotheses (H_3) and (H_6), we have

$$\int_0^t \int_0^1 -\theta^2 F_0''(\theta)\theta_t dx d\tau = \int_0^1 \int_0^t \frac{d}{d\tau} \left(\int_0^{\theta(x,\tau)} -\eta^2 F_0''(\eta)d\eta \right) d\tau dx$$

$$= \int_0^1 \int_{\theta_0(x)}^{\theta(x,t)} -\eta^2 F_0''(\eta)d\eta dx \geq C_1 \|\theta(t)\|^2 - C_2. \tag{4.2.41}$$

Next we use integration by parts and the fact that $u_t|_{x=0,1} = 0$ to obtain

$$\left| \int_0^t \int_0^1 \theta^2 F_1'(u_x)u_{xt} dx d\tau \right|$$

$$= \left| \int_0^t \int_0^1 \left((\theta^2 F_1'(u_x)u_t)_x - 2\theta\theta_x F_1'(u_x)u_t - \theta^2 F_1''(u_x)u_{xx}u_t \right) dx d\tau \right|$$

$$\leq 2 \int_0^t \int_0^1 |\theta\theta_x F_1'(u_x)u_t| dx d\tau + \int_0^t \int_0^1 |\theta^2 F_1''(u_x)u_{xx}u_t| dx d\tau$$

$$= I_1 + I_2. \tag{4.2.42}$$

Let $\delta > 0$ be a constant to be specified later. Applying the Young inequality and invoking (4.2.34), (4.2.35), we deduce that

$$I_1 \leq \delta \int_0^t \|\theta_x\|^2 d\tau + \delta^{-1}C \int_0^t \int_0^1 \theta^2 u_t^2 dx d\tau$$

$$\leq \delta \int_0^t \|\theta_x\|^2 d\tau + C\delta^{-1} \int_0^t \|\theta\|_{L^\infty}^2 \|u_t\|^2 d\tau$$

$$\leq \delta \int_0^t \|\theta_x\|^2 d\tau + C\delta^{-1} \int_0^t \|\theta\|_{L^\infty}^2 d\tau. \tag{4.2.43}$$

By the Nirenberg inequality, the Young inequality and (4.2.34), we have

$$\|\theta\|_{L^\infty}^2 \leq \tilde{C}_1 \|\theta_x\|^{\frac{4}{3}} \|\theta\|_{L^1}^{\frac{2}{3}} + \tilde{C}_2 \|\theta\|_{L^1}^2$$

$$\leq \frac{\delta^2}{C} \|\theta_x\|^2 + C. \tag{4.2.44}$$

Thus

$$I_1 \leq \delta \int_0^t \|\theta_x\|^2 d\tau + C. \tag{4.2.45}$$

Similarly, by (4.2.34), (4.2.35) we obtain

$$I_2 \leq C \int_0^t \|\theta\|_{L^\infty}^2 \|u_{xx}\| \|u_t\| d\tau \leq C \int_0^t \|\theta\|_{L^\infty}^2 d\tau$$

$$\leq \delta \int_0^t \|\theta_x\|^2 d\tau + C. \tag{4.2.46}$$

Combining (4.2.40)–(4.2.46) and taking δ small enough, we get (4.2.38). Using the Nirenberg inequality

$$\|\theta\|_{L^\infty} \leq \tilde{C}_1 \|\theta_x\|^{\frac{1}{2}} \|\theta\|^{\frac{1}{2}} + \tilde{C}_2 \|\theta\| \tag{4.2.47}$$

and combining it with (4.2.38) yields (4.2.39). The proof is complete.   □

Now that the key estimates (4.2.38), (4.2.39) have been proved, we can proceed to derive a priori bounds for the L^∞ norm of θ and for higher-order derivatives of u.

Lemma 4.2.4 *For any $T > 0$, there exists $C_T > 0$ such that for $0 \leq t \leq T$, the following estimates hold:*

$$\|u_{xt}(t)\|^2 + \|u_{xxx}(t)\|^2 + \|\theta(t)\|_{H^1}^2 + \int_0^t \int_0^1 \theta_t^2 dx d\tau \leq C_T, \qquad (4.2.48)$$

$$\|\theta(t)\|_{L^\infty} \leq C_T, \quad \|u_t(t)\|_{L^\infty} + \|u_{xx}(t)\|_{L^\infty} \leq C_T, \qquad (4.2.49)$$

$$\int_0^t \|\theta_{xx}\|^2 d\tau \leq C_T. \qquad (4.2.50)$$

Proof. It suffices to prove (4.2.48). Once (4.2.48) has been proved, (4.2.49), (4.2.50) follow from equation (4.2.10) and the Sobolev imbedding theorem.

Multiplying (4.2.10) by θ_t and integrating with respect to x and t yields

$$\int_0^t \int_0^1 -\theta F_0''(\theta)\theta_t^2 dx d\tau - \alpha_1 \int_0^t \int_0^1 \theta F_1'(u_x)\theta_t u_{xt} dx d\tau$$

$$+ \frac{\kappa}{2}\|D\theta(t)\|^2 + \frac{\beta}{2}\theta^2|_{x=1} = \frac{\kappa}{2}\|D\theta_0\|^2 + \frac{\beta}{2}\theta_0^2|_{x=1}. \qquad (4.2.51)$$

Let $\delta > 0$ be a constant to be specified later. Then by Lemma 4.2.2 and Lemma 4.2.3, we have

$$\left| \alpha_1 \int_0^t \int_0^1 \theta F_1'(u_x)\theta_t u_{xt} dx d\tau \right|$$

$$\leq \delta \int_0^t \|\theta_t\|^2 d\tau + \frac{C}{\delta} \int_0^t \|\theta\|_{L^\infty}^2 \|u_{xt}\|^2 d\tau$$

$$\leq \delta \int_0^t \|\theta_t\|^2 d\tau + \frac{C}{\delta} \sup_{0 \leq \tau \leq t} \|u_{xt}(\tau)\|^2. \qquad (4.2.52)$$

Combining (4.2.51) with (4.2.52), using the hypothesis (H_3) and taking $\delta = \frac{\bar{C}_1}{2}$, we get

$$\int_0^t \|\theta_t\|^2 d\tau + \|\theta\|_{H^1}^2 \leq C + C \sup_{0 \leq \tau \leq t} \|u_{xt}(\tau)\|^2. \qquad (4.2.53)$$

Next we multiply (4.2.9) by u_{xxt}, integrate with respect to x and t and use integration by parts to obtain

$$\frac{1}{2}\|u_{xt}(t)\|^2 + \frac{\gamma}{2}\|D^3u(t)\|^2 - \alpha_1 \int_0^t\int_0^1 \theta F_1'(u_x)u_{xxxt}dxd\tau$$

$$-\int_0^t\int_0^1 F_2'(u_x)u_{xxxt}dxd\tau = \frac{1}{2}\|Du_1\|^2 + \frac{\gamma}{2}\|D^3u_0\|^2. \qquad (4.2.54)$$

Using the results obtained in the previous lemmas, we have

$$\left|\int_0^t\int_0^1 \theta F_1'(u_x)u_{xxxt}dxd\tau\right|$$

$$= \left|\int_0^1 \theta F_1'(u_x)u_{xxx}dx - \int_0^1 \theta_0 F_1'(Du_0)D^3u_0dx - \int_0^t\int_0^1 (\theta F_1'(u_x))_t D^3u\,dxd\tau\right|$$

$$\leq \frac{\gamma}{8}\|D^3u\|^2 + \frac{\delta}{2}\int_0^t\|\theta_t\|^2d\tau + \frac{C}{\delta}\int_0^t\|D^3u\|^2d\tau + \frac{\delta}{2}\int_0^t\|\theta\|_{L^\infty}^2\|u_{xt}\|^2d\tau + C$$

$$\leq \frac{\gamma}{8}\|D^3u\|^2 + \frac{\delta}{2}\int_0^t\|\theta_t\|^2d\tau + \frac{C}{\delta}\int_0^t\|D^3u\|^2d\tau + \frac{\delta C}{2}\sup_{0\leq\tau\leq t}\|u_{xt}(\tau)\|^2 + C. \qquad (4.2.55)$$

An analogous calculation also gives

$$\left|\int_0^t\int_0^1 F_2'(u_x)u_{xxxt}dx\right|$$

$$\leq C + \frac{\gamma}{8}\|D^3u(t)\|^2 + C\int_0^t(\|u_{xt}\|^2 + \|D^3u\|^2)d\tau. \qquad (4.2.56)$$

Combining (4.2.53)–(4.2.56) yields

$$\frac{1}{2}\|u_{xt}\|^2 + \frac{\gamma}{4}\|D^3u(t)\|^2$$

$$\leq \frac{\delta}{2}\int_0^t\|\theta_t\|^2d\tau + C\int_0^t(\|u_{xt}\|^2 + \|D^3u\|^2)d\tau + \frac{\delta C}{2}\sup_{0\leq\tau\leq t}\|u_{xt}(\tau)\|^2 + C$$

$$\leq C\delta\sup_{0\leq\tau\leq t}\|u_{xt}(\tau)\|^2 + C\int_0^t(\|u_{xt}\|^2 + \|D^3u\|^2)d\tau + C. \qquad (4.2.57)$$

Choosing δ small enough, taking the supremum with respect to t and applying the Gronwall inequality, we arrive at

$$\sup_{0\leq\tau\leq t}(\|u_{xt}(\tau)\|^2 + \|D^3u(\tau)\|^2) \leq C_T. \qquad (4.2.58)$$

Combining (4.2.53), (4.2.58) yields (4.2.48). Thus the proof of Lemma 4.2.4 is complete.   □

Having established Lemmas 4.2.2–4.2.4, we can easily get estimates of higher-order derivatives.

Lemma 4.2.5 *For any $T > 0$, there exists $C_T > 0$ such that for $0 \leq t \leq T$, the following estimates hold.*

$$\|\theta_{xx}(t)\|^2 + \|D^4 u(t)\|^2 + \|u_{xxt}(t)\|^2 + \int_0^t \|\theta_{xt}\|^2 d\tau + \int_0^t \theta_t^2|_{x=1} d\tau \leq C_T, \quad (4.2.59)$$

$$\|\theta_t(t)\|^2 + \|u_{tt}(t)\|^2 \leq C_T, \quad (4.2.60)$$

$$\int_0^t \|\theta_{xxx}\|^2 d\tau \leq C_T. \quad (4.2.61)$$

Proof. It can be easily seen that (4.2.60), (4.2.61) follow from (4.2.59) and equations (4.2.9), (4.2.10). Therefore, it suffices to prove (4.2.59). We can write equation (4.2.9) as

$$u_{tt} + \gamma D^4 u = \tilde{f} \quad (4.2.62)$$

with

$$\tilde{f} = (\alpha_1 \theta F_1'(Du) + F_2'(Du))_x . \quad (4.2.63)$$

Then the usual energy method gives

$$\frac{1}{2}\|D^2 u_t(t)\|^2 + \frac{\gamma}{2}\|D^4 u\|^2 = \frac{1}{2}\|D^2 u_1\|^2 + \frac{\gamma}{2}\|D^4 u_0\|^2 + \int_0^1 \int_0^t \tilde{f}_t D^4 u \, dx \, d\tau. \quad (4.2.64)$$

Observe that

$$\tilde{f}_t = \alpha_1 \theta_{xt} F_1'(Du) + \alpha_1 \theta F_1''(Du) u_{xxt} + F_2''(Du) u_{xxt} + h \quad (4.2.65)$$

where

$$h = \alpha_1 \theta_x F_1''(Du) u_{xt} + \alpha_1 \theta_t F_1''(Du) u_{xx} + \alpha_1 \theta F_1'''(Du) u_{xt} u_{xx} + F_2'''(Du) u_{xt} u_{xx}. \quad (4.2.66)$$

By virtue of the previous estimates,

$$\|h(t)\| \leq C_T. \quad (4.2.67)$$

It turns out from (4.2.64) and the Young inequality that

$$\frac{1}{2}\|u_{xxt}(t)\|^2 + \frac{\gamma}{2}\|D^4u\|^2$$

$$\leq \delta \int_0^t \|\theta_{xt}\|^2 d\tau + C\left(1 + \int_0^t (\|u_{xxt}\|^2 + \|D^4u\|^2)d\tau\right). \tag{4.2.68}$$

Applying the Gronwall inequality to (4.2.68) yields

$$\|u_{xxt}(t)\|^2 + \|D^4u(t)\|^2 \leq C(1 + \delta \int_0^t \|\theta_{xt}\|^2 d\tau). \tag{4.2.69}$$

Differentiating equation (4.2.10) with respect to t, multiplying the resulting equation by θ_t and integrating over $\Omega \times [0,t]$ yields

$$\frac{1}{2}\int_0^1 c_0(\theta)\theta_t^2 dx + \frac{1}{2}\int_0^t \int_0^1 c_0'(\theta)\theta_t^3 dx d\tau$$

$$-\alpha_1 \int_0^t \int_0^1 \left(\theta_t F_1'(Du)Du_t + \theta F_1''(Du)(Du_t)^2 + \theta F_1'(Du)u_{xtt}\right)\theta_t dx d\tau$$

$$+\frac{\kappa}{2}\int_0^t \|\theta_{xt}\|^2 d\tau + \frac{\beta}{2}\int_0^t \theta_t^2|_{x=1} d\tau \leq C. \tag{4.2.70}$$

By the previous estimates established in Lemma 4.2.2–Lemma 4.2.4 and the Nirenberg inequality, we have

$$\left|\frac{1}{2}\int_0^t \int_0^1 c_0'(\theta)\theta_t^3 dx d\tau\right| \leq C\int_0^t \|\theta_t\|_{L^3}^3 d\tau$$

$$\leq C\left(\int_0^t \|\theta_{xt}\|^{\frac{1}{2}}\|\theta_t\|^{\frac{5}{2}} d\tau + \int_0^t \|\theta_t\|^3 d\tau\right)$$

$$\leq C\left((\int_0^t \|\theta_{xt}\|^2 d\tau)^{\frac{1}{4}}(\int_0^t \|\theta_t\|^{\frac{10}{3}} d\tau)^{\frac{3}{4}} + \int_0^t \|\theta_t\|^3 d\tau\right)$$

$$\leq C\left(\max_{0\leq\tau\leq t}\|\theta_t(\tau)\|(\int_0^t \|\theta_{xt}\|^2 d\tau)^{\frac{1}{4}} + \max_{0\leq\tau\leq t}\|\theta_t(\tau)\|\right)$$

$$\leq \delta \max_{0\leq\tau\leq t}\|\theta_t(\tau)\|^2 + \delta \int_0^t \|\theta_{xt}\|^2 d\tau + C_\delta, \tag{4.2.71}$$

$$\left| \alpha_1 \int_0^t \int_0^1 \theta_t^2 F_1'(Du) Du_t \, dx d\tau \right| \le C \int_0^t \|\theta_t\|_{L^4}^2 \|u_{xt}\| d\tau$$

$$\le C \int_0^t \|\theta_t\|_{L^4}^2 d\tau \le C \left(\int_0^t \|\theta_{xt}\|^{\frac{1}{2}} \|\theta_t\|^{\frac{3}{2}} d\tau + \int_0^t \|\theta_t\|^2 d\tau \right)$$

$$\le C \left(\int_0^t \|\theta_{xt}\|^2 d\tau \right)^{\frac{1}{4}} + C \le \delta \int_0^t \|\theta_{xt}\|^2 d\tau + C_\delta, \qquad (4.2.72)$$

and

$$\left| \alpha_1 \int_0^t \int_0^1 \theta F_1''(Du)(Du_t)^2 \theta_t \, dx d\tau \right| \le C \int_0^t \|\theta_t\|_{L^\infty} \|u_{xt}\|^2 d\tau$$

$$\le C \int_0^t \|\theta_t\|_{L^\infty} d\tau \le \delta \int_0^t \|\theta_t\|_{H^1}^2 d\tau + C_\delta. \qquad (4.2.73)$$

By integration by parts, we get

$$\left| \alpha_1 \int_0^t \int_0^1 \theta F_1'(Du) u_{xtt} \theta_t \, dx d\tau \right|$$

$$= \left| \alpha_1 \int_0^t \int_0^1 \left(\theta F_1'(Du) u_{tt} \theta_{xt} + F_1'(Du) \theta_x u_{tt} \theta_t + \theta F_1''(Du) D^2 u u_{tt} \theta_t \right) dx d\tau \right|$$

$$\le C \int_0^t \|u_{tt}\| \|\theta_{xt}\| d\tau + C \int_0^t \|\theta_x\| \|u_{tt}\| \|\theta_t\|_{L^\infty} d\tau$$

$$+ C \int_0^t \|u_{xx}\|_{L^\infty} \|u_{tt}\| \|\theta_t\| d\tau \le C \max_{0 \le \tau \le t} \|u_{tt}\| \left(\int_0^t \|\theta_{xt}\|^2 d\tau \right)^{\frac{1}{2}}$$

$$+ C \max_{0 \le \tau \le t} \|u_{tt}\| \left(\int_0^t \|\theta_{xt}\|^2 d\tau \right)^{\frac{1}{4}} + C \max_{0 \le \tau \le t} \|u_{tt}\|$$

$$\le C\delta \int_0^t \|\theta_{xt}\|^2 d\tau + C \qquad (4.2.74)$$

where (4.2.69) and equation (4.2.9) have been used.

Combining (4.2.70)–(4.2.74), taking δ small enough and taking (4.2.69) into account yields (4.2.59). The proof is complete. □

Lemma 4.2.6 *For any $T > 0$, there exists $C_T > 0$ such that for $0 \le t \le T$, the* *following estimate holds.*

$$\|D^5 u(t)\|^2 + \|D^3 u_t(t)\|^2 + \|D u_{tt}(t)\|^2 + \|\theta_t\|_{H^1}^2 + \|D^3 \theta(t)\|^2$$

$$+ \int_0^t (\|\theta_{tt}\|^2 + \|\theta_{xxt}\|^2 + \|D^4\theta\|^2) d\tau \leq C_T. \tag{4.2.75}$$

The proof is essentially the same as before. Therefore, we can leave it to the reader. □

Having established uniform a priori estimates, the global existence and uniqueness follows from the continuation argument. To prove the Hölder continuity of functions $D^4 u, D^2 u_t, u_{tt}, \theta_t$ and θ_{xx}, we need the following interpolation Lemma (see Ladyzenskaja, Solonnikov & Uralceva [1], Chapter 2):

Lemma 4.2.7 *Let* $w(x,t) : Q_T \mapsto I\!R$ *be a function and constants* $\alpha, \beta \in (0,1]$, K_1, $K_2 > 0$ *such that*

$$\max_{0 \leq x \leq 1} |w(x,t) - w(x,\tau)| \leq K_1 |t - \tau|^\alpha, \ \forall t, \tau \in [0,T], \tag{4.2.76}$$

$$\max_{0 \leq t \leq T} |w_x(x,t) - w_x(y,t)| \leq K_2 |x - y|^\beta, \ \forall x, y \in [0,1]. \tag{4.2.77}$$

Then there is a constant $K_3 > 0$ *such that* $\forall t, \tau \in [0,T]$,

$$\max_{0 \leq x \leq 1} |w_x(x,t) - w_x(x,\tau)| \leq K_3 |t - \tau|^{\frac{\alpha\beta}{1+\beta}} \tag{4.2.78}$$

which together with (4.2.77) implies that $w_x \in C^{\delta, \frac{\delta}{2}}$ *with* $\delta = 2\min(\frac{\beta}{2}, \frac{\alpha\beta}{1+\beta})$.

Taking what has been proved in Lemmas 4.2.2–4.2.6 into account, we deduce that u is uniformly Lipschitz continuous in t. By the Schwartz inequality, $D^i u$ ($0 \leq i \leq 4$) are uniformly Hölder continuous in x with exponent $\frac{1}{2}$. Thus successively applying Lemma 4.2.7, we deduce that $D^i u$ ($i = 0, \cdots, 4$) are uniformly Hölder continuous in Q_T. Then by equation (4.2.9) we also deduce the Hölder continuity of u_{tt}. The same argument yields the Hölder continuity of $D u_t, D^2 u_t, \theta_{xx}, \theta_t$. The proof of Theorem 4.2.1 is complete. □

Remark 4.2.2 *For mathematical results on Fremond's model in three space dimension, we refer to* Colli, Fremond & Visintin [1], Hoffmann, Niezgodka & Zheng [1], Colli [1–2], Colli & Sprekels [1]. *We also refer the reader to* Dafermos [1], Dafermos & Hsiao [1], Niezgodka, Zheng & Sprekles [1], Niezgodka & Sprekels [1], Chen & Hoffmann [1] *for the case of thermal and viscous materials in which the viscosity plays a crucial role in the analysis.*

4.3　The Coupled Cahn–Hilliard Equations

In this section we are concerned with a system of nonlinear partial differential equations, in short, the coupled Cahn–Hilliard equations, which consist of a fourth-order quasilinear parabolic equation and a second-order quasilinear parabolic equation.

This system was recently derived by Penrose & Fife [1] and also by Alt & Pawlow [1] to describe the non-isothermal phase separation of a two-component system. The derivation by Penrose and Fife was based on the idea that the value of the entropy functional cannot decrease along solutions paths, which is in agreement with what one expects from the second law of thermodynamics. It turns out that the conserved order parameter φ, which is the concentration of one component in a binary system, and the absolute temperature θ satisfy a system of nonlinear parabolic equations that we call the coupled Cahn–Hilliard equations:

$$\frac{\partial \varphi}{\partial t} = M_1 \Delta \left(-K_1 \Delta \varphi + \frac{\frac{\partial f}{\partial \varphi}}{\theta} \right), \tag{4.3.1}$$

$$\frac{\partial e}{\partial t} = -M_2 \, \Delta \left(\frac{1}{\theta} \right) \tag{4.3.2}$$

where M_1, M_2, K_1 are positive constants, $e = e(\varphi, \theta)$ is the internal energy and $f = f(\varphi, \theta)$ is the free energy density such that

$$e = f - \theta \frac{\partial f}{\partial \theta}. \tag{4.3.3}$$

On the other hand, Alt & Pawlow [1] recently also proposed a mathematical model of the dynamics of non-isothermal phase separation. The model extends the Cahn–Hilliard approach to the non-isothermal situation and the construction is based on the Landau–Ginzberg free energy functional and formalism of non-equilibrium thermodynamics. However, as observed by Shen & Zheng [3], under certain circumstances these two models are essentially the same.

In this section we consider equations (4.3.1), (4.3.2) in one space dimension. Following Penrose and Fife, we take

$$f(\varphi, \theta) = -C_V(\theta \ln \theta + \delta) - \frac{a}{2}\varphi^2 - b\varphi + c - \theta s_o(\varphi) \tag{4.3.4}$$

where $C_V, a > 0, \delta \geq 0$, and b, c are given constants and $s_o(\varphi)$ is a concave function. Typically, we can take

$$s_o(\varphi) = -\frac{\varphi^4}{4}. \tag{4.3.5}$$

However, the function satisfying the general assumption (A) in Section 4.1.2 is also allowed.

It turns out that equations (4.3.1), (4.3.2) become

$$\varphi_t = M_1 D^2 \left(-K_1 D^2 \varphi + \varphi^3 - \frac{a\varphi + b}{\theta} \right), \tag{4.3.6}$$

$$C_V \theta_t - (a\varphi + b)\varphi_t = -M_2 D^2 \left(\frac{1}{\theta} \right) \tag{4.3.7}$$

where, as usual, the notation $D^i = \frac{\partial^i}{\partial x^i}$ is used. We imposed the following boundary conditions and initial conditions :

$$(D\varphi) \mid_{x=0,1} = (D^3\varphi) \mid_{x=0,1} = (D\theta) \mid_{x=0,1} = 0, \tag{4.3.8}$$

$$\varphi \mid_{t=0} = \varphi_0(x), \quad \theta \mid_{t=0} = \theta_0(x) > 0. \tag{4.3.9}$$

The boundary conditions (4.3.8) physically mean that there is no mass and heat flux through the boundary.

In this section we are concerned with the global existence and uniqueness of the solution. The main mathematical difficulties in proving the global existence and uniqueness come from several sources. As for the phase-field equations in Section 4.1.2, they come from the appearance of the term $\frac{1}{\theta}$, which may become singular, and also come from the high nonlinearity. Furthermore, equation (4.3.6) for the order parameter φ is a fourth-order parabolic equation instead of a second-order parabolic equation. It turns out that the maximum principle does not hold and at the very beginning we have a relatively weaker energy estimate ($\varphi_t \in L^2([0,T]; H^{-1})$) instead of $\varphi_t \in L^2([0,T]; L^2)$ for the phase-field equations. Again the key issue is to get uniform L^∞ bounds of θ and $\frac{1}{\theta}$. Now that it is a one-dimensional problem, we can use a different technique from that in Section 4.1.2. To overcome the difficulty caused by the weaker energy estimate of φ, we establish the following lemma (see Shen & Zheng [3]):

Lemma 4.3.1 *Suppose $y(t)$ and $h(t)$ are nonnegative functions, $y'(t)$ is locally integrable on $(0, +\infty)$ and $y(t), h(t)$ satisfy*

$$\frac{dy}{dt} \leq A_1 y^2 + A_2 + h(t), \quad \forall t \geq 0, \tag{4.3.10}$$

$$\int_0^T y(\tau)d\tau \leq A_3, \quad \int_0^T h(\tau)d\tau \leq A_4, \quad \forall T > 0 \tag{4.3.11}$$

with A_i ($i = 1, \cdots, 4$) being nonnegative constants independent of t, T. Then for any $r > 0$,

$$y(t+r) \leq \left(\frac{A_3}{r} + A_2 r + A_4\right) e^{A_1 A_3}, \quad \forall t \geq 0. \tag{4.3.12}$$

Moreover,

$$\lim_{t \to +\infty} y(t) = 0. \tag{4.3.13}$$

Proof. The proof of the first part is similar to that of the Uniform Gronwall Lemma (see Lemma 1.1 in R.Temam [1], p. 89). Assume $0 \leq t \leq s \leq t + r$ with any given $r > 0$. We multiply (4.3.10) by $exp(-\int_t^s A_1 y(\tau)d\tau)$ to obtain that

$$\frac{d}{ds}\left[y(s)e^{-\int_t^s A_1 y(\tau)d\tau}\right] \leq (A_2 + h(s))e^{-\int_t^s A_1 y(\tau)d\tau} \leq A_2 + h(s). \tag{4.3.14}$$

Then integration between s and $t + r$ gives

$$y(t+r) \leq y(s)e^{\int_s^{t+r} A_1 y(\tau)d\tau} + (A_2 r + A_4)e^{\int_t^{t+r} A_1 y(\tau)d\tau} \leq (y(s) + A_2 r + A_4)e^{A_1 A_3}. \tag{4.3.15}$$

Intergation of this inequality, with respect to s between t and $t + r$, gives (4.3.12). By (4.3.10) and (4.3.12), we have

$$\frac{dy}{dt} \leq A_1\left[\left(\frac{A_3}{r} + A_2 r + A_4\right)e^{A_1 A_3}\right]^2 + A_2 + h(t) = A_r + h(t), \quad \forall \ t \geq r, \tag{4.3.16}$$

where

$$A_r \overset{\text{def}}{=} A_1 \left[\left(\frac{A_3}{r} + A_2 r + A_4 \right) e^{A_1 A_3} \right]^2 + A_2. \tag{4.3.17}$$

To prove (4.3.13), we use a contradiction argument. Suppose it is not true. Then there exists a monotone increasing sequence $\{t_n\}$ and a constant $a > 0$ such that

$$t_n \geq r + \frac{a}{4A_r}, \quad t_{n+1} \geq t_n + \frac{a}{4A_r}, \quad \forall \ n \in I\!N, \tag{4.3.18}$$

$$\lim_{n \to +\infty} t_n = +\infty, \tag{4.3.19}$$

$$y(t_n) \geq \frac{a}{2} > 0, \quad \forall \ n \in I\!N. \tag{4.3.20}$$

On the other hand, we deduce from (4.3.16) that

$$y(t_n) - y(t) \leq A_r(t_n - t) + \int_t^{t_n} h(\tau) d\tau, \quad as \ \ t_n - \frac{a}{4A_r} \leq t < t_n. \tag{4.3.21}$$

Combining (4.3.20) and (4.3.21) yields

$$\frac{a}{2} - y(t) \leq y(t_n) - y(t) \leq \frac{a}{4} + \int_{t_n - \frac{a}{4A_r}}^{t_n} h(\tau) d\tau, \quad as \ \ t_n - \frac{a}{4A_r} \leq t < t_n. \tag{4.3.22}$$

Therefore,

$$y(t) + \int_{t_n - \frac{a}{4A_r}}^{t_n} h(\tau) d\tau \geq \frac{a}{4}, \quad as \ \ t_n - \frac{a}{4A_r} \leq t < t_n. \tag{4.3.23}$$

Let

$$n_T = \max\{n | n \in I\!N, \ r + \frac{a}{4A_r} \leq t_n \leq T\}. \tag{4.3.24}$$

Thus,

$$\lim_{T \to +\infty} n_T = +\infty. \tag{4.3.25}$$

It turns out from (4.3.23) that for any $T > 0$,

$$A_3 + \frac{aA_4}{4A_r} \geq \int_0^T y(\tau) d\tau + \frac{a}{4A_r} \int_0^T h(\tau) d\tau$$

$$\geq \sum_{1 \leq n \leq n_T} \left(\int_{t_n - \frac{a}{4A_r}}^{t_n} y(\tau) d\tau + \frac{a}{4A_r} \int_{t_n - \frac{a}{4A_r}}^{t_n} h(\tau) d\tau \right) \geq \frac{a^2}{16A_r} n_T, \tag{4.3.26}$$

a contradiction. Thus the proof is complete.   □

Remark 4.3.1 *This lemma plays an important role in deriving uniform a priori estimates of the solution to problem (4.3.6)–(4.3.9) and has many applications to other problems.*

The main results in this section are as follows (see Shen & Zheng [3]).

Theorem 4.3.1 *Suppose $\theta_0(x) \in H^3, \varphi_0(x) \in H^5$ satisfying the compatibility conditions: $(D\theta_0)|_{x=0,1} = (D\varphi_0)|_{x=0,1} = (D^3\varphi_0)|_{x=0,1} = 0$ and suppose $\theta_0(x) > 0$ in $[0,1]$. Then problem (4.3.6)–(4.3.9) admits a unique global classical solution $(\varphi(x,t), \theta(x,t))$ such that*

1.

$$\varphi \in C(I\!\!R^+; H^5) \bigcap C^1(I\!\!R^+; H^1), \quad \varphi_t \in L^2(I\!\!R^+; H^3);$$

$$\theta \in C(I\!\!R^+; H^3), \quad D\theta \in L^2(I\!\!R^+; H^3), \quad \theta_t \in C(I\!\!R^+; H^1) \bigcap L^2(I\!\!R^+; H^2);$$

$$\theta(x,t) > 0, \forall (x,t) \in [0,1] \times I\!\!R^+.$$

2. *Moreover, as time goes to infinity,*

$$\|\varphi_t\|_{H^1} \to 0, \quad \|D(-K_1 D^2\varphi + \frac{\frac{\partial f}{\partial \varphi}}{\theta})\|_{H^2} \to 0, \tag{4.3.27}$$

$$\|D\theta\|_{H^2} \to 0 \quad \|D(\frac{1}{\theta})\|_{H^2} \to 0, \quad \|\frac{1}{\theta_t}\|_{H^1} \to 0, \quad \|\theta_t\|_{H^1} \to 0. \tag{4.3.28}$$

Remark 4.3.2 *The smoothness requirement of the initial data in Theorem 4.3.1 is just for simplicity of exposition and can be significantly weakened to $\varphi_0 \in H^2, \theta_0 \in H^1$ using the smoothing property of the parabolic operators and the bootstrap argument. The detailed proof of this point will be given in Section 5.2. Multiplicity of equilibria and the convergence of (φ, θ) to an equilibrium, as time goes to infinity, will also be discussed in Chapter 5.*

We refer to Alt & Pawlow [1] for the existence of a weak solution in $\Omega \times (0, T)$ with $\Omega \subset I\!\!R^n, n \leq 3$ and with a different boundary condition for the temperature. We also refer to Kenmochi & Niezgodka [1–2] for the mathematical results in this direction.

Proof of Theorem 4.3.1. In what follows, without loss of generality, we always assume that $C_V = M_1 = 1$. Let

$$u = \frac{1}{\theta}. \tag{4.3.29}$$

Instead of problem (4.3.6)–(4.3.9), we study the following equivalent problem:

$$\varphi_t = D^2\left(-K_1 D^2\varphi + \varphi^3 - (a\varphi + b)u\right), \tag{4.3.30}$$

$$u_t + (a\varphi + b)u^2\varphi_t = M_2 u^2 D^2 u, \tag{4.3.31}$$

$$(Du)|_{x=0,1} = (D\varphi)|_{x=0,1} = (D^3\varphi)|_{x=0,1} = 0, \tag{4.3.32}$$

$$\varphi\,|_{t=0} = \varphi_0(x), \quad u\,|_{t=0} = u_0(x) = \frac{1}{\theta_0(x)} > 0. \tag{4.3.33}$$

The whole proof of Theorem 4.3.1 consists of several lemmas. We first establish the following local existence and uniqueness lemma.

Lemma 4.3.2 *Suppose $\varphi_0(x) \in H^5, u_0(x) \in H^3$ satisfying the compatibility conditions: $(Du_0)|_{x=0,1} = (D\varphi_0)|_{x=0,1} = (D^3\varphi_0)|_{x=0,1} = 0$, and $u_0(x) > 0$. Then there exists a positive constant t^* depending only on $\min\limits_{0 \le x \le 1} u_0(x), \|u_0\|_{H^3}, \|\varphi_0\|_{H^5}$ such that problem (4.3.30)–(4.3.33) admits a unique solution $(\varphi(x,t), u(x,t))$ in $[0,1] \times [0, t^*]$:*
$\varphi \in C([0, t^*]; H^5) \cap C^1([0, t^*]; H^1), \ \varphi_t \in L^2([0, t^*]; H^3), \ u \in C([0, t^*]; H^3) \cap L^2([0, t^*];$
$H^4), \ u_t \in C([0, t^*]; H^1) \cap L^2([0, t^*]; H^2), u(x,t) > 0, \forall(x,t) \in [0,1] \times [0, t^*].$
Moreover,

$$\|\varphi(t)\|_{H^5}^2 \le C_1, \quad \|u(t)\|_{H^3}^2 \le 2\|u_o\|_{H^3}^2, \ \ 0 \le t \le t^* \tag{4.3.34}$$

where C_1 depends only on $\|\varphi_0\|_{H^5}$ and $\|u_0\|_{H^3}$.

Sketch of the proof. As before , we use the contraction mapping theorem to prove this lemma. Let

$$
X_h(C_1, C_2) = \left\{ (\varphi, u) \; \middle| \;
\begin{aligned}
&\varphi \in C([0, h]; H^5) \cap C^1([0, h]; H^1), \varphi_t \in L^2([0, h]; H^3), \\
&u \in C([0, h]; H^3) \cap L^2([0, h]; H^4), \\
&u_t \in C([0, h]; H^1) \cap L^2([0, h]; H^2), \\
&(D\varphi)|_{x=0,1} = (D^3\varphi)|_{x=0,1} = (Du)|_{x=0,1} = 0, \\
&\varphi|_{t=0} = \varphi_0(x), \; u|_{t=0} = u_0(x), \\
&u(x, t) \geq \tfrac{1}{2} \min_{0 \leq x \leq 1} u_0(x) > 0, \\
&\|\varphi(t)\|_{H^5}^2 \leq C_1, \; \|\varphi_t\|_{H^1}^2 \leq 2C_1, \|u(t)\|_{H^3}^2 \leq 2\|u_0\|_{H^3}^2, \\
&\|u_t(t)\|_{H^1}^2 \leq C_2, \int_0^h \|u_t\|_{H^2} d\tau \leq C_2.
\end{aligned}
\right\},
$$

where C_1, C_2 are positive constants to be specified later.

For $(\tilde{\varphi}, \tilde{u}) \in X_h(C_1, C_2)$, we consider the following linear auxiliary problems for φ and u, respectively.

$$
\varphi_t + K_1 D^4 \varphi = \tilde{f}, \tag{4.3.35}
$$

$$
(D\varphi)|_{x=0,1} = (D^3\varphi)|_{x=0,1} = 0, \tag{4.3.36}
$$

$$
\varphi|_{t=0} = \varphi_0(x) \tag{4.3.37}
$$

with

$$
\tilde{f}(x, t) = D^2(\tilde{\varphi}^3 - (a\tilde{\varphi} + b)\tilde{u}) \tag{4.3.38}
$$

and

$$
u_t - \tilde{a} D^2 u + \tilde{g} u = 0, \tag{4.3.39}
$$

$$
(Du)|_{x=0,1} = 0, \tag{4.3.40}
$$

$$
u|_{t=0} = u_0(x) \tag{4.3.41}
$$

with

$$
\tilde{a} = M_2 \tilde{u}^2, \quad \tilde{g} = (a\tilde{\varphi} + b)\tilde{u}\tilde{\varphi}_t. \tag{4.3.42}
$$

We denote by $\mathcal{N}$ the mapping from $(\tilde{\varphi}, \tilde{u})$ to (φ, u). Applying the standard energy method yields that there exist constants C_1, C_2 and a positive t^* depending only on $\|u_0\|_{H^3}, \|\varphi_0\|_{H^5}, \min\limits_{0 \leq x \leq 1} u_0(x)$ such that the mapping $\mathcal{N}$ maps $X_{t^*}(C_1, C_2)$ into itself and is a strict contraction with the corresponding norm. Therefore, the assertion of Lemma 4.3.2 follows. Since the proof is quite standard and similar to those in the previous sections, we leave the details to the reader. $\square$

To complete the proof of Theorem 4.3.1, we now proceed to get uniform a priori estimates of the solution. Let

$$z = -K_1 D^2 \varphi + \varphi^3 - (a\varphi + b)u \tag{4.3.43}$$

which usually is called the chemical potential. In the following, we denote by $C_i, i \in \mathbb{N}$, positive constants depending only on the initial data.

Lemma 4.3.3 *For any $t > 0$, the following estimates hold*

$$\|\varphi\|_{H^1} \leq C_3, \tag{4.3.44}$$

$$\int_0^t (\|Dz\|^2 + \|Du\|^2) d\tau \leq C_3, \tag{4.3.45}$$

$$0 < C_4 \leq \int_0^1 \frac{1}{u} dx \leq C_5 < \infty. \tag{4.3.46}$$

Proof. Let

$$F = -\int_0^1 [\frac{1}{4}\varphi^4 + \ln u + \frac{K_1}{2}(D\varphi)^2] dx \tag{4.3.47}$$

which serves as the entropy functional. Multiplying (4.3.30) by z, (4.3.31) by u^{-1}, then integrating with respect to x and adding together, we get

$$-\frac{dF}{dt} + \|Dz\|^2 + M_2\|Du\|^2 = 0. \tag{4.3.48}$$

Integrating (4.3.48) with respect to t yields

$$\int_0^1 [\frac{K_1}{2}(D\varphi)^2 + \frac{1}{4}\varphi^4 + \ln u] dx + \int_0^t [\|Dz\|^2 + M_2\|Du\|^2] d\tau$$

$$= \int_0^1 [\frac{K_1}{2}(D\varphi_o)^2 + \frac{1}{4}\varphi_o^4 + \ln u_o] dx \stackrel{\text{def}}{=} \tilde{C}_6. \tag{4.3.49}$$

Although u is positive, $\int_0^1 \ln u dx$ is not necessarily positive. To get rid of this difficulty, multiplying (4.3.31) by u^{-2} and integrating with respect to x and t, we obtain

$$\frac{a}{2}\|\varphi\|^2 + b\int_0^1 \varphi dx - \int_0^1 \frac{1}{u} dx = \frac{a}{2}\|\varphi_0\|^2 + b\int_0^1 \varphi_0 dx - \int_0^1 \frac{1}{u_0} dx \overset{\text{def}}{=} \tilde{C}_7. \quad (4.3.50)$$

By Young's inequality $a \leq \frac{\varepsilon}{2}a^2 + \frac{1}{2\varepsilon}$, we have

$$\int_0^1 \frac{1}{u} dx = -\tilde{C}_7 + \frac{a}{2}\|\varphi\|^2 + b\int_0^1 \varphi dx \leq C_8 + \frac{1}{16}\int_0^1 \varphi^4 dx. \quad (4.3.51)$$

Combining (4.3.49) with (4.3.51) yields

$$\int_0^1 [\frac{K_1}{2}(D\varphi)^2 + \frac{1}{4}\varphi^4] dx + \int_0^t [\|Dz\|^2 + M_2\|Du\|^2] d\tau$$

$$= \tilde{C}_6 + \int_0^1 \ln \frac{1}{u} dx$$

$$\leq C_6 + \int_0^1 \frac{1}{u} dx \leq C_6 + C_8 + \frac{1}{16}\int_0^1 \varphi^4 dx. \quad (4.3.52)$$

Thus (4.3.44) and (4.3.45) follow. Therefore, it follows from (4.3.51) (4.3.52) that

$$\int_0^1 \ln u dx \leq C_9, \quad \int_0^1 \frac{1}{u} dx \leq C_5. \quad (4.3.53)$$

It remains to bound $\int_0^1 \frac{1}{u} dx$ from below. For this purpose, we apply the Jensen inequality to the convex function $-\ln y$ to obtain

$$-\ln \left(\int_0^1 \frac{1}{u} dx\right) \leq -\int_0^1 \ln \frac{1}{u} dx \leq C_9. \quad (4.3.54)$$

$$\int_0^1 \frac{1}{u} dx \geq e^{-C_9} \overset{\text{def}}{=} C_4 > 0. \quad (4.3.55)$$

Thus the proof is complete. $\qquad\qquad\qquad\qquad\qquad\qquad\qquad\qquad\qquad\qquad\qquad\quad \square$

Notice that estimate (4.3.46) only gives the L^1 norm bound of θ. Our task is to get the L^∞ norm estimate of θ (and u) and also the estimates of higher-order derivatives. A similar situation has already occurred in Section 4.1 and Section 4.2. For this one-dimensional problem, we now use a different technique to get the L^∞ norm estimate of u. For any $t > 0$, let

$$m(t) = \min_{0 \leq x \leq 1} u(x,t), \quad M(t) = \max_{0 \leq x \leq 1} u(x,t). \quad (4.3.56)$$

It follows from (4.3.46) that

$$m(t) \leq \frac{1}{C_4}. \tag{4.3.57}$$

For any $t > 0$, let $x^*(t)$ and $x_*(t)$ be the points in which $u(x,t)$ achieves its maximum and minimum, respectively. Then we have

$$M(t) - m(t) = \int_{x_*}^{x^*} (Du)dx \leq \int_0^1 |Du|dx \leq \|Du\|. \tag{4.3.58}$$

Combining (4.3.58) with (4.3.57) yields

$$M(t) \leq m(t) + \|Du\| \leq \frac{1}{C_4} + \|Du\|. \tag{4.3.59}$$

Lemma 4.3.4 *For any $t \geq 0$, the following estimates hold.*

$$\|Du(t)\| \leq C_{10}, \quad \|Dz(t)\| \leq C_{10}, \tag{4.3.60}$$

$$\int_0^t \|D\varphi_t\|^2 d\tau \leq C_{10}, \quad \int_0^t \left\|\frac{u_t}{u}\right\|^2 d\tau \leq C_{10}, \quad \int_0^t \|\varphi\varphi_t\|^2 d\tau \leq C_{10}, \tag{4.3.61}$$

$$\int_0^t \|u_t\|^2 d\tau \leq C_{10}, \quad \int_0^t \|\varphi_t\|^2 d\tau \leq C_{10}, \tag{4.3.62}$$

$$\|\varphi(t)\|_{H^3} \leq C_{10}, \quad \|u(t)\|_{L^\infty} \leq C_{10}. \tag{4.3.63}$$

Moreover,

$$\lim_{t \to +\infty} (\|Du\|^2 + \|Dz\|^2) = 0. \tag{4.3.64}$$

Proof. First, we prove (4.3.60) for $t \geq t^*$ with t^* a positive constant indicated in Lemma 4.3.2. Differentiating (4.3.43) with respect to t, then multiplying it by $-\varphi_t = -D^2 z$, we obtain

$$-D^2 z z_t = \left(K_1 D^2 \varphi_t - 3\varphi^2 \varphi_t + (a\varphi + b)u_t + a\varphi_t u\right) \varphi_t. \tag{4.3.65}$$

Multiplying (4.3.31) by $\dfrac{u_t}{u^2}$, we get

$$\frac{u_t^2}{u^2} + (a\varphi + b)\varphi_t u_t = M_2 D^2 u u_t. \tag{4.3.66}$$

Adding (4.3.65), (4.3.66) together, then integrating with respect to x yields

$$\frac{1}{2}\frac{d}{dt}\left[M_2\|Du\|^2 + \|Dz\|^2\right] + K_1\|D\varphi_t\|^2 + 3\int_0^1 \varphi^2\varphi_t^2 dx + \left\|\frac{u_t}{u}\right\|^2$$

$$= \int_0^1 au\varphi_t^2 dx. \tag{4.3.67}$$

It turns out from $\varphi_t = D^2 z$ and the boundary conditions $(D\varphi)\mid_{x=0,1} = (Dz)\mid_{x=0,1} = 0$
that

$$\|\varphi_t\|^2 = \int_0^1 (D^2 z)^2 dx = -\int_0^1 (D(D^2 z))Dz dx \le \|D\varphi_t\|\|Dz\|. \tag{4.3.68}$$

By (4.3.59), we can estimate the superlinear term on the right-hand side of (4.3.67)
as follows:

$$\int_0^1 au\varphi_t^2 dx \le aM(t)\|\varphi_t\|^2 \le \left(a\|Du\| + \frac{a}{C_4}\right)\|\varphi_t\|^2. \tag{4.3.69}$$

Combining (4.3.67)–(4.3.69) yields

$$\frac{1}{2}\frac{d}{dt}\left[M_2\|Du\|^2 + \|Dz\|^2\right] + K_1\|D\varphi_t\|^2 + 3\int_0^1 \varphi^2\varphi_t^2 dx + \left\|\frac{u_t}{u}\right\|^2$$

$$\le a\|D\varphi_t\|\|Dz\|\|Du\| + \frac{a}{C_4}\|D\varphi_t\|\|Dz\|$$

$$\le \left(\frac{a}{C_4} + a\right)\frac{\varepsilon}{2}\|D\varphi_t\|^2 + \frac{a}{C_4}\frac{1}{2\varepsilon}\|Dz\|^2 + \frac{a}{2\varepsilon}\|Du\|^2\|Dz\|^2. \tag{4.3.70}$$

Taking $\varepsilon = \dfrac{K_1 C_4}{a(1+C_4)}$ and applying Young's inequality, we get

$$\frac{d}{dt}\left[M_2\|Du\|^2 + \|Dz\|^2\right] + K_1\|D\varphi_t\|^2 + 6\int_0^1 \varphi^2\varphi_t^2 dx + 2\left\|\frac{u_t}{u}\right\|^2$$

$$\le C_{11}\left(M_2\|Du\|^2 + \|Dz\|^2\right)^2 + C_{12}. \tag{4.3.71}$$

Let $y(t) = M_2\|Du\|^2 + \|Dz\|^2$. Applying Lemma 4.3.1 yields (4.3.60) for $t \ge t^*$ and
(4.3.64). Combining it with the local existence results yields (4.3.60) for $t \ge 0$. Integrating (4.3.70) with respect to t and applying (4.3.60) and (4.3.45), we obtain
(4.3.61). The estimates (4.3.62) follow from (4.3.59)–(4.3.60) and (4.3.44). The estimates in (4.3.63) follow easily from (4.3.59), (4.3.43) and the elliptic regularity
estimates.   □

Lemma 4.3.5 *For $t \geq 0$, the following estimates hold:*

$$\|\varphi_t(t)\| \leq C_{13}, \quad \|\varphi_t(t)\| \to 0, \quad as \ t \to +\infty, \qquad (4.3.72)$$

$$\int_0^t \|D^2 \varphi_t(\tau)\|^2 d\tau \leq C_{13}. \qquad (4.3.73)$$

Proof. Differentiating (4.3.30) with respect to t, then multiplying it by φ_t, integrating with respect to x and integrating by parts yield:

$$\frac{1}{2} \frac{d}{dt} \|\varphi_t\|^2 + K_1 \|D^2 \varphi_t\|^2 + 3 \|\varphi D \varphi_t\|^2 = 3 \int_0^1 \varphi_t^2 (D\varphi)^2 dx$$

$$+ 3 \int_0^1 \varphi \varphi_t^2 D^2 \varphi \, dx - \int_0^1 (au_t \varphi + au\varphi_t + bu_t) D^2 \varphi_t dx. \qquad (4.3.74)$$

It turns out from Young's inequality and (4.3.63) that

$$\left| \int_0^1 (au_t \varphi + bu_t) D^2 \varphi_t dx \right| \leq \frac{\varepsilon}{2} \|D^2 \varphi_t\|^2 + C_\varepsilon \|u_t\|^2, \qquad (4.3.75)$$

$$\left| \int_0^1 au\varphi_t D^2 \varphi_t dx \right| \leq \frac{\varepsilon}{2} \|D^2 \varphi_t\|^2 + C_\varepsilon \|\varphi_t\|^2, \qquad (4.3.76)$$

$$\left| 3 \int_0^1 \varphi_t^2 |D\varphi|^2 dx \right| \leq C \|\varphi_t\|^2 \qquad (4.3.77)$$

with C_ε being a constant depending on ε.

Choosing $\varepsilon = \frac{K_1}{2}$, then using (4.3.62) and applying Lemma 4.3.1 with $y(t) = \|\varphi_t\|^2$, we obtain (4.3.72). The estimate (4.3.73) follows from integrating (4.3.74) with respect to t. $\qquad \square$

The following lemma gives the crucial L^∞ norm estimate of θ.

Lemma 4.3.6 *For $t \geq 0$, the following estimate holds:*

$$\|\theta(t)\|_{L^\infty} \leq C_{14}. \qquad (4.3.78)$$

Proof. It follows from (4.3.7) with $C_V = 1$ that $\theta = \frac{1}{u}$ also satisfies

$$\theta^3 \theta_t - M_2 \theta D^2 \theta + 2M_2 (D\theta)^2 = (a\varphi + b)\theta^3 \varphi_t. \qquad (4.3.79)$$

Integrating with respect to x yields

$$\frac{1}{4}\frac{d}{dt}\int_0^1 \theta^4 dx + 3M_2\|D\theta\|^2 = \int_0^1 (a\varphi + b)\theta^3 \varphi_t dx. \tag{4.3.80}$$

By the Nirenberg inequality, we have

$$\|\theta\|_{L^4}^4 \leq \tilde{C}_1 \|D\theta\|^2 \|\theta\|_{L^1}^2 + \tilde{C}_2 \|\theta\|_{L^1}^4. \tag{4.3.81}$$

Then we deduce from (4.3.80), (4.3.81), (4.3.46) and the Young inequality that

$$\frac{d}{dt}\|\theta\|_{L^4}^4 + C_{15}\|\theta\|_{L^4}^4 \leq C_{16} + \varepsilon\|\theta\|_{L^4}^4 + C_\varepsilon\|\varphi_t\|_{L^4}^4. \tag{4.3.82}$$

Choosing $\varepsilon = \frac{C_{15}}{2}$ and appltying Nirenberg's inequality again yields

$$\frac{d}{dt}\|\theta\|_{L^4}^4 + \frac{C_{15}}{2}\|\theta\|_{L^4}^4 \leq C_{16} + C_{17}(\|D\varphi_t\|\|\varphi_t\|^3 + \|\varphi_t\|^4). \tag{4.3.83}$$

By (4.3.72), (4.3.61), (4.3.62) and the Gronwall inequality, we deduce from (4.3.83) that

$$\|\theta(t)\|_{L^4}^4 \leq C_{18}, \quad for \ t \geq 0. \tag{4.3.84}$$

For $t \geq 0$, let $\tilde{x}(t)$ be the point in which θ achieves its minimum. Then

$$\theta(x,t) - \theta(\tilde{x},t) = \int_{\tilde{x}}^x D\theta dx. \tag{4.3.85}$$

By (4.3.46), (4.3.60) and (4.3.85), we have

$$\|\theta(t)\|_{L^\infty} \leq \theta(\tilde{x},t) + \int_0^1 |D\theta| dx \leq \int_0^1 \theta dx + \int_0^1 |D\theta| dx$$

$$\leq C_5 + \int_0^1 |D(\frac{1}{u})| dx = C_5 + \int_0^1 \left|\frac{Du}{u^2}\right| dx$$

$$\leq C_5 + \left(\int_0^1 |Du|^2 dx\right)^{\frac{1}{2}} \left(\int_0^1 \theta^4 dx\right)^{\frac{1}{2}} \leq C_{14}. \tag{4.3.86}$$

Thus the proof is complete. □

After getting the uniform boundedness of the L^∞ norm of u and θ, we can easily proceed to obtain the estimates of higher-order derivatives.

Lemma 4.3.7 *For $t \geq 0$, the following estimates hold:*

$$\|u_t(t)\| \leq C_{19}, \ \|D\varphi_t(t)\| \leq C_{19}, \tag{4.3.87}$$

$$\int_0^t \|Dz_t\|^2 d\tau \leq C_{19}, \ \int_0^t \|Du_t\|^2 d\tau \leq C_{19}, \tag{4.3.88}$$

$$\|\varphi(t)\|_{H^5} \leq C_{19}, \ \|\varphi_t\|_{L^\infty} \leq C_{19}. \tag{4.3.89}$$

Proof. We first differentiate (4.3.43) with respect to t, then multiply it by $D^2 z_t = \varphi_{tt}$. We also multiply (4.3.31) by u^{-2} and differentiate the resulting equation with respect to t, then multiply it by u_t. Finally, adding them together and integrating with respect to x and t, we obtain

$$\int_0^t \|Dz_t\|^2 d\tau + \frac{K_1}{2}\|D\varphi_t(t)\|^2 + M_2 \int_0^t \|Du_t\|^2 d\tau + \frac{1}{2}\left\|\frac{u_t}{u}(t)\right\|^2$$

$$= C_{20} + \int_0^t \int_0^1 \varphi_{tt}G\,dx\,d\tau + \int_0^t \int_0^1 \frac{u_t^3}{u^3}dx\,d\tau + \int_0^t \int_0^1 a\varphi_t^2 u_t\,dx\,d\tau \tag{4.3.90}$$

where

$$G = (-3\varphi^2 + au)\varphi_t. \tag{4.3.91}$$

Now we estimate the last three terms on the right hand side of (4.3.90). Using integration by parts, applying the Nirenberg inequality and Lemma 4.3.4, we have

$$\left|\int_0^t \int_0^1 \varphi_{tt}G\,dx\,d\tau\right| = \left|\int_0^t \int_0^1 D^2 z_t G\,dx\,d\tau\right| = \left|\int_0^t \int_0^1 Dz_t DG\,dx\,d\tau\right|$$

$$\leq \frac{1}{2}\int_0^t \|Dz_t\|^2 d\tau + C_{21}\int_0^t \left(\|\varphi_t\|^2 + \|D\varphi_t\|^2\right)d\tau + C_{21}\int_0^t \|Du\|^2\|\varphi_t\|_{L^\infty}^2 d\tau$$

$$\leq \frac{1}{2}\int_0^t \|Dz_t\|^2 d\tau + C_{22} + C_{22}\int_0^t \left(\|\varphi_t\|\|D\varphi_t\| + \|\varphi_t\|^2\right)d\tau$$

$$\leq \frac{1}{2}\int_0^t \|Dz_t\|^2 d\tau + C_{23}, \tag{4.3.92}$$

$$\left|\int_0^t \int_0^1 \frac{u_t^3}{u^3}dx\,d\tau\right| = \left|\int_0^t \int_0^1 u_t^3 \theta^3\,dx\,d\tau\right|$$

$$\le C_{24} \int_0^t \|u_t\|_{L^3}^3 d\tau \le C_{25} \int_0^t \left(\|Du_t\|^{\frac{1}{2}} \|u_t\|^{\frac{5}{2}} + \|u_t\|^3 \right) d\tau$$

$$\le C_{25} \left(\int_0^t \|Du_t\|^2 d\tau \right)^{\frac{1}{4}} \left(\int_0^t \|u_t\|^{\frac{10}{3}} d\tau \right)^{\frac{3}{4}} + C_{25} \max_{0\le\tau\le t} \|u_t\| \int_0^t \|u_t\|^2 d\tau$$

$$\le C_{25} \max_{0\le\tau\le t} \|u_t\| \left(\int_0^t \|Du_t\|^2 d\tau \right)^{\frac{1}{4}} \left(\int_0^t \|u_t\|^2 d\tau \right)^{\frac{3}{4}} + C_{25} \max_{0\le\tau\le t} \|u_t\| \int_0^t \|u_t\|^2 d\tau$$

$$\le \varepsilon \max_{0\le\tau\le t} \|u_t\|^2 + \varepsilon \int_0^t \|Du_t\|^2 d\tau + C_{26} \tag{4.3.93}$$

with ε being a positive constant to be specified later.

$$\left| \int_0^t \int_0^1 a\varphi_t^2 u_t \, dx d\tau \right| \le a \int_0^t \|u_t\| \|\varphi_t\|_{L^4}^2 d\tau$$

$$\le C_{27} \int_0^t \left(\|D\varphi_t\|^{\frac{1}{2}} \|\varphi_t\|^{\frac{3}{2}} + \|\varphi_t\|^2 \right) \|u_t\| d\tau$$

$$\le C_{27} \max_{0\le\tau\le t} \|u_t\| \left(\left(\int_0^t \|D\varphi_t\|^2 d\tau \right)^{\frac{1}{4}} \left(\int_0^t \|\varphi_t\|^2 d\tau \right)^{\frac{3}{4}} + \int_0^t \|\varphi_t\|^2 d\tau \right)$$

$$\le \varepsilon \max_{0\le\tau\le t} \|u_t\|^2 + C_{28}. \tag{4.3.94}$$

It follows from (4.3.63) that

$$\left\| \frac{u_t}{u} \right\|^2 \ge \frac{1}{C_{10}^2} \|u_t\|^2. \tag{4.3.95}$$

Combining (4.3.90), (4.3.92)–(4.3.95), choosing $\varepsilon = \min(\frac{M_2}{2}, \frac{1}{8C_{10}^2})$ and taking the supremum with respect to t in (4.3.90), we arrive at (4.3.87), (4.3.88). The estimates of (4.3.89) follow from the elliptic regularity theorem and the Sobolev imbedding theorem.   □

Remark 4.3.3 *We can also deduce from the proof of Lemma 4.3.7 and Lemma 4.3.1 with $y(t) = \frac{K_1}{2} \|D\varphi_t\|^2 + \frac{1}{2} \|\frac{u_t}{u}\|^2$ that*

$$\|D\varphi_t(t)\| \to 0, \quad \|u_t(t)\| \to 0, \text{ as } t \to \infty. \tag{4.3.96}$$

Lemma 4.3.8 *For $t \geq 0$, the following estimates hold:*

$$\|u(t)\|_{H^3} \leq C_{29}, \quad \|Du_t(t)\| \leq C_{29}, \tag{4.3.97}$$

$$\|Du_t(t)\| \to 0, \quad as \ t \to \infty, \tag{4.3.98}$$

$$\int_0^t \|Du\|_{H^3}^2 d\tau \leq C_{29}. \tag{4.3.99}$$

Proof. It easily follows from (4.3.62), (4.3.73), (4.3.89) that for $t \geq 0$,

$$\int_0^t \|D^3 \varphi_t\|^2 d\tau \leq C_{30}, \quad \int_0^t \|Du\|_{H^2}^2 d\tau \leq C_{30}. \tag{4.3.100}$$

Owing to the uniform estimates of the L^∞ norm of u and θ, equation (4.3.31) is a nondegenerate second-order parabolic equation

$$u_t - M_2 u^2 D^2 u = g \tag{4.3.101}$$

with

$$g = -(a\varphi + b)u^2 \varphi_t. \tag{4.3.102}$$

The previous estimates in Lemmas 4.3.2–4.3.7 show that $D^3 g$ and $D^3 u$ are bounded in $L^2(\mathbb{R}^+; L^2)$. Acting with D^3 on (4.3.101), then multiplying the resulting equation by $D^3 u$ and integrating over $(0, 1)$, we obtain

$$\frac{1}{2} \frac{d}{dt} \|D^3 u\|^2 + M_2 \int_0^1 u^2 |D^4 u|^2 dx$$

$$= -M_2 \int_0^1 D^4 u (4u Du D^3 u + 2|Du|^2 D^2 u + 2u|D^2 u|^2) dx$$

$$+ \int_0^1 D^3 u D^3 g \, dx. \tag{4.3.103}$$

It can be seen from the previous estimates in Lemmas 4.3.2–4.3.7 and (4.3.101) that

$$\|D^2 u(t)\| \leq C_{31}, \quad \forall t \geq 0. \tag{4.3.104}$$

Thus it follows from (4.3.103) and the Nirenberg inequality that

$$\frac{1}{2}\frac{d}{dt}\|D^3u\|^2 + C_{32}\|D^4u\|^2$$

$$\leq \varepsilon\|D^4u\|^2 + C_{33}\|D^3u\|^2 + C_{33} + C_{33}\|D^3u\|\|D^2u\|^3 + \frac{1}{2}\|D^3u\|^2 + \frac{1}{2}\|D^3g\|^2$$

$$\leq \varepsilon\|D^4u\|^2 + C_{34}\|D^3u\|^4 + C_{34} + \frac{1}{2}\|D^3g\|^2. \tag{4.3.105}$$

Choosing $\varepsilon = \frac{C_{32}}{2}$ and applying Lemma 4.3.1 with $y(t) = \|D^3u\|^2$ yields

$$\|D^3u(t)\| \rightarrow 0, \quad as \ t \rightarrow \infty, \tag{4.3.106}$$

$$\int_0^t \|D^4u\|^2 d\tau \leq C_{35}, \quad \forall t \geq 0. \tag{4.3.107}$$

Thus (4.3.97)–(4.3.99) easily follow. The proof is complete. □

Owing to Lemma 4.3.2 and the uniform a priori estimates established in Lemmas 4.3.3–4.3.8, we deduce the global existence and uniqueness of smooth solutions and (4.3.27), (4.3.28). Thus the proof of Theorem 4.3.1 is complete. □

Chapter 5

Stationary Problems and

Asymptotic Behaviour

In this chapter we are concerned with the asymptotic behaviour of solutions to the phase-field equations and coupled Cahn–Hilliard equations as $t \to \infty$.

In Section 5.1 we will consider an unusual type of nonlinear elliptic boundary value problem with nonlocal terms and constraints arising from the study of stationary problems of the coupled Cahn–Hilliard equations and also of the phase-field equations. A new technique is introduced to deal with problems in one space dimension. The problems are reduced to finding intersection points of two analytic functions of one variable. It turns out that in one space dimension the conclusion about the precise number of solutions or at most a countable number of solutions is obtained.

In Section 5.2 we will combine the results obtained in Chapter 4 and in Section 5.1 to conclude the convergence of solution of the corresponding evolution equations to an equilibrium as $t \to \infty$ in one space dimension.

The final section of this chapter is devoted to the study of infinite-dimensional dynamical systems associated with the phase-field equations. The results on the existence of a global attractor of finite Hausdorff dimension and the results on the existence of inertial manifolds and inertial sets are presented. Since the phase-field equations are not diagonal parabolic system. It turns out that the existing theory about existence of inertial manifolds stated in Temam [1] is not directly applicable.

The theory developed by Sell & You [1] about the existence of inertial manifolds requires some information concerning the distribution of the spectrum which, in our case, is not easy to verify. We use the so-called symmetrizer technique to convert the problem to one with a self-adjoint operator being the major linear part. Then the results on the existence of inertial manifolds are obtained by using the theory stated in Temam [1].

5.1   Nonlinear BVP with Nonlocal Terms and Constraints

Let us look at the coupled Cahn–Hilliard equations first (see Shen & Zheng [3] and Novick-Cohen & Zheng [1]). It easily follows from (4.3.6)–(4.3.9) that $\int_0^1 \varphi dx$ and $\frac{a}{2}\|\varphi\|^2 + b\int_0^1 \varphi dx - \int_0^1 \theta dx$ are conserved for all $t \geq 0$. The first one is simply the conservation of mass. Then the corresponding stationary problem to (4.3.6)–(4.3.9) reads

$$-K_1 D^2 \psi + \psi^3 - \frac{a\psi}{\theta} = \sigma, \tag{5.1.1}$$

$$\bar{\theta} = Const., \tag{5.1.2}$$

$$D\psi|_{x=0,1} = 0, \tag{5.1.3}$$

$$\int_0^1 \left(\bar{\theta} - \frac{a}{2}\psi^2 + b\psi \right) dx = \int_0^1 \left(\theta_0 - \frac{a}{2}\varphi_0^2 + b\varphi_0 \right) dx, \tag{5.1.4}$$

$$\int_0^1 \psi dx = \int_0^1 \varphi_0 dx \tag{5.1.5}$$

where σ is a constant to be determined.

We look for a function $\psi(x)$ and constants $\bar{\theta} > 0$ and σ such that (5.1.1)–(5.1.5) are satisfied. Let

$$M_0 = \int_0^1 \varphi_0 dx, \ M_1 = \int_0^1 \theta_0 dx, \ M_2 = \frac{a}{2}\int_0^1 \varphi_0^2 dx, \ M_3 = \frac{M_1 - M_2}{a}. \tag{5.1.6}$$

By (5.1.2), (5.1.4) and (5.1.5), we have

$$\bar{\theta} = \frac{a}{2}\|\psi\|^2 + M_1 - M_2. \tag{5.1.7}$$

Substituting (5.1.7) into (5.1.1) and using (5.1.2), we obtain

$$- K_1 D^2 \psi + \psi^3 - \frac{a}{\frac{a}{2}\|\psi\|^2 + M_1 - M_2} \psi = \sigma \tag{5.1.8}$$

with

$$\sigma = \int_0^1 \psi^3 dx - \frac{a M_0}{\frac{a}{2}\|\psi\|^2 + M_1 - M_2}. \tag{5.1.9}$$

Then problem (5.1.1)–(5.1.5) is equivalent to solving the problem (P): (5.1.8)–(5.1.9), (5.1.3) and (5.1.5). This is a nonlinear boundary value problem with nonlocal terms and constraints. A similar situation also occurs for the phase-field equations (see Elliott & Zheng [2] and Luckhaus & Zheng [1]).

It is easy to see that problem (P) always has a trivial solution:

$$\psi = M_0, \quad \bar\theta = \frac{a M_0^2}{2} + M_1 - M_2, \quad \sigma = \psi^3 - \frac{a}{\theta}\psi. \tag{5.1.10}$$

We are interested in the existence and multiplicity of nontrivial solutions.

In what follows we use the 'time map' and the method developed in Zheng [13], Luckhaus & Zheng [1] and Novick-Cohen & Peletier [1] to discuss the existence and multiplicity of nontrivial solutions to Problem (P). The basic idea of the method is to convert the problem to finding intersection points of two analytic functions of one variable. We refer to Novick-Cohen & Zheng [1] for more results about the bifurcation. We also refer to Grinfeld & Novick-Cohen [1] for related results on the Cahn–Hilliard equation.

The main results in this section are as follows.

Theorem 5.1.1 (i) *If $M_3 \geq \dfrac{1}{\pi^2}$, then problem (5.1.1)–(5.1.5) has no nontrivial solutions.*

(ii) *If $M_3 \leq -\dfrac{M_0^2}{2}$, then problem (5.1.1)–(5.1.5) must have an infinitely countable number of nontrivial solutions.*

(iii) *If $-\dfrac{M_0^2}{2} < M_3 < \dfrac{1}{\pi^2}$, then problem (5.1.1)–(5.1.5) has at most an infinitely countable number of nontrivial solutions. More precisely, there is $M_3^* \in [-\dfrac{M_0^2}{2},$*

$\frac{1}{\pi^2}$) depending on M_0 such that when $M_3 \leq M_3^*$, there are an infinitely countable

number of nontrivial solutions; when $M_3^* < M_3 < \frac{1}{\pi^2}$, there are only a finite

number of nontrivial solutions.

Remark 5.1.1 *In the special case $M_0 = 0$, Theorem 5.1.1 has been proved in Shen
and Zheng [3]:*

(i) *If $M_3 \geq \frac{1}{\pi^2}$, then problem (5.1.1)–(5.1.5) has no nontrivial solutions.*

(ii) *If $0 < M_3 < \frac{1}{\pi^2}$, then problem (5.1.1)–(5.1.5) has exactly $2k_0$ nontrivial solutions,
where $k_0 \in \mathbb{N}$ with $\frac{1}{\pi\sqrt{M_3}} - 1 \leq k_0 < \frac{1}{\pi\sqrt{M_3}}$.*

(iii) *If $M_3 \leq 0$, then problem (5.1.1)–(5.1.5) has an infinitely countable number of
solutions.*

Proof of Theorem 5.1.1. Without loss of generality, we always assume that $K_1 = 1$.
In order to eleminate the nonlocal term appearing in the equation, we make the
following change of variables (see also Luckhaus & Zheng [1]):

$$L = \sqrt{\frac{a}{\theta}}, \quad x' = Lx, \quad u(x') = \frac{\psi(x)}{L} \tag{5.1.11}$$

and in the following we still denote x' by x. Then problem (5.1.1)–(5.1.5) is reduced
to

$$-D^2u + u^3 - u = \sigma, \quad x \in (0, L) \tag{5.1.12}$$

$$Du|_{x=0,L} = 0, \tag{5.1.13}$$

$$\int_0^L u\,dx = M_0, \tag{5.1.14}$$

$$\frac{a}{L^2} = \bar{\theta} = aM_3 + \frac{aL}{2}\int_0^L u^2 dx. \tag{5.1.15}$$

where a, M_0, M_3 are given constants; σ, L are constants to be determined along with
the solution u.

Notice that (5.1.12)–(5.1.14) is similar to the stationary problem for the Cahn–
Hilliard equation (see Carr, Gurtin & Slemrod [1] and Zheng [7]). However, in the

present case the constant L is not given and has to be determined along with the solution u and another constant σ. On the other hand, we have the additional constraint (5.1.15). Let

$$f(u; \sigma) = u^3 - u - \sigma \tag{5.1.16}$$

and

$$F(u; \sigma, b) = \frac{1}{4}u^4 - \frac{1}{2}u^2 - \sigma u - b. \tag{5.1.17}$$

For $-\frac{2\sqrt{3}}{9} < \sigma < \frac{2\sqrt{3}}{9}$, $f(u; \sigma)$ has three roots. We denote the three roots of $f(u; \sigma) = 0$ by $w_0(\sigma), w_1(\sigma), w_2(\sigma)$ with

$$w_0(\sigma) \leq w_1(\sigma) \leq w_2(\sigma). \tag{5.1.18}$$

Notice that, as indicated in Carr, Gurtin & Slemlod [1] and Zheng [7], in the (σ, b) plane, $\Delta = (\sigma, b)$ is defined in the admissible region Σ which is bounded by

$$\begin{cases} \partial_1\Sigma = \{(\sigma, b) \mid \ b = \Phi_\sigma(w_0(\sigma)), \ 0 \leq \sigma < \frac{2}{3\sqrt{3}}\}, \\ \partial_2\Sigma = \{(\sigma, b) \mid \ b = \Phi_\sigma(w_2(\sigma)), \ -\frac{2}{3\sqrt{3}} < \sigma \leq 0\}, \\ \partial_3\Sigma = \{(\sigma, b) \mid \ b = \Phi_\sigma(w_1(\sigma)), \ -\frac{2}{3\sqrt{3}} \leq \sigma \leq \frac{2}{3\sqrt{3}}\} \end{cases} \tag{5.1.19}$$

where

$$\Phi_\sigma(w) = \frac{1}{4}w^4 - \frac{1}{2}w^2 - \sigma w. \tag{5.1.20}$$

When $\Delta = (\sigma, b) \in \Sigma, F(u; \sigma, b) = 0$ has four roots. We denote the four roots of $F(\sigma, b) = 0$ by $u_0(\sigma, b), u_1(\sigma, b), u_2(\sigma, b), u_3(\sigma, b)$ such that

$$u_0 \leq w_0 \leq u_1 \leq w_1 \leq u_2 \leq w_2 \leq u_3. \tag{5.1.21}$$

$$F(u; \sigma, b) > 0, \quad only \ for \ u_1 < u < u_2. \tag{5.1.22}$$

Multiplying (5.1.12) by Du and then integrating with respect to x yields the first integral:

$$-\frac{1}{2}(Du)^2 + F(u; \sigma, b) = 0. \tag{5.1.23}$$

Thus,

$$dx = \pm\frac{du}{\sqrt{2F(u; \sigma, b)}}. \tag{5.1.24}$$

It follows from (5.1.22)–(5.1.23) that the solution u of problem (5.1.12)–(5.1.15) must stay in $[u_1, u_2]$ and when $u = u_1$ (or u_2), $Du = 0$. On the other hand, since $(Du)|_{x=0,L} = 0$, u must take u_1 (or u_2) on the boundary. Let

$$I_0(\sigma, b) = \int_{u_1}^{u_2} \frac{ds}{\sqrt{2F(s; \sigma, b)}} \tag{5.1.25}$$

which, as in Schaaf [1], is called the 'time map'. If we think of x as the time variable, then by (5.1.24) and (5.1.25), I_0 indicates the time which a solution needs to proceed in the phase plane from $(u, u') = (u_1, 0)$ to the next intersection point $(u, u') = (u_2, 0)$. Thus, any solution u must consist of k ($k \in \mathbb{N}$) pieces of monotone branches ranging from u_1 (or u_2) to u_2 (u_1). Therefore,

$$k I_0(\sigma, b) = L. \tag{5.1.26}$$

We define

$$I_1(\sigma, b) = \int_{u_1}^{u_2} \frac{s\,ds}{\sqrt{2F(s; \sigma, b)}} \tag{5.1.27}$$

and

$$I_2(\sigma, b) = \int_{u_1}^{u_2} \frac{s^2\,ds}{\sqrt{2F(s; \sigma, b)}}. \tag{5.1.28}$$

Then it follows from (5.1.24) that (5.1.14) and (5.1.15) turn out to be

$$k I_1(\sigma, b) = M_0, \tag{5.1.29}$$

$$\frac{a}{L^2} = M_1 - M_2 + \frac{aLk I_2(\sigma, b)}{2}. \tag{5.1.30}$$

Substituting L of (5.1.26) into (5.1.30) yields

$$\frac{1}{k^2 I_0^2} - \frac{k^2 I_0 I_2}{2} = M_3. \tag{5.1.31}$$

Thus problem (5.1.12)–(5.1.15) is reduced to finding $(\sigma, b) \in \Sigma$ and $k \in \mathbb{N}$ such that (5.1.29), (5.1.31) are satisfied.

It is known (see Schaaf [1], Shen & Zheng [3]) that $I_i(\sigma, b)$ ($i = 0, 1, 2$) are analytic functions of $\Delta = (\sigma, b)$ in Σ. Moreover, we have the following limiting behaviour.

Lemma 5.1.1

$$\lim_{\Delta \to \partial_3 \Sigma} I_0 = \frac{\pi}{\sqrt{1 - 3w_1^2(\sigma)}}, \quad \lim_{\Delta \to \partial_i \Sigma} I_0 = \infty, \quad (i = 1, 2), \qquad (5.1.32)$$

$$\lim_{\Delta \to \partial_3 \Sigma} I_1 = \frac{\pi w_1(\sigma)}{\sqrt{1 - 3w_1^2(\sigma)}}, \qquad (5.1.33)$$

$$\lim_{\Delta \to \partial_3 \Sigma} I_2 = \frac{\pi w_1^2}{\sqrt{1 - 3w_1^2(\sigma)}}, \quad \lim_{\Delta \to \partial_i \Sigma} I_2 = \infty, \quad (i = 1, 2). \qquad (5.1.34)$$

The method used in this section is the following: we first establish some lemmas to explore the monotone properties of the functions $I_i, (i = 0, 1, 2)$. Then we show that b can be uniquely solved as a function of σ from equation (5.1.29) and (5.1.31), respectively. Hence, we are led to finding the intersection points of two curves given by these analytic functions of the variable σ. Thus the conclusion of at most an infinitely countable number of solutions follows. By a more detailed study of these analytic functions, we can draw more precise conclusion about the multiplicity of solutions.

In what follows we prove the monotone properties of the functions I_i $(i = 0, 1, 2)$. Let

$$\Sigma^+ = \{(\sigma, b) \in \Sigma, \ \sigma > 0\}, \qquad (5.1.35)$$

$$\Sigma^- = \{(\sigma, b) \in \Sigma, \ \sigma < 0\}. \qquad (5.1.36)$$

Then we have

Lemma 5.1.2 *For $\Delta \in \Sigma$, we have*

(i)

$$\begin{cases} I_1 < 0, & \sigma > 0, \\ I_1 = 0, & \sigma = 0, \\ I_1 > 0, & \sigma < 0. \end{cases} \qquad (5.1.37)$$

(ii)

$$\frac{\partial I_1}{\partial \sigma} < 0. \qquad (5.1.38)$$

(iii)

$$
\begin{cases}
\dfrac{\partial I_1}{\partial b} > 0, & \sigma > 0, \\[2mm]
\dfrac{\partial I_1}{\partial b} = 0, & \sigma = 0, \\[2mm]
\dfrac{\partial I_1}{\partial b} < 0, & \sigma < 0.
\end{cases}
\tag{5.1.39}
$$

Proof. That $I_1 = 0$ as $\sigma = 0$ has been proved in Zheng [7]. We now proceed to prove (ii) which also implies (i). In what follows we use the notation in Novick-Cohn & Peletier [1].

After changing the integral variable, we can rewrite I_1 as

$$
I_1 = \frac{1}{\sqrt{2}} \left(\int_0^1 \frac{(u_1 - w_1)x + w_1}{\sqrt{(1-x)Q_1}} \, dx + \int_0^1 \frac{(u_2 - w_1)x + w_1}{\sqrt{(1-x)Q_2}} \, dx \right)
\tag{5.1.40}
$$

where

$$
Q_i = -\frac{1}{4} u_i^2 (1 + x + x^2 + x^3) - \frac{1}{2} u_i w_1 (1 + x + x^2 - x^3)
$$
$$
- \frac{1}{4} w_1^2 (3 + 3x - 3x^2 + x^3) + \frac{1}{2}(1 + x), \quad (i = 1, 2).
\tag{5.1.41}
$$

A straightforward calculation gives

$$
\frac{\partial I_1}{\partial \sigma} = \frac{1}{\sqrt{2}} \left(\int_0^1 \frac{(u_{1\sigma} - w_{1\sigma})x + w_{1\sigma}}{\sqrt{(1-x)Q_1}} \, dx + \int_0^1 \frac{(u_{2\sigma} - w_{1\sigma})x + w_{1\sigma}}{\sqrt{(1-x)Q_2}} \, dx \right.
$$

$$
+ \frac{1}{4} \int_0^1 \frac{((u_1 - w_1)x + w_1)u_{1\sigma}q_1}{\sqrt{(1-x)Q_1^3}} \, dx + \frac{1}{4} \int_0^1 \frac{((u_2 - w_1)x + w_1)u_{2\sigma}q_2}{\sqrt{(1-x)Q_2^3}} \, dx
$$

$$
\left. + \frac{1}{4} \int_0^1 \frac{((u_1 - w_1)x + w_1)w_{1\sigma}r_1}{\sqrt{(1-x)Q_1^3}} \, dx + \frac{1}{4} \int_0^1 \frac{((u_2 - w_1)x + w_1)w_{1\sigma}r_2}{\sqrt{(1-x)Q_2^3}} \, dx \right)
$$

$$
\stackrel{\text{def}}{=} \sum_{i=1}^{6} A_i
\tag{5.1.42}
$$

where the subscript σ denotes the partial derivative with respect to σ and

$$
\begin{cases}
q_i = u_i(1 + x + x^2 + x^3) + w_1(1 + x + x^2 - x^3), & (i = 1, 2), \\[2mm]
r_i = u_i(1 + x + x^2 - x^3) + w_1(3 + 3x - 3x^2 + x^3), & (i = 1, 2).
\end{cases}
\tag{5.1.43}
$$

It suffices to give the proof for $\Delta \in \Sigma^+$. The situation for $\Delta \in \Sigma^-$ can be similarly treated. The result in the case $\sigma = 0$ follows from continuity. As proved in Novick-Cohen & Peletier [1], a straightforward calculation yields for $\sigma > 0$

$$u_1 + |u_2| < 0, \quad Q_1 < Q_2, \quad w_{1\sigma} < 0, \quad u_{1\sigma} + |u_{2\sigma}| < 0. \tag{5.1.44}$$

$$w_1 < 0, \quad q_1 + |q_2| < 0, \quad q_1 u_{1\sigma} + q_2 u_{2\sigma} > 0, \tag{5.1.45}$$

$$(u_1 - w_1)x + w_1 < 0, \quad (u_1 - w_1)x + |(u_2 - w_1)x + w_1| < 0. \tag{5.1.46}$$

$$r_1 < 0, \quad (r_1 + r_2)w_{1\sigma} < 0, \quad w_{1\sigma} < 0. \tag{5.1.47}$$

Now we can easily deduce from (5.1.44), (5.1.45) and (5.1.47) that

$$A_1 + A_2 < 0, \tag{5.1.48}$$

$$A_3 + A_4 < 0 \tag{5.1.49}$$

and

$$A_5 + A_6 < 0, \tag{5.1.50}$$

respectively. Thus (ii) follows.

The conclusion of (iii) can be derived directly in the same manner and actually follows from Lemma 5.1, Lemma 7.1 and Lemma 7.2 in the paper by Novick-Cohen & Peletier. $\qquad\square$

Lemma 5.1.3 *For $\Delta \in \Sigma$, we have*

(i)

$$\frac{\partial I_0}{\partial b} < 0. \tag{5.1.51}$$

(ii)

$$\begin{cases} \dfrac{\partial I_0}{\partial \sigma} > 0, \quad \sigma > 0, \\[2mm] \dfrac{\partial I_0}{\partial \sigma} = 0, \quad \sigma = 0, \\[2mm] \dfrac{\partial I_0}{\partial \sigma} < 0, \quad \sigma < 0. \end{cases} \tag{5.1.52}$$

Proof. The proof can be carried out in the same manner as in the previous lemma. We refer to Novick-Cohen & Peletier [1] for the details.   □

We now consider the function I_2.

Lemma 5.1.4 *For $\Delta \in \Sigma$, we have*

(i)

$$\frac{\partial I_2}{\partial b} < 0, \quad for \ \ \Delta \in \Sigma. \tag{5.1.53}$$

(ii)

$$\begin{cases} \frac{\partial I_2}{\partial \sigma} > 0, \quad \sigma > 0, \\\\ \frac{\partial I_2}{\partial \sigma} = 0, \quad \sigma = 0, \\\\ \frac{\partial I_2}{\partial \sigma} < 0, \quad \sigma < 0. \end{cases} \tag{5.1.54}$$

Proof. Define

$$n(\sigma, b) = \frac{I_2}{I_0} \tag{5.1.55}$$

Then as proved in Grinfeld & Novick-Cohen [1] , Lemma 3 with b corresponding to $-p$ in their notation,

$$\frac{\partial n}{\partial b} < 0. \tag{5.1.56}$$

Since I_0 and n are positive, it follows from (5.1.51) and (5.1.56) that

$$\frac{\partial I_2}{\partial b} = \frac{\partial I_0}{\partial b} n + I_0 \frac{\partial n}{\partial b} < 0. \tag{5.1.57}$$

To prove (ii), we rewrite I_2 as

$$I_2 = \frac{1}{\sqrt{2}} \left(\int_0^1 \frac{((u_1 - w_1)x + w_1)^2}{\sqrt{(1-x)Q_1}} dx + \int_0^1 \frac{((u_2 - w_1)x + w_1)^2}{\sqrt{(1-x)Q_2}} dx \right). \tag{5.1.58}$$

Therefore,

$$\frac{\partial I_2}{\partial \sigma} = \frac{1}{\sqrt{2}} \left(\int_0^1 \frac{2((u_1 - w_1)x + w_1)((u_{1\sigma} - w_{1\sigma})x + w_{1\sigma})}{\sqrt{(1-x)Q_1}} dx \right.$$

$$+ \int_0^1 \frac{2((u_2 - w_1)x + w_1)((u_{2\sigma} - w_{1\sigma})x + w_{1\sigma})}{\sqrt{(1-x)}Q_2} dx$$

$$+ \frac{1}{4} \left(\int_0^1 \frac{((u_1 - w_1)x + w_1)^2 u_{1\sigma} q_1}{\sqrt{(1-x)}Q_1^3} dx + \int_0^1 \frac{((u_2 - w_1)x + w_1)^2 u_{2\sigma} q_2}{\sqrt{(1-x)}Q_2^3} dx \right.$$

$$\left. + \int_0^1 \frac{((u_1 - w_1)x + w_1)^2 w_{1\sigma} r_1}{\sqrt{(1-x)}Q_1^3} dx + \int_0^1 \frac{((u_2 - w_1)x + w_1)^2 w_{1\sigma} r_2}{\sqrt{(1-x)}Q_2^3} dx \right) \right)$$

$$\stackrel{\text{def}}{=} \sum_{i=1}^6 B_i. \tag{5.1.59}$$

For $\Delta \in \Sigma^+$, as proved in Novick-Cohen & Peletier [1],

$$((u_{1\sigma} - w_{1\sigma})x + w_{1\sigma}) + |(u_{2\sigma} - w_{1\sigma})x + w_{1\sigma}| < 0 \tag{5.1.60}$$

which together with (5.1.45)–(5.1.47) gives

$$B_1 + B_2 > 0. \tag{5.1.61}$$

It follows from (5.1.44)–(5.1.46) that

$$B_3 + B_4 > 0. \tag{5.1.62}$$

Finally, combining (5.1.44) with (5.1.45)–(5.1.47) yields

$$B_5 + B_6 > 0. \tag{5.1.63}$$

Thus when $\Delta \in \Sigma^+$,

$$\frac{\partial I_2}{\partial \sigma} > 0. \tag{5.1.64}$$

Similarly, we can prove that when $\Delta \in \Sigma^-$,

$$\frac{\partial I_2}{\partial \sigma} < 0. \tag{5.1.65}$$

That $\frac{\partial I_2}{\partial \sigma} = 0$, as $\sigma = 0$ is deduced from the continuity of $\frac{\partial I_2}{\partial \sigma}$. □

Having established these lemmas, we now proceed to complete the proof of Theorem 5.1.1 by solving equations (5.1.29) for $k \in I\!N$. As mentioned before, the case

$M_0 = 0$ has been proved in Shen & Zheng [3]. Therefore, without loss of generality, we assume that $M_0 < 0$.

Thus by Lemma 5.1.2, for any $k \in IN$ we must have a unique analytic function $b = b_k(\sigma)$ such that $(\sigma, b_k(\sigma)) \in \Sigma^+$ and

$$I_1(\sigma, b_k(\sigma)) = \frac{M_0}{k}. \tag{5.1.66}$$

Therefore, along the curve $b = b_k(\sigma)$,

$$\frac{\partial I_1}{\partial b} \frac{\partial b}{\partial \sigma} + \frac{\partial I_1}{\partial \sigma} = 0, \tag{5.1.67}$$

$$\frac{\partial b}{\partial \sigma} = -\frac{\frac{\partial I_1}{\partial \sigma}}{\frac{\partial I_1}{\partial b}} > 0, \quad for \quad 0 < \sigma < \frac{2}{3\sqrt{3}}. \tag{5.1.68}$$

By Lemma 5.1.1, the curve defined by (5.1.66) must intersect

$$\partial_3 \Sigma^+ = \{(\sigma, b) | \ b = \Phi_\sigma(w_1(\sigma)), \ 0 \le \sigma \le \frac{2}{3\sqrt{3}}\}. \tag{5.1.69}$$

The coordinates of this intersection point can be determined as follows. By Lemma 5.1.1, as $\Delta \to \partial_3 \Sigma^+$,

$$I_1 \to \frac{\pi w_1(\sigma)}{\sqrt{1 - 3w_1^2(\sigma)}}. \tag{5.1.70}$$

By (5.1.44), $\frac{\pi w_1(\sigma)}{\sqrt{1 - 3w_1^2(\sigma)}}$ is a strict monotone decreasing function of σ and it vanishes at $\sigma = 0$. Therefore, there is a unique σ^* such that

$$\frac{\pi w_1(\sigma^*)}{\sqrt{1 - 3w_1^2(\sigma^*)}} = \frac{M_0}{k}. \tag{5.1.71}$$

It turns out from (5.1.16), (5.1.20) that the coordinates of the intersection point are

$$(w_1^3(\sigma^*) - w_1(\sigma^*), \ -\frac{3}{4}w_1^4(\sigma^*) + \frac{1}{2}w_1^2(\sigma^*)). \tag{5.1.72}$$

By (5.1.37), the curve given by the function $b = b_k(\sigma)$ defined on $(0, \sigma^*]$ must start from this intersection point and approach the singular point $(0, -\frac{1}{4})$ as σ goes to zero. Moreover, as k increases, the curve $b = b_k(\sigma)$ becomes closer to the straight line $\sigma = 0$. We now look at the function

$$H_k(\sigma, b) = \frac{1}{k^2 I_0^2} - \frac{k^2 I_0 I_2}{2}. \tag{5.1.73}$$

By Lemmas 5.1.1–5.1.3,

$$H_k \to -\infty, \quad as \ \Delta \to \partial_1 \Sigma, \tag{5.1.74}$$

$$H_k \to h_k(\sigma) = \frac{1 - 3w_1^2(\sigma)}{k^2 \pi^2} - \frac{k^2 \pi^2 w_1^2(\sigma)}{2(1 - 3w_1^2(\sigma))}, \quad as \ \Delta \to \partial_3 \Sigma^+. \tag{5.1.75}$$

It is easy to see that $h_k(\sigma)$ is a monotone decreasing function of σ. On the other hand, $H_k(0, b)$ is a monotone increasing function of b. Moreover, $\frac{\partial H_k}{\partial \sigma} < 0$, $\frac{\partial H_k}{\partial b} > 0$, as $\Delta \in \Sigma^+$ and H_k achieves its maximum at $\sigma = 0, b = 0$:

$$H_k(0,0) = \frac{1}{k^2 \pi^2}. \tag{5.1.76}$$

Thus, if $M_3 \geq \frac{1}{\pi^2}$, there is no point $(\sigma, b) \in \Sigma^+$ such that

$$H_k(\sigma, b) = M_3, \tag{5.1.77}$$

which implies that the stationary problem has no nontrivial solutions.

Notice that $h_k(0) = \frac{1}{k^2 \pi^2}$ and $h_k(\sigma) \to -\infty$, as $\sigma \to \frac{2}{3\sqrt{3}}$. Therefore, when $M_3 < \frac{1}{\pi^2}$, there are some $k \in I\!N$ (only a finite number of $k < \frac{1}{\sqrt{M_3}\pi}$, as $0 < M_3 < \frac{1}{\pi^2}$; for all $k \in I\!N$ as $M_3 \leq 0$) such that

$$H_k(\sigma, b) = M_3 \tag{5.1.78}$$

has solutions $(\sigma, b) \in \Sigma^+$. By Lemmas 5.1.2–5.1.3, the curve of the solution is represented by a monotone increasing function

$$b = \tilde{b}_k(\sigma, M_3). \tag{5.1.79}$$

Moreover, the family of curves becomes closer to $b = 0$ as k increases or as M_3 increases. In what follows we consider the intersection of the curves defined by (5.1.66) and (5.1.77). It follows from (5.1.71) that

$$-\frac{M_0^2}{2} = g_k(\sigma^*) = -\frac{k^2 \pi^2 w_1^2(\sigma^*)}{2(1 - 3w_1^2(\sigma^*))}. \tag{5.1.80}$$

By (5.1.45), (5.1.47), $g_k(\sigma)$ is also a decreasing function in σ. Thus if $M_3 \leq -\frac{M_0^2}{2}$, then for any $k \in I\!N$, there is a unique pair of points $P_k, \tilde{P}_k \in \partial_3 \Sigma^+$ such that

$$h_k(\sigma_{P_k}) = M_3, \quad g_k(\sigma_{\tilde{P}_k}) = -\frac{M_0^2}{2}. \tag{5.1.81}$$

Since h_k and g_k are decreasing functions and $g_k \leq h_k$, it follows that $\tilde{P}_k$ must be on the left-hand side of P_k . Thus for any $k \in I\!N$, the curves defined by (5.1.66) and (5.1.77) must have at least one intersection point. It turns out from the analytic property of two curves that the stationary problem must exactly have an infinitely countable number of nontrivial solutions.

We now consider the case that $-\dfrac{M_0^2}{2} < M_3 < \dfrac{1}{\pi^2}$. Since the curves defined by (5.1.66) and (5.1.77) are analytic, they have at most an infinitely countable number of intersection points in Σ^+. This implies that the stationary problem has at most an infinitely countable number of nontrivial solutions.

For any M_3, $0 < M_3 < \dfrac{1}{\pi^2}$ it follows from (5.1.76) that there is only finite number of $k \in I\!N$ such that (5.1.77) has a solution $(\sigma, b) \in \Sigma^+$. More precisely, if $\dfrac{1}{(k_0 + 1)^2\pi^2} \leq M_3 < \dfrac{1}{k_0^2\pi^2}$, then for $k = 1, \cdots, k_0$, (5.1.77) defines k curves in Σ^+. It can be seen from the previous discussion that if $M_3 \neq \dfrac{1}{3M_0^2 + k^2\pi^2} - \dfrac{M_0^2}{2}$, $k = 1, \cdots, k_0$, then $P_k \neq \tilde{P}_k$. It turns out that the number of intersection points of curves (5.1.66) and (5.1.77) is finite. It follows from the monotone property of curves (5.1.79) with respect to M_3 that when $M_3 \in (0, \dfrac{1}{\pi^2})$, the stationary problem has only a finite number of nontrivial solutions.

For any $M_3 \in (-\dfrac{M_0^2}{2}, 0]$, (5.1.78) defines a family of monotone curves for all $k \in I\!N$. It follows from the monotone property of curves (5.1.79) with respect to M_3 that there is $M_3^*(M_0) \in [-\dfrac{M_0^2}{2}, 0]$ such that when $M_3 > M_3^*(M_0)$, there is only a finite number of intersection points and when $M_3 \leq M_3^*(M_0)$, there is an infinitely countable number of intersection points. We refer the reader to Novick-Cohen & Zheng [1] for a more detailed proof on this point and the bifurcation diagram. The proof of Theorem 5.1.1 is complete.   □

We now consider the Neumann initial boundary value problem for the phase-field equations of Caginalp's model (see Section 4.1.1). The corresponding stationary problem in one space dimension reads:

$$\xi^2 D^2\varphi + \frac{1}{2}(\varphi - \varphi^3) + 2\bar{u} = 0, \quad x \in [0, 1], \tag{5.1.82}$$

$$\varphi_x|_{x=0,1} = 0, \tag{5.1.83}$$

$$\bar{u} = Const., \tag{5.1.84}$$

$$\bar{u} + \frac{l}{2}\int_0^1 \varphi dx = \frac{M_0}{2} = \int_0^1 u_0 dx + \frac{l}{2}\int_0^1 \varphi_0 dx. \tag{5.1.85}$$

If we solve $\bar{u}$ from (5.1.85) and substitute it into (5.1.82), then the above problem again becomes a nonlinear boundary value problem with nonlocal term:

$$\xi^2 D^2\varphi + \frac{1}{2}(\varphi - \varphi^3) + M_0 - l\int_0^1 \varphi dx = 0, \quad x \in [0,1], \tag{5.1.86}$$

$$\varphi_x|_{x=0,1} = 0. \tag{5.1.87}$$

It is easy to verify by the variational method and the bootstrap argument that problem (5.1.86)–(5.1.87) is equivalent to finding critical points of the following functional

$$F(\varphi) = \int_0^1 \left(\frac{\xi^2}{2}|D\varphi|^2 + \frac{1}{8}\varphi^4 - \frac{1}{4}\varphi^2 - M_0\varphi\right) dx + \frac{l}{2}\left(\int_0^1 \varphi dx\right)^2 \tag{5.1.88}$$

over $H^1(0,1)$.

Since F is bounded from below, we have a minimizing sequence $\varphi_n \in H^1$. It follows from the expression for F that φ_n is bounded in H^1. Therefore, there exists a subsequence, which we still denote by φ_n, such that

$$\varphi_n \rightharpoonup \varphi, \quad weakly \ in \ H^1 \tag{5.1.89}$$

and

$$\varphi_n \rightarrow \varphi \ in \ C[0,1]. \tag{5.1.90}$$

It turns out from the lower semicontinuity of F that φ is a minimizer.

It is easy to see that problem (5.1.86), (5.1.87) has three trivial (constant) solutions which may coincide for some M_0 and l. We refer to Elliott & Zheng [2] for detailed discussion about whether and when the stationary problem has at least one nontrivial solution. We are now interested in the multiplicity of nontrivial solutions. The above variational method is applicable to the problem in higher space dimension. However, this method usually only yields at least how many solutions there are. For the one-dimensional problem, the method previously used in this section gives upper bounds on the number of solutions.

Theorem 5.1.2 *For any given M_0, l, and ξ, the stationary problem (5.1.86)–(5.1.87) has at most a finite number of nontrivial solutions.*

Proof. Let

$$x' = \frac{x}{\sqrt{2\xi}}, \quad \sigma = 4\bar{u}. \tag{5.1.91}$$

Then it turns out that

$$\frac{d^2\varphi}{dx'^2} + (\varphi - \varphi^3) + \sigma = 0, \quad x' \in [0, \frac{1}{\sqrt{2\xi}}), \tag{5.1.92}$$

$$\varphi_{x'}\big|_{x'=0,\frac{1}{\sqrt{2\xi}}} = 0, \tag{5.1.93}$$

$$\frac{\sigma}{2} + \sqrt{2}l\xi \int_0^{\frac{1}{\sqrt{2\xi}}} \varphi dx' = M_0. \tag{5.1.94}$$

In the same way as before, the problem is reduced to finding $(\sigma, b) \in \Sigma$ and $k \in \mathbb{N}$ such that

$$kI_0(\sigma, b) = \frac{1}{\sqrt{2\xi}}, \tag{5.1.95}$$

$$\frac{\sigma}{2} + \sqrt{2}l\xi k I_1(\sigma, b) = M_0. \tag{5.1.96}$$

It can be easily seen from Lemma 5.1.1 that when $\pi \geq \frac{1}{\sqrt{2\xi}}$, there is no $(\sigma, b) \in \Sigma$ such that (5.1.95) is satisfied. Therefore, when $\pi \geq \frac{1}{\sqrt{2\xi}}$, the stationary problem has no nontrivial solution. When $\pi < \frac{1}{\sqrt{2\xi}}$, there is only a finite number of $k \in \mathbb{N}$ such that (5.1.95) defines a curve in Σ. For every such k, solving b as a function of σ from (5.1.95), then substituting it into (5.1.96), we are led to finding the roots of an analytic function in σ. It turns out that the stationary problem has at most a finite number of nontrivial solutions. The proof is complete. $\square$

5.2   Convergence to Equilibrium

First let us consider the following one-dimensional phase-field equations of Caginalp's model:

$$\tau\varphi_t = \xi^2 D^2\varphi + \frac{1}{2}(\varphi - \varphi^3) + 2u, \tag{5.2.1}$$

$$u_t + \frac{l}{2}\varphi_t = KD^2\varphi, \tag{5.2.2}$$

$$D\varphi\big|_{x=0,1} = Du\big|_{x=0,1} = 0, \tag{5.2.3}$$

$$\varphi\big|_{t=0} = \varphi_0(x), \quad u\big|_{t=0} = u_0(x). \tag{5.2.4}$$

For any $(\varphi_0, u_0) \in H^1 \times L^2$, it can be easily deduced by a similar argument to that in Section 4.1.1 that problem (5.2.1)–(5.2.4) has a unique global solution $(\varphi(x,t), u(x,t))$ such that $(\varphi, u) \in C([0, +\infty); H^1) \times C([0, +\infty); L^2)$. Moreover, $\varphi, u \in C((0, +\infty); C^\infty)$ and for any $\varepsilon > 0$, the orbit $t \in [\varepsilon, \infty) \mapsto (\varphi(\cdot, t), u(\cdot, t))$ is compact in $H^1 \times L^2$. Since the dynamical system has a Liapunov functional defined by (4.1.21) which is continuous in $H^1 \times L^2$, we deduce from Theorems 1.5.1, 1.5.3 that the ω-limit set:

$$\omega(\varphi_0, u_0) = \left\{ (\psi, v) \,\middle|\, \begin{array}{l} \exists t_n, \ t_n \to \infty, \ \text{such that} \\ \varphi(x, t_n) \to \psi, \ \text{in } H^1, \ u(x, t_n) \to v \ \text{in } L^2 \end{array} \right\} \qquad (5.2.5)$$

is a compact connected subset of $H^1 \times L^2$ and it consists of equilibria. Thus combining this result with Theorem 5.1.2, we have

Theorem 5.2.1 *For any $(\varphi_0, u_0) \in H^1 \times L^2$, problem (5.2.1)–(5.2.4) admits a unique global solution $(\varphi(x,t), u(x,t))$ such that $(\varphi, u) \in C([0, +\infty); H^1) \cap C((0, \infty); C^\infty) \times C([0, +\infty); L^2) \cap C((0, +\infty); C^\infty)$. Moreover, as $t \to \infty$, $(\varphi(x,t), u(x,t))$ converges in $H^1 \times L^2$ to an equilibrium.* $\qquad \square$

We now consider the one-dimensional coupled Cahn–Hilliard equations. The following is a version of Theorem 4.3.1 with weaker assumptions on the initial data.

Theorem 5.2.2 *For any $(\varphi_0, \theta_0) \in H^2 \times H^1$, $\theta_0 > 0$ in $[0, 1]$ satisfying the compatibility condition, problem (4.3.6)–(4.3.9) admits a unique global solution $(\varphi(x,t), \theta(x,t))$ $\in C([0, \infty); H^2) \cap C((0, \infty); C^\infty) \times C([0, \infty); H^1) \cap C((0, \infty); C^\infty), \theta(x,t) > 0, \forall (x,t) \in [0, 1] \times \mathrm{I\!R}^+$. Moreover, as time goes to infinity, in addition to (4.3.27), (4.3.28), (φ, θ) converges to one of the equilibria.*

Proof. First we point out that for any $\varphi_0 \in H^2, \theta_0 \in H^1, \theta_0 > 0$ in $[0, 1]$ we can deduce from a similar argument as those in Section 4.3 that problem (4.3.6)–(4.3.9) admits a unique local solution (φ, θ) such that $\varphi \in C([0, t^*]; H^2) \cap L^2([0, t^*]; H^4), \varphi_t \in L^2([0, t^*]; L^2), \theta \in C([0, t^*]; H^1) \cap L^2([0, t^*]; H^2), \theta_t \in L^2([0, t^*]; L^2)$. We use the density argument and Lemma 4.3.1 to prove the present theorem. It suffices to prove that for any small $\delta > 0$, $\|\varphi(\delta)\|_{H^5}$ and $\|\theta(\delta)\|_{H^3}$ can be bounded by a constant depending only on $\|\varphi_0\|_{H^2}, \|\theta_0\|_{H^1}$ and δ. To this end, we apply Lemma 4.3.1 successively to reconsider Lemma 4.3.4–Lemma 4.3.8.

Indeed, applying Lemma 4.3.1 to (4.3.71) yields (see (4.3.12))

$$\|Du(t)\|^2 + \|Dz(t)\|^2 \leq C_\delta, \quad t \geq \frac{\delta}{4} \tag{5.2.6}$$

with C_δ being a positive constant depending only on $\|\varphi_0\|_{H^2}, \|\theta_0\|_{H^1}$ and δ. Thus integrating (4.3.71) with respect to t from $\frac{\delta}{4}$ to t yields

$$\int_{\frac{\delta}{4}}^t \|D\varphi_t\|^2 \, d\tau \leq C_\delta, \quad \int_{\frac{\delta}{4}}^t \left\|\frac{u_t}{u}\right\|^2 \, d\tau \leq C_\delta, \tag{5.2.7}$$

$$\|\varphi(t)\|_{H^3} \leq C_\delta, \quad \|u(t)\|_{L^\infty} \leq C_\delta, \quad t \geq \frac{\delta}{4}. \tag{5.2.8}$$

Then apply Lemma 4.3.1 again to (4.3.74) in the proof of Lemma 4.3.5 to obtain

$$\|\varphi_t(t)\| \leq C_\delta, \quad t \geq \frac{\delta}{2} \tag{5.2.9}$$

and also

$$\int_{\frac{\delta}{2}}^t \|D^2\varphi_t\|^2 \, d\tau \leq C_\delta. \tag{5.2.10}$$

It turns out from the proof of Lemma 4.3.6 that

$$\|\theta(t)\|_{L^\infty} \leq C_\delta, \quad t \geq \frac{\delta}{2}. \tag{5.2.11}$$

Combining the local existence result with (5.2.6) and (5.2.11) yields

$$\|\theta(t)\|_{H^1} \leq C, \quad \|u(t)\|_{H^1} \leq C, \quad t \geq 0. \tag{5.2.12}$$

In the same manner as above, we obtain from the proof of Lemma 4.3.7 and Lemma 4.3.8, that

$$\|u_t(t)\| \leq C_\delta, \quad ,\|D\varphi_t(t)\| \leq C_\delta, \quad t \geq \frac{3\delta}{4}, \tag{5.2.13}$$

$$\|\varphi(t)\|_{H^5} \leq C_\delta, \quad \|\varphi_t(t)\|_{L^\infty} \leq C_\delta, \quad t \geq \frac{3\delta}{4}, \tag{5.2.14}$$

$$\int_{\frac{3\delta}{4}}^t \left(\|Dz_t\|^2 + \|Du_t\|^2\right) \, d\tau \leq C_\delta, \tag{5.2.15}$$

$$\|u(t)\|_{H^3} \leq C_\delta, \quad \|Du_t(t)\| \leq C_\delta, \quad t \geq \delta. \tag{5.2.16}$$

Therefore, it follows that for any $\varphi_0 \in H^2, \theta_0 \in H^1, \theta_0 > 0$ in $[0,1]$, problem (4.3.6)–(4.3.9) admits a unique global solution. Moreover, $0 < C_1 \leq u(x,t) \leq C_2$, $0 < C_3 \leq \theta(x,t) \leq C_4$ and the orbit $t \in [\varepsilon, +\infty) \rightarrow (\varphi(\cdot,t), \theta(\cdot,t))$ is compact in $H^2 \times H^1$. Let

$$X = \left\{(\varphi, \theta) \,\middle|\, (\varphi, \theta) \in H^2 \times H^1, \, \theta > 0 \text{ in } [0,1]\right\}. \tag{5.2.17}$$

Then the semigroup $S(t)$ defined by the problem (4.3.6)–(4.3.9) is continuous on the metric space X. Let

$$\omega(\varphi_0, \theta_0) = \left\{(\psi, v) \,\middle|\, \begin{array}{l} \exists t_n, \ t_n \rightarrow \infty, \ such \ that \\ \varphi(x, t_n) \rightarrow \psi, \ in \ H^2, \theta(x, t_n) \rightarrow v \ in \ H^1. \end{array}\right\}. \tag{5.2.18}$$

Then following the proof of Proposition 2.1 in Dafermos [2] (p. 106), in the same manner we deduce that the ω-limit set $\omega(\varphi_0, \theta_0)$ is a compact, connected subset of X. Furthermore, it is positively invariant under $S(t)$. The functional $V = -F$ with F defined by (4.3.47) is continuous on X and $V(S(t)(\varphi_0, \theta_0)) \leq V((\varphi_0, \theta_0))$ for any $(\varphi_0, \theta_0) \in X$ and $t \geq 0$. Therefore, by Proposition 2.2 in Dafermos [2] (p. 107) we conclude that V is constant on $\omega(\varphi_0, \theta_0)$. Furthermore, it follows from (4.3.48), (4.3.50) and

$$\int_0^1 \varphi dx = \int_0^1 \varphi_0 dx \tag{5.2.19}$$

that the ω-limit set $\omega(\varphi_0, \theta_0)$ consists of equilibria. Combining it with Theorem 5.1.1 yields that $\omega(\varphi_0, \theta_0)$ is a single point which implies that as time goes to infinity, the solution (φ, θ) converges to an equilibrium. Thus the proof is complete. $\quad\square$

5.3 Global Attractors and Inertial Manifolds

In this section we will establish results on the existence of a global attractor, an inertial manifold and an inertial set concerning the initial boundary value problem for the phase-field equations:

$$\tau \varphi_t = \xi^2 \Delta \varphi + \frac{1}{2}(\varphi - \varphi^3) + 2u, \tag{5.3.1}$$

$$u_t + \frac{l}{2}\varphi_t = K \Delta u, \tag{5.3.2}$$

$$\varphi|_\Gamma = u|_\Gamma = 0, \tag{5.3.3}$$

$$\varphi|_{t=0} = \varphi_0(x), \quad u|_{t=0} = u_0(x) \tag{5.3.4}$$

where Γ is a smooth boundary of a bounded domain $\Omega \subseteq \mathbb{R}^n$ with $n \leq 3$. In order to overcome the difficulty that the system is not diagonal and so does not fit the framework in Temam [1] to produce existence of inertial manifolds, we use a tool, the so called 'symmetrizer', motivated by the results in Bates & Fife [1]. We believe that this technique may be applicable to some other equations. As proved in Section 4.1.1, for any $(\varphi_0,\, u_0) \in H_o^1 \times L^2$; $H^2 \cap H_o^1 \times H_o^1$, respectively, problem (5.3.1)–(5.3.4) has a unique global solution $(\varphi, u) \in C([0, +\infty); H_o^1) \times C([0, +\infty); L^2)$, $C([0, +\infty); H^2 \cap H_o^1)$ $\times C([0, +\infty); H_o^1)$, respectively, and the nonlinear semigroup $S(t)$ associated with the system (5.3.1)–(5.3.4) is a dynamical system. Furthermore, we have

Theorem 5.3.1 *The semigroup $S(t)$ associated with the system (5.3.1)–(5.3.4) possesses a global (maximal) attractor $\mathcal{M}$ which is bounded in $H^m \times H^m$, for any $m \in \mathbb{N}$, compact and connected in $H_o^1 \times L^2$; $H^2 \cap H_o^1 \times H_o^1$, respectively and attracts all the bounded sets of $H_o^1 \times L^2$; $H^2 \cap H_o^1 \times H_o^1$, respectively.*

Proof. In order to apply Theorem 1.5.1 to obtain the existence of a global attractor, by Theorem 4.1.1, and Remark 4.1.2 it suffices to verify that there exists an absorbing ball.

It is easy to see from the expression for $E(t)$ in (4.1.21) that the boundedness of $E(t)$ from above implies the boundedness of $\|\varphi\|_{H^1}^2 + \|u\|^2$ from above. Therefore, we need only to prove that $\limsup\limits_{t \to \infty} E(t) \leq C$ with a constant $C > 0$ independent of φ_0 and u_0. Multiplying (5.3.1) by φ, then integrating with respect to x yields

$$\int_\Omega \left(\xi^2 |\nabla \varphi|^2 + \frac{1}{2}(\varphi^4 - \varphi^2) + \tau \varphi_t \varphi - 2 \varphi u \right) dx = 0. \tag{5.3.5}$$

By the Young inequality, we easily get from (5.3.5) that

$$\int_\Omega \left(\xi^2 |\nabla \varphi|^2 + \frac{1}{4}\varphi^4 - \frac{1}{2}\varphi^2 \right) dx \leq 2\tau \|\varphi_t\|^2 + \varepsilon \|u\|^2 + C_\varepsilon \tag{5.3.6}$$

with ε being an arbitrary positive constant and $C_\varepsilon > 0$ a constant depending only on ε and τ. By the Poincarè inequality, we have

$$\|u\|^2 \leq C \|\nabla u\|^2 \tag{5.3.7}$$

with $C > 0$ depending only on the domain Ω. Dividing (5.3.6) by 2 and choosing $\varepsilon = \frac{4K}{Cl}$, then adding up with (4.1.22) yields

$$\frac{dE}{dt} + \alpha E(t) \leq C' \tag{5.3.8}$$

with $\alpha = min(1, \frac{K}{C})$ and $C' > 0$ depending only on Ω, τ, l and K. It follows from (5.3.8) that

$$E(t) \leq E(0)e^{-\alpha t} + C'. \tag{5.3.9}$$

Notice that $E(0)$ is bounded if $\|\varphi_0\|^2_{H^1} + \|u_0\|^2$ is bounded. Therefore, the inequality (5.3.9) implies the existence of an absorbing set in $H^1_o \times L^2$. The existence of an absorbing ball in $H^2 \cap H^1_o \times H^1_o$ can be easily proved by combining the above argument and (4.1.33)–(4.1.34). We leave the details to the reader. The proof of Theorem 5.3.1 is complete. □

Remark 5.3.1 *As can be seen from Section 4.1.1, the global attractor actually is also bounded in $H^{2m} \times H^{2m}$ for any $m \in IN$.*

Remark 5.3.2 *Since the system has a Liapunov functional $E(t)$, by Theorem 1.5.4, the global attractor $\mathcal{M}$ can be expressed as*

$$\mathcal{M} = \mathcal{M}_+(\mathcal{E}) \tag{5.3.10}$$

where $\mathcal{E}$ is the set of equilibria.

Remark 5.3.3 *As can be seen from the book by Temam [1], it is quite routine to prove that the global attractor is of finite fractal dimension and finite Hausdorff dimension with upper bound estimates depending on the coefficients and Ω. We leave the details to the reader.*

We now discuss the existence of an inertial manifold. Since the phase-field equations are not a diagonal parabolic system, if we put them into the abstract framework of the first-order evolution equation

$$\frac{du}{dt} + \mathcal{A}u + F(u) = 0, \tag{5.3.11}$$

by substracting (5.3.2) from (5.3.1) times $-\frac{l}{2\tau}$, then the corresponding operator $\mathcal{A}$ is not self-adjoint. But the existing theory for inertial manifolds (see Temam [1]) usually requires that $\mathcal{A}$ be a self-adjoint operator. We refer to Sell & You [1] for the theory on non-self-adjoint operators. However, the theory requires some information about the distribution of the complex eigenvalues of $\mathcal{A}$ which is, in our case, not easy to verify. In what follows we use a 'symmetrizer' technique similar to that in Bates & Fife [1] to reduce the problem to one with $\mathcal{A}$ being self-adjoint.

Dividing (5.3.1) by τ, we obtain

$$\varphi_t = \frac{\xi^2}{\tau}\Delta\varphi + \frac{1}{\tau}\left(\frac{1}{2}(\varphi - \varphi^3) + 2u\right). \tag{5.3.12}$$

Multiplying (5.3.12) by $-\frac{l}{2}$, then adding up with (5.3.2) yields

$$u_t = K\Delta u - \frac{l\xi^2}{2\tau}\Delta\varphi - \frac{l}{2\tau}\left(\frac{1}{2}(\varphi - \varphi^3) + 2u\right). \tag{5.3.13}$$

Since $-\Delta$ defined on $H^2 \cap H_o^1 \subset L^2$ is a positive definite operator, as shown in Chapter 1, we can write $-\Delta$ as

$$-\Delta = A^2 \tag{5.3.14}$$

where A is a self-adjoint positive definite operator. It can be given explicitly by

$$Au = \sum_{n=1}^{\infty} \lambda_n^{\frac{1}{2}}(u, u_n)u_n, \quad \forall u \in D(A) = H_o^1 \tag{5.3.15}$$

with u_n being normalized eigenfunctions of $-\Delta$ associated with eigenvalues λ_n, and (u, u_n) being the inner product in L^2. Also

$$A^{-1}u = \sum_{n=1}^{\infty} \lambda_n^{-\frac{1}{2}}(u, u_n)u_n. \tag{5.3.16}$$

Let

$$a = \frac{2}{\sqrt{l}\,\xi}, \quad e = aA^{-1}u. \tag{5.3.17}$$

Then (5.3.12) becomes

$$\varphi_t = \frac{\xi^2}{\tau}\Delta\varphi + \frac{\sqrt{l}\,\xi}{\tau}Ae + f_1(\varphi) \tag{5.3.18}$$

with

$$f_1 = \frac{1}{2\tau}(\varphi - \varphi^3). \tag{5.3.19}$$

Acting with aA^{-1} on equation (5.3.13) yields

$$e_t = K\Delta e + \frac{\sqrt{l}\,\xi}{\tau}A\varphi + f_2(\varphi) - \frac{l}{\tau}e \tag{5.3.20}$$

with

$$f_2(\varphi) = \frac{-\sqrt{l}}{2\tau\xi} A^{-1}(\varphi - \varphi^3).\tag{5.3.21}$$

Then the system (5.3.18), (5.3.20) can be written as

$$\frac{dU}{dt} + \mathcal{A}U = F(U)\tag{5.3.22}$$

with

$$U = \begin{pmatrix} \varphi \\ e \end{pmatrix}, \quad \mathcal{A} = \begin{pmatrix} -\frac{\xi^2}{\tau} & -\frac{\sqrt{l}\xi}{\tau}A \\ -\frac{\sqrt{l}\xi}{\tau}A & -K\Delta + \frac{l}{\tau} \end{pmatrix}, \quad F = \begin{pmatrix} f_1 \\ f_2 \end{pmatrix}\tag{5.3.23}$$

and initial condition

$$U|_{t=0} = U_0 \equiv \begin{pmatrix} \varphi_0 \\ aA^{-1}u_0 \end{pmatrix}.\tag{5.3.24}$$

It is easy to see that the system (5.3.1)–(5.3.4) is equivalent to the system (5.3.22)–(5.3.24) in the sense that if $(\varphi, u) \in C([0, +\infty); H_o^1) \times C([0, +\infty); L^2)$ $(C([0, +\infty); H^2 \cap H_o^1) \times C([0, +\infty); H_o^1))$ is a solution to the system (5.3.1)–(5.3.4), then $(\varphi, e) \in C([0, +\infty); H_o^1) \times C([0, +\infty); H_o^1)$ $(C([0, +\infty); H^2 \cap H_o^1) \times C([0, +\infty); H^2 \cap H_o^1))$ is a solution to the system (5.3.22)–(5.3.24), and vice versa.

Theorem 5.3.1 shows, by the equivalence of the two systems mentioned above, that the semigroup $\tilde{S}(t)$ associated with the system (5.3.22)–(5.3.24) possesses a global (maximal) attractor which is bounded in $H^m \times H^m$, for any $m \in \mathbb{N}$, compact and connected in $H_o^1 \times H_o^1$ $(H^2 \cap H_o^1 \times H^2 \cap H_o^1)$ and attracts all the bounded sets of $H_o^1 \times H_o^1$ $(H^2 \cap H_o^1 \times H^2 \cap H_o^1)$. Consider $\mathcal{A}$ as an operator with domain $D(\mathcal{A}) = H^2 \cap H_o^1 \times H^2 \cap H_o^1$ in $L^2 \times L^2$. Then it is easy to see that $\mathcal{A}$ is self-adjoint. Furthermore, if we denote by $(\,,\,)$ the inner product in $L^2 \times L^2$ and also in L^2 when no confusion occurs, then we have

$$(\mathcal{A}U, U)$$

$$= \left(-\frac{\xi^2}{\tau}\Delta\varphi, \varphi\right) - \frac{2\sqrt{l}\xi}{\tau}(A\varphi, e) + \left(-K\Delta e + \frac{l}{\tau}e, e\right)$$

$$= \frac{\xi^2}{\tau}\sum_{n=1}^{\infty} \lambda_n(\varphi, u_n)^2 - \frac{2\sqrt{l}\xi}{\tau}\sum_{n=1}^{\infty} \lambda_n^{\frac{1}{2}}(\varphi, u_n)(e, u_n)$$

$$+K \sum_{n=1}^{\infty} \lambda_n(e, u_n)^2 + \frac{l}{\tau} \sum_{n=1}^{\infty}(e, u_n)^2. \tag{5.3.25}$$

Since

$$\frac{2\sqrt{l}\xi}{\tau}\lambda_n^{\frac{1}{2}}(\varphi, u_n)(e, u_n) \leq \frac{\xi^2(1-\varepsilon)}{\tau}\lambda_n(\varphi, u_n)^2 + \frac{l}{\tau}\frac{1}{1-\varepsilon}(e, u_n)^2 \tag{5.3.26}$$

with small $\varepsilon > 0$, it follows from (5.3.25), (5.3.26) that

$$(AU, U) \geq \frac{\xi^2\varepsilon}{\tau}\sum_{n=1}^{\infty}\lambda_n(\varphi, u_n)^2 + \sum_{n=1}^{\infty}\left(K\lambda_n - \frac{\varepsilon}{1-\varepsilon}\frac{l}{\tau}\right)(e, u_n)^2. \tag{5.3.27}$$

Taking ε small enough so that $K\lambda_1 - \dfrac{\varepsilon}{1-\varepsilon}\dfrac{l}{\tau} > 0$, we deduce from (5.3.27) that $\mathcal{A}$ is a positive definite operator.

Theorem 5.3.2 *Suppose $n = 1$, $(\Omega = (0, L))$ or $n = 2$ and $\Omega = (0, L) \times (0, L)$. Then system (5.3.22)–(5.3.24) possesses an inertial manifold in $D(\mathcal{A}^{\frac{1}{2}}) = H_o^1 \times H_o^1, (n = 1)$ or in $D(\mathcal{A}) = H^2 \cap H_o^1 \times H^2 \cap H_o^1, (n = 2)$. This implies that the system (5.3.1)– (5.3.4) has an inertial manifold in $H_o^1 \times L^2$ $(n = 1)$ or in $H^2 \cap H_o^1 \times H_o^1$ $(n = 2)$.*

Proof. First consider the case $n = 1$.

In order to apply Theorem 1.5.6 to obtain the existence of an inertial manifold, by Theorem 4.1.1, it remains to prove that F defined in (5.3.23) is a bounded mapping from $D(\mathcal{A}^{\frac{1}{2}})$ into $D(\mathcal{A}^{\frac{1}{2}})$, $(\alpha = \frac{1}{2}, \gamma = 0)$ and F is locally Lipschitz and, also, to prove that the spectral gap condition is satisfied. By Sobolev's imbedding theorem, it is easy to see that f_1, f_2 are bounded mappings from H_o^1 to H_o^1 and f_1 is locally Lipschitz from H_o^1 to H_o^1. To prove that f_2 is also locally Lipschitz, since A^{-1} is a bounded operator from L^2 to H_o^1, we need only to consider the term $A^{-1}(\varphi^3)$. By (5.3.16), we have

$$\|A^{-1}(\varphi_1^3) - A^{-1}(\varphi_2^3)\|_{H^1} = \|\varphi_1^3 - \varphi_2^3\| \leq C_M\|\varphi_1 - \varphi_2\|_{H^1},$$

$$if \quad \|\varphi_1\|_{H^1} \leq M, \ \|\varphi_2\|_{H^1} \leq M \tag{5.3.28}$$

with C_M being a constant depending on M.

In what follows we verify the spectral gap condition. For $\Omega = (0, L)$, we look for the eigenvalue λ and the associated eigenfunction $(\varphi, e) \in H^2 \cap H_o^1 \times H^2 \cap H_o^1$ such that

$$\begin{pmatrix} -\frac{\xi^2}{\tau}\Delta & -\frac{\sqrt{l}\xi}{\tau}A \\ -\frac{\sqrt{l}\xi}{\tau}A & -K\Delta + \frac{l}{\tau} \end{pmatrix} \begin{pmatrix} \varphi \\ e \end{pmatrix} = \lambda \begin{pmatrix} \varphi \\ e \end{pmatrix}. \tag{5.3.29}$$

We rewrite the equations separately:

$$-\frac{\xi^2}{\tau}\Delta\varphi - \frac{\sqrt{l}\xi}{\tau}Ae = \lambda\varphi, \tag{5.3.30}$$

$$-\frac{\sqrt{l}\xi}{\tau}A\varphi - K\Delta e + \frac{l}{\tau}e = \lambda e. \tag{5.3.31}$$

Acting with A on (5.3.30), using (5.3.14), replacing $A\varphi$ by the one in (5.3.31), we get

$$\frac{K\xi}{\sqrt{l}}\Delta^2 e + \frac{K\tau + \xi^2}{\sqrt{l}\xi}\lambda\Delta e + \left(\frac{\lambda^2\tau}{\sqrt{l}\xi} - \frac{\sqrt{l}}{\xi}\lambda\right)e = 0. \tag{5.3.32}$$

The normalized eigenfunctions, which are also the eigenfunctions of $-\Delta$ on $H^2 \cap H_o^1$, are

$$e_n = \sqrt{\frac{2}{L}}\sin\frac{n\pi x}{L}, \quad n = 1, 2, \cdots. \tag{5.3.33}$$

The corresponding eigenvalues $\lambda = \tilde{\lambda}_n$ satisfy

$$\left(\frac{\lambda^2\tau}{\sqrt{l}\xi} - \frac{\sqrt{l}}{\xi}\lambda\right) - \lambda\left(\frac{K\tau + \xi^2}{\sqrt{l}\xi}\right)\mu_n + \frac{K\xi}{\sqrt{l}}\mu_n^2 = 0 \tag{5.3.34}$$

where $\{\mu_n\}$ are the eigenvalues of $-\Delta$ on $H^2 \cap H_o^1$, which in this case are given by $\mu_n = \left(\frac{n\pi}{L}\right)^2, \quad n = 1, 2, \cdots$.

Thus the eigenvalues $\tilde{\lambda}_n$ are given by the two forms:

$$\tilde{\lambda}_n^+ = \tilde{a}\mu_n + \tilde{b} + \sqrt{(\tilde{a}\mu_n + \tilde{b})^2 - \tilde{c}^2\mu_n^2}, \tag{5.3.35}$$

$$\tilde{\lambda}_n^- = \tilde{a}\mu_n + \tilde{b} - \sqrt{(\tilde{a}\mu_n + \tilde{b})^2 - \tilde{c}^2\mu_n^2}, \tag{5.3.36}$$

where $\tilde{a} = \frac{K\tau + \xi^2}{2\tau}$, $\tilde{b} = \frac{l}{2\tau}$, $\tilde{c} = \sqrt{\frac{K}{\tau}}\xi$.

Noting that $\tilde{a}^2 \geq \tilde{c}^2$, we find

$$d_n^+ \stackrel{\text{def}}{=} \tilde{\lambda}_{n+1}^+ - \tilde{\lambda}_n^+ = (\mu_{n+1} - \mu_n)[a + \alpha_n], \tag{5.3.37}$$

$$d_n^- \stackrel{\text{def}}{=} \tilde{\lambda}_{n+1}^- - \tilde{\lambda}_n^- = (\mu_{n+1} - \mu_n)[a - \alpha_n], \tag{5.3.38}$$

where $[x]$ denotes the integer part of x and $\alpha_n \to \sqrt{a^2 - c^2}$ as $n \to \infty$.

It follows that $d_n^+ \geq d_n^- \to \infty$ because $\mu_{n+1} - \mu_n \to \infty$ as $n \to \infty$. For fixed N, we define the gap at $\tilde{\lambda}_N^-$ to be the maximum of $(\tilde{\lambda}_N^- - \mu)$ and $(\nu - \tilde{\lambda}_N^-)$ where μ (ν) is the largest (smallest) eigenvalue of $\mathcal{A}$ less (greater) than $\tilde{\lambda}_N^-$.

Let $\widetilde{K} = \widetilde{K}(N)$ be defined by

$$\tilde{\lambda}_{\widetilde{K}}^+ \leq \tilde{\lambda}_N^- \leq \tilde{\lambda}_{\widetilde{K}+1}^+. \tag{5.3.39}$$

Then

$$either \quad \mu = \tilde{\lambda}_{\widetilde{K}}^+ \quad or \quad \mu = \tilde{\lambda}_{N-1}^-, \tag{5.3.40}$$

and

$$either \quad \nu = \tilde{\lambda}_{\widetilde{K}+1}^+ \quad or \quad \nu = \tilde{\lambda}_{N+1}^-. \tag{5.3.41}$$

It follows that the gap at $\tilde{\lambda}_N^-$ is at least

$$d_N = \min(d_N^-, d_{N-1}^-, \frac{1}{2}d_{\widetilde{K}}^+). \tag{5.3.42}$$

Clearly, $\widetilde{K} = \widetilde{K}(N) \to \infty$ and hence $d_N \to \infty$ as $N \to \infty$. Thus the spectral gap conditions (1.5.26) and (1.5.27) are satisfied and the existence of an inertial manifold follows.

We now consider the case $n = 2, \Omega = (0, L) \times (0, L)$. In order to apply Theorem 1.5.6 to obtain the existence of an inertial manifold, by Remark 4.1.2, it remains to prove that F is a bounded mapping from $D(\mathcal{A})$ into $D(\mathcal{A})$ ($\alpha = 1, \gamma = 0$), F is also locally Lipschitz and the spectral gap condition is satisfied. By Sobolev's imbedding theorem, H^2 ($n = 2$) is continuously imbedded in $C(\bar{\Omega})$. Therefore, F is a bounded mapping from $D(\mathcal{A})$ into $D(\mathcal{A}), (\alpha = 1, \gamma = 0)$. Using the same argument as before yields that F is also locally Lipschitz. The spectrum $\tilde{\lambda}_n$ is still given by (5.3.35), (5.3.36), but μ_n is now the nth eigenvalue of $-\Delta$ with domain $H^2 \cap H_o^1$. These eigenvalues have the form

$$\left(\frac{\pi}{L}\right)^2 (i^2 + j^2) \quad with\ i\ and\ j\ integers. \tag{5.3.43}$$

and a result in number theory (see Richards [1]) then implies the existence of a constant $\beta > 0$ such that

$$\mu_{n+1} - \mu_n > \beta \log n, \quad as\ n \to \infty. \tag{5.3.44}$$

As before, (1.5.26) and the spectral gap condition (1.5.27) are satisfied and, therefore, the proof is complete. □

Remark 5.3.4 *If we have the Neumann boundary condition for φ and the Dirichlet boundary condition for u instead of both Dirichlet boundary conditions, then the theorem on the existence of an absorbing set and the global attractor still hold. But the above symmetrized method fails and the existence of an inertial manifold remains open.*

We can see from the above that the spectral gap condition imposed severe restrictions on the domain for the existence of an inertial manifold. In what follows we prove that for the system (5.3.1)–(5.3.4) and for a general smooth bounded domain $\Omega \subset \mathbb{R}^n (n \leq 3)$, there exists an inertial set, which is a notion, as shows in Chapter 1, recently introduced by A. Eden et al.

Let B_R be the absorbing ball of radius R in $H_o^1 \times L^2$ given by Theorem 5.3.1 for the system (5.3.1)–(5.3.4). Accordingly, there is an absorbing ball B_{R_0} of radius R_0 in $H_o^1 \times H_o^1$ for the system (5.3.22)–(5.3.23). Let $t_1 > 0$ be a constant such that for any $(\varphi_0, e_0) \in V = H_o^1 \times H_o^1$, with $\|(\varphi_0, e_0)\|_V \leq R_0$, $\|\tilde{S}(t)(\varphi_0, e_0)\|_V \leq R_0$ for $t \geq t_1$. Let

$$B = \overline{\bigcup_{t \geq t_1}^{\infty} \tilde{S}(t)(\varphi_0, e_0)}, \quad for \ \|(\varphi_0, e_0)\|_V \leq R_0. \tag{5.3.45}$$

Then B is an invariant set of the nonlinear semigroup $\tilde{S}(t)$. Moreover, as proved in Theorem 4.1.1, B is also bounded in $H^{2m} \times H^{2m}$ for all $m \in \mathbb{N}$.

We now have

Theorem 5.3.3 *Let $\Omega \subseteq \mathbb{R}^n (n \leq 3)$ be a bounded domain with smooth boundary Γ. Then the system (5.3.22)–(5.3.24) (accordingly the system (5.3.1)–(5.3.4)) has an inertial set $\widetilde{\mathcal{M}}$ in $H^2 \cap H_o^1 \times H^2 \cap H_o^1$ (accordingly, an inertial set $\mathcal{M}$ in $H^2 \cap H_o^1 \times H_o^1$). Moreover, the estimates (1.5.30), (1.5.31) hold.*

Proof. To apply Theorem 1.5.7 , we have to verify that $\left(\{\tilde{S}(t)\}_{t \geq 0}, B\right)$ satisfies the squeezing property. As indicated before, $\tilde{S}(t)$ is a nonlinear semigroup on $H_o^1 \times H_o^1$ associated with the system (5.3.22)–(5.3.24). Let U and $\bar{U}$ be two solutions of (5.3.22)–(5.3.23) and

$$\bar{V} = U - \bar{U}. \tag{5.3.46}$$

Then $\bar{V}$ satisfies

$$\frac{d\bar{V}}{dt} + \mathcal{A}\bar{V} = F(U) - F(\bar{U}), \qquad (5.3.47)$$

$$\bar{V}(0) = \bar{V}_0. \qquad (5.3.48)$$

The self-adjoint positive definite operator $\mathcal{A} : D(\mathcal{A}) = H^2 \cap H_o^1 \times H^2 \cap H_o^1 \mapsto L^2 \times L^2$, as shown before, has relabelled eigenvalues λ_n $(n = 1, 2, \cdots)$ satisfying

$$\lambda_n \to \infty, \quad as \ n \to \infty. \qquad (5.3.49)$$

Let $\bar{V}_n$ be the corresponding eigenvector functions, i.e.,

$$\mathcal{A}\bar{V}_n = \lambda_n \bar{V}_n. \qquad (5.3.50)$$

Let

$$H_N = span\{\bar{V}_1, \cdots, \bar{V}_N\} \qquad (5.3.51)$$

and

$$P_N : \ H = L^2 \times L^2 \mapsto H_N, \qquad (5.3.52)$$

the orthogonal projection onto H_N, and

$$Q_N = I - P_N. \qquad (5.3.53)$$

Let

$$W = Q_N \bar{V}. \qquad (5.3.54)$$

Then by (5.3.47), (5.3.48) we have

$$\frac{dW}{dt} + \mathcal{A}W = Q_N(F(U) - F(\bar{U})), \qquad (5.3.55)$$

$$W(0) = Q_N \bar{V}_0. \qquad (5.3.56)$$

Let $V = D(\mathcal{A}^{\frac{1}{2}}) = H_o^1 \times H_o^1$. Then multiplying (5.3.55) by $(\mathcal{A}W)^T$ and integrating with respect to x yields

$$\frac{1}{2}\frac{d}{dt}\|W\|_V^2 + \|\mathcal{A}W\|_H^2 \leq \|\mathcal{A}W\|_H \|Q_N(F(U) - F(\bar{U}))\|_H$$

$$\leq \frac{1}{2}\|\mathcal{A}W\|_H^2 + \frac{1}{2}\|Q_N(F(U) - F(\bar{U}))\|_H^2$$

$$\leq \frac{1}{2}\|\mathcal{A}W\|_H^2 + \frac{1}{2\lambda_{N+1}}\|F(U) - F(\bar{U})\|_V^2. \qquad (5.3.57)$$

When $U(0), \bar{U}(0) \in B$, by Theorem 4.1.1 we have

$$\|U(t)\|_E \leq C, \quad \|\bar{U}(t)\|_E \leq C, \ \forall t \geq 0 \tag{5.3.58}$$

where $E = H^2 \times H^2$ and $C > 0$ is a constant depending on B.

From the expression for F and the Sobolev imbedding theorem, we have

$$\|F(U) - F(\bar{U})\|_V \leq \tilde{C}\|U - \bar{U}\|_V = \tilde{C}\|\bar{V}\|_V \tag{5.3.59}$$

with $\tilde{C} > 0$ a constant depending only on B. Therefore, it follows from (5.3.57) that

$$\frac{1}{2}\frac{d}{dt}\|W\|_V^2 + \frac{\lambda_{N+1}}{2}\|W\|_V^2$$

$$\leq \frac{1}{2}\frac{d}{dt}\|W\|_V^2 + \frac{1}{2}\|AW\|_H^2 \leq \frac{\tilde{C}^2}{2\lambda_{N+1}}\|\bar{V}\|_V^2. \tag{5.3.60}$$

Applying Gronwall's inequality to (5.3.60) yields

$$\|W(t)\|_V^2 \leq e^{-\lambda_{N+1}t}\|W(0)\|_V^2 + \frac{\tilde{C}^2}{\lambda_{N+1}}\int_0^t \|\bar{V}\|_V^2 d\tau. \tag{5.3.61}$$

On the other hand, we deduce from (5.3.47) that

$$\frac{1}{2}\frac{d}{dt}\|\bar{V}\|_V^2 + \|A\bar{V}\|_H^2 \leq \|A\bar{V}\|_H\|F(U) - F(\bar{U})\|_H$$

$$\leq \frac{1}{2}\|A\bar{V}\|_H^2 + \frac{1}{2}\|F(U) - F(\bar{U})\|_H^2 \leq \frac{1}{2}\|A\bar{V}\|_H^2 + \frac{C_1}{2}\|\bar{V}\|_V^2 \tag{5.3.62}$$

with $C_1 > 0$ being a constant depending on B. Applying Gronwall's inequality to (5.3.62) yields

$$\|\bar{V}(t)\|_V^2 \leq e^{C_1 t}\|\bar{V}(0)\|_V^2 \tag{5.3.63}$$

and

$$\int_0^t \|\bar{V}\|_V^2 d\tau \leq \frac{1}{C_1}e^{C_1 t}\|\bar{V}(0)\|_V^2. \tag{5.3.64}$$

Inserting (5.3.64) into (5.3.61) yields

$$\|W(t)\|_V^2 \leq e^{-\lambda_{N+1}t}\|W(0)\|_V^2 + \frac{\tilde{C}^2}{C_1\lambda_{N+1}}e^{C_1 t}\|\bar{V}(0)\|_V^2$$

$$\leq \left(e^{-\lambda_{N+1}t} + \frac{\tilde{C}^2}{C_1\lambda_{N+1}}e^{C_1 t}\right)\|\bar{V}(0)\|_V^2. \tag{5.3.65}$$

We can deduce the squeezing property from (5.3.65). Indeed, we choose

$$t_* = \frac{8 \ln 2}{\lambda_1} \tag{5.3.66}$$

and we choose N_0 such that when $N \geq N_0$,

$$\lambda_{N+1} \geq \frac{256 \tilde{C}^2}{C_1} e^{C_1 t_*}. \tag{5.3.67}$$

On the other hand, if

$$\|P_{N_0} \bar{V}(t_*)\|_V \leq \|Q_{N_0} \bar{V}(t_*)\|_V, \tag{5.3.68}$$

then

$$\|\bar{V}(t_*)\|_V^2 \leq 2\|Q_{N_0} \bar{V}(t_*)\|_V^2. \tag{5.3.69}$$

Combining (5.3.65) with (5.3.69) yields

$$\|\bar{V}(t_*)\|_V^2 \leq \frac{1}{64} \|\bar{V}(0)\|^2, \tag{5.3.70}$$

i.e.,

$$\|\bar{V}(t_*)\|_V \leq \frac{1}{8} \|\bar{V}(0)\|_V \tag{5.3.71}$$

which implies the squeezing property. The fact that $\tilde{S}(t_*)$ is Lipschitz on B follows from (5.3.63) and (5.3.66) with Lipschitz constant $L = e^{\frac{C_1 t_*}{2}} = 2^{\frac{4C_1}{\lambda_1}}$. The proof of Theorem 5.3.3 is complete. $\qquad \square$

Bibliography

R.A. Adams

[1] Sobolev Spaces, Academic Press, New York, 1975.

N. Alikakos

[1] L^p bounds of solutions of reaction–diffusion equations, Comm. PDE, Vol. 4, No. 8 (1979), 827–868.

H.W. Alt and I. Pawlow

[1] A mathematical model of dynamics of non-isothermal phase separation, Physica D, Vol. 59, No. 4 (1992), 389–416.

[2] Existence of solutions for non-isothermal phase separation, Advances in Math. Sci. and Appl., Vol. 2, No. 2 (1992), 319–409.

H. Amann

[1] Existence and regularity for semilinear parabolic evolution equations, Ann. Scuola Norm. Sup. Pisa, Ser. IV, 11 (1984), 593–676.

[2] Global existence for semilinear parabolic systems, J. Reine Angew. Math., 360 (1985), 47–83.

[3] Parabolic evolution equations with nonlinear boundary conditions, in F.E. Browder, editor, Nonlinear Functional Analysis and Its Applications, pp. 17–27, Proc. Symp. Pure Math. 45, Amer. Math. Soc., Providence, R. I., 1986.

[4] Quasilinear evolution equations and parabolic systems, Trans. Amer. Math. Soc. 293 (1986), 191–227.

[5] Quasilinear parabolic systems under nonlinear boundary conditions, Arch. Rat. Mech. Anal., 92 (1986), 153–192.

[6] Dynamic theory of quasilinear parabolic equations–I, Abstract evolution equations, Nonlinear Analysis, 12 (1988), 895–919.

231

[7] Parabolic evolution equations and nonlinear boundary conditions, J. Diff. Eqs., 72 (1988), 201–269.

[8] Dynamic theory of quasilinear parabolic systems–III, global existence, Math. Z., 202 (1989), 219–250; Erratum 205 (1990), 331.

[9] Nonhomogeneous linear and quasilinear elliptic and parabolic boundary value problems, H.J. Schmeisser, H. Triebel eds, Function Spaces, Differential Operator and Nonlinear Analysis, Teubner-Texte zur Math. 133, 9–126, Stuttgart-Leipzig, 1993.

A.V. Babin and M.I. Visik

[1] Regular attractors of semigroups of evolutionary equations, J. Math. Pures Appl., 62 (1983), 441–491.

J.M. Ball

[1] Finite time blow-up in nonlinear problems, "Nonlinear Evolution Equations" edited by M.G. Crandall, p.189–206. Academic Press, New York, 1977.

[2] Remarks on blow-up and nonexistence theorems for nonlinear evolution equations, Quarterly J. Math. Oxford, 28 (1977), 473–486.

P. Bates and P. Fife

[1] Spectral comparison principles for the Cahn–Hilliard and phase-field equations, and time scales for coarsening, Physica D, 43 (1990), 335–348.

P. Bates and Songmu Zheng

[1] Inertial manifolds and inertial sets for the phase-field equations, Journal of Dynamics and Differential Equations, Vol. 4, No. 2 (1992), 375–398.

E.F. Beckenbach and R. Bellman

[1] Inequalities, Springer–Verlag, New York, 1961.

R. Bellman

[1] On the existence and boundedness of solutions of nonlinear partial differential equations of parabolic type, Trans. Amer. Math. Soc., 64 (1948), 21–44.

D. Brochet, X. Chen and D. Hilhorst

[1] Finite dimensional exponential attractor for the phase field model, to appear in Applicable Analysis.

D. Brochet and D. Hilhorst

[1] Universal attractor and inertial sets for the phase field model, Appl. Math. Letter, Vol. 4, No. 6 (1991), 59–62.

J.A. Burns, Zhuangyi Liu and Songmu Zheng

[1] On the energy decay of a linear thermoelastic bar, J. Math. Anal. Appl. Vol. 179, No. 2 (1993), 574–591.

G. Caginalp

[1] An analysis of a phase field model of a free boundary, Arch. Rat. Mech. Anal., 92 (1986), 205–245.

J. Carr, M.E. Gurtin and M. Slemrod

[1] Structured phase transitions on a finite interval, Arch. Rat. Mech. Anal., 86 (1984), 317–351.

Z. Chen and K.H. Hoffmann

[1] On a one-dimensional nonlinear thermoviscoelastic model for structural phase transitions in shape memory alloys, J. Diff. Eqs., 112 (1994), 325–350.

P. Colli

[1] Global existence for a second-order thermo-mechanical model of shape memory alloys, to appear in J. Math. Anal. Appl.

[2] An existence result for a thermomechanical model of shape memory alloys, Advances in Mathematical Sciences and Applications, Vol. 1, No. 1 (1992), 83–97.

P. Colli, M. Fremond and A. Visintin

[1] Shape memory alloys, a continuum mechanics model, Quarterly Applied Mathematics, 48 (1990), 31–47.

P. Colli and J. Sprekels

[1] Global existence for a three-dimensional model for the thermo-mechanical evolution of shape memory alloys, to appear in Nonlinear Analysis.

R. Courant and D. Hilbert

[1] Methods of Mathematical Physics, Vol. 1, Intersciences, New York, 1953.

C.M. Dafermos

[1] On the existence and the asymptotic stability of solutions to the equations of linear thermoelasticity, Arch. Rat. Mech. Anal., Vol. 29 (1968), 241–271.

[2] Asymptotic stability in viscoelasticity, Arch. Rat. Mech. Anal., Vol. 37 (1970), 297–308.

[3] Asymptotic behavior of solutions of evolution equations, "Nonlinear Evolution Equations", edited by M.G. Crandall, p.103–124. Academic Press, New York, 1977.

[4] Conservation laws with dissipation, in Nonlinear Phenomena in Mathematical Sciences, V. Lakshmikantham, ed., Academic Press, New York, 1981.

[5] Global smooth solutions to the initial boundary value problem for the equations of one-dimensional nonlinear thermoviscoelasticity, SIAM, J. Math. Anal., 13 (1982), 397–408.

C.M. Dafermos and L. Hsiao

[1] Global smooth thermomechanical process in one-dimensional nonlinear thermo-viscoelasticity, Nonlinear Analysis, 6 (1982), 435–454.

L. Delaey, R.V. Krishnan, H. Tas and H. Warlimont

[1] Thermoelasticity, pseudoelasticity and the memory effects associated with martensitic transformations, J. Mat. Sci., 9 (1974), 1521–1555.

Y. Ebihara and T. Nanbu

[1] On the global solution of some nonlinear parabolic equations, II, Funkcialaj Ekvacioj, 20 (1977), 273–286.

A. Eden, C. Foias, B. Nicolaenko and R. Temam

[1] Ensembles inertiels pour des equations d'evolution dissipatives, C. R. Acas. Sci. Paris, 310, Ser.1 (1990), 559–562.

[2] Inertial sets for dissipative evolution equations, to appear in Appl. Math. Letter.

D. Eidus

[1] Asymptotical expansions of solutions of linear parabolic equations as $t \to \infty$, J. Math. Anal. Appl., 130 (1988), 155–170.

C.M. Elliott and Songmu Zheng

[1] On the Cahn–Hilliard equation, Arch. Rat. Mech. Anal. Vol. 96, No. 4 (1986), 339–357.

[2] Global existence and stability of solutions to the phase field equations, Free Boundary Problems, edited by K.H. Hoffmann and J. Sprekels, International Series of Numerical Mathematics, Vol. 95, p. 46–58. Birkhaus–Verlag, Basel, 1990.

F. Falk

[1] Ginzberg–Landau theory of static domain walls in shape memory alloys, Physica B, 51 (1983), 177–185.

[2] Ginzberg–Landau theory and solitary waves in shape memory alloys, Physica B, 54 (1984), 159–167.

C. Foias, G. Sell and R. Temam

[1] Inertial manifolds for nonlinear evolution equations, J. Diff. Eqs., 73 (1988), 309–353.

A. Friedman

[1] Partial Differential Equations of Parabolic Type, Prentice-Hall, Englewood Cliffs, New Jewsey, 1964.

[2] Remarks on nonlinear parabolic equations, Proc. Symp. Appl. Math., 17 (1965), 3–23.

[3] Partial Differential Equations, Holt, Rinehart and Winston, New York, 1969.

K.O. Friedrichs

[1] Symmetric hyperbolic linear differential equations, Comm. Pure Appl. Math., 7 (1954), 345–392.

[2] Symmetric positive linear differential equations, Comm. Pure Appl. Math., 11 (1958), 333–418.

H. Fujita

[1] On the blowing up of solutions of the Cauchy problem for $u_t = \Delta u + u^{1+\alpha}$, J. Fac. Sci. Univ. Tokyo, sect. 1, 13 (1966), 109–124.

[2] On some nonexistence and nonuniqueness theorems for nonlinear parabolic equations, Proc. Symp. Pure Math., AMS, 18 (1970), 105–113.

J.S. Gibson, I.G. Rosen and G. Tao

[1] Approximation in control of thermoelastic systems, SIAM J. Control and Optimization, Vol. 30, No. 5 (1992), 1163–1189.

R.T. Glassey

[1] Blow-up theorems for nonlinear wave equations, Math. Z., 132 (1973), 183–302.

[2] Finite-time blow-up for solutions of nonlinear wave equations, Math. Z., 177 (1981), 323–340.

M. Grinfeld and A. Novick–Cohen

[1] Counting stationary solutions of the Cahn–Hilliard equation by transversality arguments, to appear in Proc. Royal Soc. of Edinburgh, Sec. A.

J.K. Hale

[1] Asymptotic Behavior of Dissipative Systems, Mathematical Surveys and Monographs 25, AMS, 1988.

S.W. Hansen

[1] Exponential energy decay in a linear thermoelastic rod, J. Math. Anal. Appl., Vol. 167 (1992), 429–442.

K. Hayakawa

[1] On nonexistence of global solutions of some semilinear parabolic equations, Proc. Japan Acad., 49 (1973), 503–505.

D. Henry

[1] Geometric Theory of Semilinear Parabolic Equations, Lecture Notes in Math. 840, Springer–Verlag, New York, 1981.

K.H. Hoffmann, M. Niezgodka and Songmu Zheng

[1] Existence and uniqueness of global solutions to an extended model of the dynamical developments in shape memory alloys, Nonlinear Analysis, Vol. 15, No. 10 (1990), 977–990.

K.H. Hoffmann and Songmu Zheng

[1] Uniqueness for structural phase transitions in shape memory alloys, Math. Methods in Applied Sci., Vol. 10 (1988), 145–151.

W. Horn, J. Sprekels and Songmu Zheng

[1] Global existence of smooth solutions to the Penrose–Fife model for Ising ferromagnets, to appear in Advances in Mathematical Sciences and Applications.

W.J. Hrusa and M.A. Tarabek

[1] On smooth solutions of the Cauchy problem in one-dimensional nonlinear thermoelasticity, Q. Appl. Math., 47 (1989), 631–644.

F.L. Huang

[1] Characteristic condition for exponential stability of linear dynamical systems in Hilbert spaces, Annals of Differential Equations, Vol. 1, No. 1 (1985), 43–56.

S. Jiang

[1] Global solution of the Neumann problem in one-dimensional thermoelasticity, to appear in Nonlinear Analysis.

S. Kaplan

[1] On the growth of solutions of quasilinear parabolic equations, Comm. Pure Appl. Math., 16 (1963), 327–330.

T.Kato

[1] Linear evolution equations of "hyperbolic" type, J. Fac. Sci. Univ. Tokyo, 17 (1970), 241–258.

[2] Linear evolution equations of "hyperbolic" type, II, J. Math. Soc. Japan, 25 (1973), 648–666.

[3] The Cauchy problem for quasilinear symmetric hyperbolic systems, Arch. Rat. Mech. Anal., 58 (1975), 181–205.

[4] Quasilinear equations of evolution, with applications to partial differential equations, in "Spectral Theory and Differential Equations", Lecture Notes in Math. No. 448 (1975), 25–70.

S. Kawashima

[1] Systems of a Hyperbolic–Parabolic Composite Type, with Applications to the Equations of Magnetohydrodynamics, Ph.D. Thesis, Kyoto Univ., Dec. 1983.

[2] Large-time behavior of solutions to hyperbolic–parabolic systems of conservation laws and applications, Proc. Roy. Soc. Edinburgh Sect. A 106 (1987), 169–194.

N. Kenmochi and M. Niezgodka

[1] Nonlinear system for non-isothermal diffusive phase separation, to appear in J. Math. Anal. Appl.

[2] Large time behavior of a nonlinear system for phase separation, pp. 12–22, in Progress in PDE: the Metz surveys 2, Pitman Research Notes Math. Ser. 296, Longman, New York, 1993.

N. Kenmochi, M. Niezgodka and Songmu Zheng

[1] Global attractor of a non-isothermal model for phase separation, to appear.

J.U. Kim

[1] Global existence of solutions of the equations of one-dimensional thermoviscoelasticity with initial data in BV and L^1, Ann. Scu. Norm. Sup. Pisa, 3 (1983), 357–427.

[2] Solutions to the equations of one-dimensional viscoelasticity in *BV*, SIAM J. Math. Anal., Vol. 14, No. 4 (1983), 684–695.

[3] On the energy decay of a linear thermoelastic bar and plate, SIAM J. Math. Anal., Vol. 23 (1992), 889–899.

S. Klainerman

[1] Long-time behavior of solutions to nonlinear evolution equations, Arch. Rat. Mech. Anal. 78 (1982), 73–98.

K. Kobayashi et al.

[1] On the blowing up problem for semilinear heat equations, J. Math. Soc. Japan, 29 (1977), 407–424.

N.V. Krylov

[1] On estimates of derivatives of solutions for nonlinear parabolic equations, Dokl. USSR, 274 (1984), 23–26.

O.A. Ladyzenskaja, V.A. Solonnikov and N.N. Uralceva

[1] Linear and Quasilinear Equations of Parabolic Type, Translations of Mathematical Monographs, Vol. 23, AMS, Rhode Island, 1968.

J. Lagnese

[1] Boundary Stabilization of Thin Plates, Studies in Applied Mathematics Vol. 10, SIAM, Philadelphia, 1989.

Ph. Laurencot

[1] Solutions to a Penrose–Fife model of phase-field type, J. Math. Anal. Appl., Vol. 185, No. 2 (1994), 262–274.

P.D. Lax and R.S. Phillips

[1] Local boundary conditions for dissipative symmetric linear differential operators, Comm. Pure Appl. Math., 13 (1960), 427–455.

H.A. Levine

[1] Some nonexistence and instability theorems for formally parabolic equations of
    the form $Pu_t = -Au + F(u)$, Arch. Rat. Mech. Anal., 51 (1973), 371–386.

[2] Instability and nonexistence of global solutions to nonlinear wave equations of
    the form $Pu_{tt} = -Au + F(u)$, Trans. Amer. Math. Soc., 192 (1974), 1–21.

Li, Dening

[1] The nonlinear initial boundary value problems for quasilinear hyperbolic sys-
    tems, Chinese Annals of Mathematics, 7B, 2 (1986), 147–159.

[2] The general initial boundary value problems for linear hyperbolic–parabolic cou-
    pled systems, Chinese Annals of Mathematics, 7B, 4 (1986), 408–424.

[3] The nonlinear initial boundary value problems and the existence of multi-dimen-
    sional shock wave for quasilinear hyperbolic–parabolic coupled systems, Chinese
    Annals of Mathematics, 8B, 2 (1987), 252–280.

J.L. Lions and E. Magenes

[1] Nonhomogeneous Boundary Value Problems and Applications, Springer–Verlag,
    New York, 1972.

Li, Tatsien and Chen, Yunmei

[1] Global Classical Solutions for Nonlinear Evolution Equations, Pitman Mono-
    graphs and Surveys in Pure and Applied Mathematics, Vol. 45, Longman Sci-
    entific and Technical, London, 1992.

Li, Tatsien, Yu, Wentzu and Shen, Weixi

[1] Boundary value problems and free boundary problems for quasilinear hyperbo-
    lic-parabolic coupled systems, MRC Tech. Summary Report 2273, Univ. of
    Wisconsin-Madison, 1981.

[2] Cauchy problem for quasilinear hyperbolic–parabolic coupled systems, Acta
    Mathematicae Applicatae Sinica, Vol. 1, No. 4 (1981), 321–338.

[3] Second initial boundary value problems for quasilinear hyperbolic–parabolic
    coupled systems, Chinese Annals of Mathematics, 2A, (1) (1981), 65–90.

[4] First initial boundary value problems for quasilinear hyperbolic–parabolic cou-
    pled systems, Chinese Annals of Mathematics, 5B, (1) (1984), 77–90.

[5] Initial boundary value problems with interface for quasilinear hyperbolic-parabolic coupled systems, Chinese Annals of Mathematics, 5B, (3) (1984), 339–356.

Liu, T-P.

[1] Nonlinear Stability of Shock Waves for Viscous Conservation Laws, American Mathematical Society, Providence, 1985.

Zhuangyi Liu and Songmu Zheng

[1] Exponential stability of semigroup associated with thermoelastic system, Q. Appl. Math., No. 3 (1993), 535–545.

[2] Uniform exponential stability and approximation in control of thermoelastic system, SIAM J. Control and Optimization, Vol. 32, No. 5 (1994), 1226–1246.

[3] On the exponential stability of linear viscoelasticity and thermoviscoelasticity, to appear in Q. Appl. Math.

[4] Exponential energy decay of the Euler–Bernoulli beam with shear or thermal diffusion, to appear in J. Math. Anal. Appl.

[5] Exponential stability of the Kirchhoff plate with thermal or viscoelastic damping, to appear in Q. Appl. Math.

S. Luckhaus and Songmu Zheng

[1] A nonlinear boundary value problem involving nonlocal term, Nonlinear Analysis, Vol. 22, No. 2 (1994), 129–135.

F.J. Massey III

[1] Abstract evolution equations and the mixed problem for symmetric hyperbolic systems, Transaction of the American Mathematical Society, 168 (1972), 165–188.

A. Matsumura

[1] Global existence and asymptotics of the solutions of the second order quasilinear hyperbolic equations with the first order dissipation, Publ. RIMS, Kyoto Univ., 13 (1977), 349–379.

[2] Initial Value Problems for Some Quasilinear Partial Differential Equations in Mathematical Physics, Ph.D. Thesis, Kyoto University, June 1980.

[3] An energy method for the equations of motion of compressible viscous heat-conductive fluids, MRC Technical Summary Report 2194, Univ. of Wisconsin-Madison, 1981.

A. Matsumura and T. Nishida

[1] The initial value problem for the equations of motion of compressible viscous and heat-conductive fluids, Proc. Jap. Acad. Ser. A, 55 (1979), 337–341.

[2] The initial value problem for the equations of motion of viscous and heat-conductive gases, J.Math. Kyoto Univ., 20 (1980), 67–104.

[3] Initial boundary value problems for the equations of motion of compressible viscous fluids, Contemporary Mathematics, 17 (1983), 109–116.

[4] Initial boundary value problems for the equations of motion of compressible viscous and heat-conductive fluids, Comm. Math. Phys., 89 (1983), 445–464.

M. Nakao

[1] L^p-estimates of solutions of some nonlinear degenerate diffusion equations, J. Math. Soc. Japan Vol. 37, No. 1 (1985), 41–63.

M. Niezgodka and J. Sprekels

[1] Existence of solutions for a mathematical model of structural phase transitions in shape memory alloys, Math. Meth. Appl. Sci., 10 (1988), 197–223.

M. Niezgodka, Songmu Zheng and J. Sprekels

[1] Global solutions to a model of structural phase transitions in shape memory alloys, J. Math. Anal. Appl., 130 (1988), 39–54.

L. Nirenberg

[1] On elliptic partial differential equations, Annali della Scoula Norm. Sup. Pisa, 13 (1959), 115–162.

[2] Topics in Nonlinear Functional Analysis, Courant Institute of Mathematical Sciences, 1974.

A. Novick-Cohen and L.A. Peletier

[1] Steady states of the one-dimensional Cahn–Hilliard equation, to appear in Quarterly Appl. Math.

A. Novick-Cohen and Songmu Zheng

[1] The Penrose–Fife type equations: Counting the one-dimensional stationary solutions, to appear in Proc. Royal Soc. of Edinburgh, series A.

L.E. Payne

[1] Improperly Posed Problems in Partial Differential Equations, SIAM Regional Conference Series in Applied Math., Vol. 22, 1975.

L.E. Payne and D.H. Sattinger

[1] Saddle points and instability of nonlinear hyperbolic equations, Israel J. Math., 22 (1975), 273–303.

A. Pazy

[1] Semigroups of Linear Operators and Applications to Partial Differential Equations, Appl. Math. Sci. Vol. 44, Springer–Verlag, New York, 1983.

O. Penrose and P. Fife

[1] Thermodynamically consistent models of phase-field type for the kinetics of phase transitions, Physica D, 43 (1990), 44–62.

G. Ponce

[1] Global existence of solutions to a class of nonlinear evolution equations, Nonlinear Analysis, 9 (1985), 71–92.

G. Ponce and R. Racke

[1] Global existence of small solutions to the initial value problem for nonlinear thermoelasticity, J. Diff. Eqs., 87 (1990), 70–83.

R. Racke

[1] Lectures on Nonlinear Evolution Equations, Aspects of Mathematics Vol. 19, VIEWEG, Bonn, 1992.

[2] On the Cauchy problem in nonlinear 3-d-thermoelasticity, Math. Z., 203 (1990), 649–682.

[3] Blow-up in nonlinear three-dimensional thermoelasticity, Math. Meth. Appl. Sci., 12 (1990), 267–273.

R. Racke, Y. Shibata and Songmu Zheng

[1] Global solvability and exponential stability in one-dimensional nonlinear thermoelasticity, Q. Appl. Math., No. 4 (1993), 751–763.

R. Racke and Songmu Zheng

[1] Global existence of solutions to a fully nonlinear fourth order parabolic equation in exterior domains, Nonlinear Analysis, Vol. 17 (1991), 1027–1038.

J. Rauch and F.J. Massey III

[1] Differentiability of solutions to hyperbolic initial boundary value problems, Transaction of the American Mathematical Society, Vol. 189 (1974), 303–318.

M. Reed and B. Simon

[1] Methods of Modern Mathematical Physics, Academic Press, New York, 1978.

J.E.M. Revira

[1] Energy decay rate in linear thermoelasticity, Funkcial Ekvac., Vol. 35 (1992), 19–30.

J. Richards

[1] On the gap between numbers which are the sum of two squares, Adv. Math. 46 (1982), 1–2.

D.H. Sattinger

[1] Stability of nonlinear hyperbolic equations, Arch. Rat. Mech. Anal., Vol. 28 (1968), 226–244.

[2] On global solution of nonlinear hyperbolic equations, Arch. Rat. Mech. Anal., Vol. 30, No. 3 (1968), 148–172.

R. Schaaf

[1] Global behavior of solution branches for some Neumann problems depending on one or several parameters, J. Reine Angew. Math., Vol. 346 (1984), 1–31.

M.E. Schonbek

[1] Decay of solutions to parabolic conservation laws, Comm. PDE, 7 (1980), 449–473.

I. Segal

[1] Nonlinear semigroups , Ann. Math., 78 (1963), 339–364.

G. Sell and Y. You

[1] Inertial manifolds: the non-self-adjoint case, Journal of Differential Equations, Vol. 96, No. 2 (1992), 203–255.

J. Shatah

[1] Global existence of small solutions to nonlinear evolution equations, Journal of Differential Equations, 46 (1982), 409–425.

Weixi Shen (Shen, Weixi) and Songmu Zheng (Zheng, Songmu)

[1] Nonlocal initial boundary value problem for nonlinear parabolic equations, Journal of Fudan University, 1 (1985), 47–57.

[2] Global smooth solutions to the equations of one-dimensional thermoelasticity with dissipation boundary conditions, Chinese Annals of Mathematics, 7B, (3) (1986), 303–317.

[3] On the coupled Cahn-Hilliard equations, Comm. PDE, Vol. 18, No. 3-4 (1993), 701–727.

Y. Shibata

[1] On a local existence theorem for some quasilinear hyperbolic–parabolic coupled systems with Neumann type boundary condition, Manuscript (1989).

[2] Neumann problem for one-dimensional nonlinear thermoelasticity, SFB 256 Preprint No. 145, Univ. Bonn 1990.

Y. Shibata and Songmu Zheng

[1] On some nonlinear hyperbolic systems with damping boundary conditions, Nonlinear Analysis, Vol. 17, No. 3 (1991), 233–266.

Y. Shizuta and S. Kawashima

[1] Systems of equations of hyperbolic–parabolic type with applications to the discrete Boltzmann equation, Hokkaido Math. J., 4 (1985), 249–275.

M. Slemrod

[1] Global existence, uniqueness and asymptotic stability of classical smooth solutions in one-dimensional nonlinear thermoelasticity, Arch. Rat. Mech. Anal.,
41 (1981), 110–133.

J. Smoller

[1] Shock Waves and Reaction–Diffusion Equations, Grundlehren math.
Wissenschaften 258, Springer–Verlag, New York, 1983.

J. Sprekels and Songmu Zheng

[1] Global solutions to the equations of a Ginzberg–Landau theory for structural
phase transitions in shape memory alloys, Physica D, 39 (1989), 59–76.

[2] Global smooth solutions to a thermodynamically consistent model of phase-field
type in higher space dimensions, J. Math. Anal. Appl., Vol. 176, No. 1 (1993),
200–223.

[3] Optimal control problem for a thermodynamically consistent model of phase-
field type for phase transitions, Advances in Mathematical Sciences and Applications, Vol. 1, No. 1 (1992), 113–125.

B. Straughan

[1] Further global nonexistence theorems for abstract nonlinear wave equations,
Proc. Amer. Math. Soc., 48 (1975), 381–390.

J.C. Strikwerda

[1] Initial boundary value problems for incompletely parabolic systems, Comm.
Pure Appl. Math., 30 (1977), 797–822.

A. Tani

[1] The existence and uniqueness of the solution of equations describing compressible viscous fluid flow in a domain, Proc. Japan Acad., 52 (1976), 334–337.

[2] On the first initial boundary value problem of compressible viscous fluid motion, Publ. RIMS, Kyoto Univ., 13 (1977), 193–253.

R. Temam

[1] Infinite-Dimensional Dynamical Systems in Mechanics and Physics, Applied Mathematical Sciences 68, Springer–Verlag, New York, 1988.

M. Tsutsumi

[1] Existence and nonexistence of global solutions for nonlinear parabolic equations, Publ. RIMS, Kyoto Univ., 8 (1972–73), 211–229.

A.I. Vol'pert and S.I. Hudjaev

[1] On the Cauchy problem for composite systems of nonlinear differential equations, Math. USSR Sbornik, 16 (1972), 517–544.

F.B. Weissler

[1] Existence and nonexistence of global solutions for a semilinear heat equation, Israel J. Math., 38 (1981), 29–40.

Jiongmin Yong and Songmu Zheng

[1] Feedback stabilization and optimal control for the Cahn–Hilliard equation, Nonlinear Analysis, Vol. 17, No. 5 (1991), 431–444.

[2] Feedback stabilization for the phase-field equations, Applicable Analysis, Vol. 42 (1991), 59–68.

K. Yosida

[1] Functional Analysis, Springer–Verlag, New York, 1965.

Y. Zeng

[1] L^1 asymptotic behavior of compressible, isentropic, viscous 1-D flow, Communications on Pure and Applied Mathematics, Vol. XLVII, No. 8 (1994), 1053–1082.

Songmu Zheng (Zheng, Songmu)

[1] Differentiability of solution to initial boundary value problem for first order symmetric hyperbolic system with characteristic boundary, Journal of Fudan University, No. 3 (1982), 331–340.

[2] Initial boundary value problem for quasilinear hyperbolic–parabolic coupled system in higher space dimensions, Chinese Annals of Mathematics, 4B, (4) (1983), 439–458.

[3] Global solutions and applications to a class of quasilinear hyperbolic–parabolic coupled system, Scientia Sinica, 17 (1984), 177–186.

[4] Global smooth solutions to quasilinear first order symmetric hyperbolic system with dissipation term, Chinese Annals of Mathematics, 5A, (4) (1984), 447–454.

[5] Global solutions to the Cauchy problems for quasilinear hyperbolic–parabolic coupled system in higher dimensions, Chinese Annals of Mathematics, 5A, (6) (1984), 681–690.

[6] Remarks on global existence for nonlinear parabolic equations, Nonlinear Analysis, 1 (1986), 107–114.

[7] Asymptotic behavior of solution to the Cahn–Hilliard equation, Applicable Analysis, 3 (1986), 165–184.

[8] Global solutions to the Cauchy problem of nonlinear thermoelastic equation with dissipation, Chinese Annals of Mathematics, 8B, (2) (1987), 142–155.

[9] Global solutions to the second initial boundary value problem for fully nonlinear parabolic equations, Acta Mathematica Sinica, No. 3 (1987), 237–246.

[10] Global existence of solution to initial boundary value problem for nonlinear parabolic equations, Acta Mathematica Scintia, 4 (1987), 361–374.

[11] Global smooth solutions to the Cauchy problem of one-dimensional viscoelastic system, Journal of Fudan University, 2 (1988), 233–237.

[12] Global solutions to the thermomechanical equations with nonconvex Laudan–Ginzberg free energy, Zeitschrift fur Angewandte Mathematik und Physik, Vol. 40 (1989), 111–127.

[13] Global existence for a thermodynamically consistent model of phase field type, Journal of Differential and Integral Equations, Vol. 5, No. 2 (1992), 241–253.

Songmu Zheng (Zheng, Songmu) and Yunmei Chen (Chen, Yunmei)

[1] Global existence for nonlinear parabolic equations, Chinese Annals of Mathematics, 7B, (1) (1986), 57–73.

[2] On the blowing-up of solutions to the initial boundary value problems for nonlinear evolution equations, Journal of Fudan University, 1 (1987), 9–27.

Songmu Zheng (Zheng, Songmu) and Weixi Shen (Shen, Weixi)

[1] Global existence of initial boundary value problem for Fitzhugh–Nagumo equations, Sciences Bulletin, 9 (1984), 16–20.

[2] Global solutions to the Cauchy problem of a class of quasilinear hyperbolic–parabolic coupled systems, Scientia Sinica, 4A (1987), 357–372.

[3] Global solutions to the Cauchy problem of the equations of one-dimensional thermoviscoelasticity, Journal of Partial Differential Equations, Vol. 2, No. 2 (1989), 26–38.

Index

Milton Keynes UK
Ingram Content Group UK Ltd.
UKHW020313111024
449327UK00040B/906